STRUCTURE AND ASSEMBLY

THE CYTOSKELETON
A Multi-Volume Treatise, Volume 1

STRUCTURE AND ASSEMBLY

Editors: JOHN E. HESKETH
Rowett Research Institute
Aberdeen, Scotland

IAN F. PRYME
Department of Biochemistry and
Molecular Biology
University of Bergen
Bergen, Norway

VOLUME 1 • 1995

 JAI PRESS INC.

Greenwich, Connecticut   *London, England*

CONTENTS

LIST OF CONTRIBUTORS

Jesús Avila

Centro de Biología Molecular
"Severo Ochoa"
Universidad Autonoma de Madrid
Cantoblanco, Madrid, Spain

W.H. Goldmann

Department of Biophysics
Technical University of Munich
Garching, Germany

G. Isenberg

Department of Biophysics
Technical University of Munich
Garching, Germany

Sutherland K. Maciver

Division of Structural Studies
MRC Laboratory of Molecular Biology
Cambridge, England

Javier Díaz Nido

Centro de Biología Molecular "Severo
Ochoa"
Universidad Autonoma de Madrid
Cantoblanco, Madrid, Spain

Verena Niggli

Pathology Institute
University of Bern
Bern, Switzerland

Robert L. Shoeman

Max-Planck-Institut für Zellbiologie
Ladenberg, Germany

Reimer Stick

Biochemistry and Cell Biology Institute
Gottingen University, Germany

Peter Traub

Max-Planck-Institut für Zellbiologie
Ladenberg, Germany

Tim J. Yen

Institute for Cancer Research
Fox Chase Cancer Centre
Philadelphia, Pennsylvania

PREFACE

Biologists have long been fascinated by cell structure and how this is related to the mechanisms involved in cell motility, particle and organelle movement, and cell division. By the late 1960s various filament structures had been observed in the cytoplasm using electron microscopy and a few people with off-beat ideas thought that nonmuscle tissues contained actin- and myosin-like proteins. From these beginnings the last 25 years has seen a dramatic surge in research interest in, and knowledge of, the filament systems which span the cytoplasm. We now know that nearly all cells contain three cytoplasmic filament systems, microfilaments, microtubules and intermediate filaments. These, as well as the plasma-membrane associated filaments make up the *cytoskeleton.*

Many protein components of the cytoskeleton have been identified, tentative functions assigned and more recently defects identified in disease states. The aim of this treatise is to bring together state-of-knowledge information on all aspects of the cytoskeleton; structural, functional and pathological. Until recently the emphasis in cytoskeleton research has been on identification of protein components and structural studies of the proteins; it was largely the domain of the protein biochemist. We have tried to redress the balance in this four volume treatise, and overall to stress functional aspects of the cytoskeleton and how this may be related to cell and tissue physiology and to disease. Research on these functional aspects of the cytoskeleton is a major thrust of cell biology and it is entering an exciting era of further expansion based on the powerful techniques of molecular biology and genetics.

It is now clear that the cytoskeleton is of fundamental importance in a wide range of cellular processes and that an appreciation of its features and functions impinges on a wide range of cell physiology, spanning research from fundamental cell biology to cancer pathology, from bones to viral infection. We believe that knowledge of the cytoskeleton will be of benefit to a large number of physiologists, biochemists and medical students as well as cell and molecular biologists; we hope that this treatise will go someway to providing this.

The first volume of the treatise deals with structural aspects of the cytoskeleton: the characteristics of the filaments and their components; the organization of the genes; motor proteins; interactions with membranes. The second volume deals with functions of the cytoskeleton in different cellular processes such as cell compartmentation and organelle transport, secretion and cell attachment. In volume III the functional theme is continued but in this case the emphasis is on the cytoskeleton in different tissues such as bone, liver and intestine. Lastly, in the fourth volume a selection of pathological situations which are related to defects in the cytoskeleton are discussed.

We would like to thank the authors for producing a series of lucid and interesting articles and especially those who delivered their manuscripts on, or close to, the deadlines. Lastly, we thank Piers Allen at JAI Press for his help during production of the treatise.

John Hesketh and Ian Pryme
Editors

INTRODUCTION

The last few years have seen a rapid increase in our knowledge of the proteins which make up the cytoskeleton: knowledge of the detailed structure of the major components such as actin and tubulin; information about the large number of associated proteins which appear to regulate assembly and interaction with other structures; and information about the genes. Increasingly genetic and molecular techniques are being used to delve further into the detailed chemistry of these proteins. It is these detailed studies which lay the base for much of the functional studies of the cytoskeleton. Therefore the first volume of this four-volume treatise is concerned largely with the structural aspects of the cytoskeletal components and regulation of cytoskeleton assembly.

Microtubules and microfilaments are simple polymers in the sense that, regardless of source, they consist of the same or very similar monomers, namely tubulin dimers or actin. However, within one cell there is a variety of different microtubules and microfilaments and this variety of form and function is produced by associated proteins which bind to the monomeric units of the filaments. The complexity of the microfilaments and their many actin-binding proteins is described by Sutherland Maciver, as is the nature of the myosin motor molecules. The microtubules appear to possess somewhat fewer monomer-binding proteins but nevertheless, as discussed by Jesús Avila and Javier Díaz Nido, microtubule-binding proteins are able to regulate the stability and assembly of microtubules. Microtubules also possess a variety of motor proteins and those involved in movements associated with the mitotic spindle are described in the third article by Tim Yen.

An anatomizing network of fine filaments is often observed underlying the cell plasma membrane and there is increasing evidence that actin interacts with membrane components. The membrane-cytoskeleton which underlies the plasma membrane is discussed in detail by Verena Niggli, whilst various biophysical approaches to the study of microfilament-lipid interactions are described by Gerhard Isenberg and Wolfgang Goldmann.

Intermediate filaments differ from other cytoskeletal components in that the basic monomeric unit is not universal in all cell types. This gives intermediate filaments a different type of complexity and a degree of tissue specificity. The properties of the various intermediate filament proteins and their close relatives, the lamins, are described in the final two articles by Peter Traub and Robert Shoeman and by Reimar Stick; these articles also highlight disease states which appear to be associated with defects in intermediate filament proteins.

John Hesketh and Ian Pryme
Aberdeen and Bergen, March 1994

MICROFILAMENT ORGANIZATION AND ACTIN-BINDING PROTEINS

Sutherland K. Maciver

The Cytoskeleton, Volume 1
Structure and Assembly, pages 1–45.
Copyright © 1995 by JAI Press Inc.
All rights of reproduction in any form reserved.
ISBN: 1-55938-687-8

I. INTRODUCTION

In most eukaryotic cell types actin is the major protein, accounting for about 15% of the total protein content. In concert with the microtubules (Avila, this volume) and intermediate filament networks (Traub, this volume), microfilaments form the cytoskeleton, a rigid yet flexible and dynamic meshwork of proteins that provides physical integrity to cells and the tissues they form. Two components of the cytoskeleton, the microtubules and the microfilaments, also provide the cell with the ability to move material within the cell and to translocate the whole cell. Both of these systems are known to employ "motor proteins" (such as kinesin, dynein and the myosins) to perform these functions, but it is possible that the dynamic behavior of these polymers themselves may contribute to some forms of motility. Microfilaments are especially dynamic and are modulated by a large number of actin binding proteins (ABPs). An astonishing number of intracellular proteins bind to actin filaments. These range from proteins whose only known function is to bind actin to those with well characterized cellular functions in addition to the ability to bind actin (Kreis & Vale, 1993; Maciver, 1994).

ABPs are traditionally classified according to their actin binding function (Weeds, 1982; Pollard & Cooper, 1986). Many types of actin filament cross-linking proteins that hold the filaments in orthogonal or bundled arrays to form viscous gels have been characterized. Other proteins exist that bind and block the ends of filaments. A smaller number of actin monomer binding proteins have been identified and these are thought to sequester a pool of unpolymerized actin within cells. This pool is presumably released to form new actin structures upon suitable stimulation such as a chemoattractant. Some ABPs cannot simply be assigned to one particular group; villin for example could be grouped with capping proteins, severing proteins and even microfilament bundling proteins.

II. ACTIN FILAMENT SEVERING PROTEINS

Actin filament severing proteins fragment filaments by mechanisms that do not require the hydrolysis of ATP. The purpose of this severing activity is probably to introduce a device whereby existing actin filament structures may be removed or remodeled to form other structures within the cell. Consistent with this role is the fact that many severing proteins are known to be controlled by products of the major signaling pathways such as calcium, pH and the phosphatidylinosides (see later) thereby allowing this severing activity to be limited spatially and temporally within the cell. Two major types of actin filament severing proteins have so far been identified. The first group is severing proteins that also cap filaments, and the sequence data available to date strongly suggest that they all are related to gelsolin. The second group is composed of small proteins (Bamburg et al., 1980) that sever at a lower rate but are typically present at a high concentration relative to the gelsolin group (Koffer et al., 1988; Bamburg & Bray, 1987). The sequence data for this group also show them to be broadly homologous to each other and distinct from the gelsolin group.

A. The Gelsolin Group

Gelsolin is the archetype of a group of ABPs that sever and cap the fast-growing, barbed end of actin filaments and that initiate the polymerization of new filaments by forming a nucleus (Yin, 1988; Weeds & Maciver, 1993) (Table 1). These different actin-binding functions are modulated by calcium, pH (Lamb et al., 1993) and phosphatidylinositol 4,5-bisphosphate (Janmey & Stossel, 1987). In humans, gelsolin is encoded by a single gene which is expressed by many cell types as a 83 kDa cytoplasmic protein (Kwiatokowski et al., 1986) and by muscle (Nodes et al., 1987; Kwiatowski et al., 1988) as a slightly larger protein incorporating an N-terminal extension that directs the protein to be secreted into plasma. This processing may be universal in the animal kingdom, since *Drosophila* also appears to generate cytoplasmic and secreted forms of gelsolin from a single gene (Heintzelman et al., 1993). Gelsolin is composed of six domains that form three distinct actin-binding sites (see Figure 1). These six domains are broadly homologous to each other and demonstrate more limited homology to profilin and perhaps Cap Z (Yu et al., 1990). Recently, the structure of domain 1 of gelsolin bound to actin has been solved by X ray crystallography (McLaughlin et al., 1993). It is generally

Table 1. The Gelsolin Group

Protein	Source	Mol. Wt. kDa	Regulated by: pH	Ca^{2+}	PIP_2
Gelsolin	Vertebrate cells & plasma	83	+	+	+
Severin	*Dictyostelium*	40	?	+	+
Fragmin	*Physarum*	42	?	+	?
Villin	Vertebrate gut microvilli	92	?	+	+
Adseverin	Adrenal medulla	74	?	+	+
Scinderin	Chromaffin cells	80	+	+	+

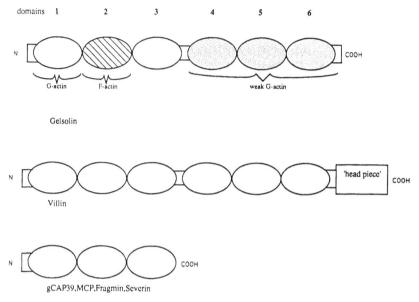

Figure 1. The domain structure of the gelsolin related ABPs. All domains are homologous except for the villin 'headpiece,' which is homologous to dematin, a filament bundling protein.

supposed that these domains arose from multiplication of an ancestral gene encoding a protein of around 15 kDa. The fragmin/severin group (which includes gCAP39 and MCP) would have arisen by triplication of this ancestral protein while gelsolin would have arisen from a duplication of this triplicate form. Villin is very similar to gelsolin with an additional F-actin binding "headpiece" that endows it with bundling activity (Arpin et al., 1988) (Figure 1). The "gene duplication" hypothesis is supported by the observation that lower eukaryotes such as *Dictyostelium* and *Physarum* possess the triplicate forms while vertebrates have the gelsolin/villin "double-triplicate" form. It has recently come to light, however, that protovillin, a newly discovered PIP$_2$ regulated, capping, nonsevering protein from *Dictyostelium* (Hofmann et al., 1992) is a villin homologue (Hofmann et al., 1993).

The function of the gelsolin group in cells can at present be rather vaguely described as regulating filament length and number in order to allow rearrangement of the actin based cytoskeleton during activities such as cell locomotion. The concentration of gelsolin in cells is correlated to the rate of cell locomotion of fibroblasts (Cunningham et al., 1991) and in myeloid cells gelsolin expression increases during differentiation as the cells become motile (Kwiatkowski, 1988). The fact that under certain situations gelsolin is enriched in ruffling membranes is in agreement with an involvement in cell locomotion (Onoda et al., 1993a), yet it has been reported that *Dictyostelium* devoid of severin is viable and able to locomote normally (André et al., 1989). Other members of the gelsolin family may of course fulfill other cellular functions. One of these, scinderin (very possibly identical to adseverin), is expected to contribute to the changes in cortical microfilament organization during stimulated secretion (Bader et al., 1986; Maekawa et al., 1989). Scinderin is a filament severing protein that becomes localized to the cortex of secretory cells just before chromaffin cells secrete (Vitale et al., 1991). This activity is controlled by Ca^{2+}, pH, acid phospholipids and protein kinase C phosphorylation (Rodriguez Del Castillo et al., 1992).

The role of gelsolin in the serum can be more definitely described. Gelsolin (present in the serum at about 2–3 µM), is thought to protect the circulation against blockage by the polymerization of actin that may leak from damaged tissues. Plasma gelsolin is synthesized by muscle tissue (Nodes et al., 1987). Vitamin D binding protein (VDBP) also inhibits actin polymerization by binding actin monomers and is produced by monocytes (Sabbatini et al., 1993) upon stimulation by lipopolysaccharides. Further light on the function of serum gelsolin may be shed by cases where a single base mutation in the human gelsolin gene (G^{654} to A^{654}) is responsible for familial amyloidosis (Finnish type). This mutation produces a protein that is proteolyzed to give a 65-kDa and a 55-kDa species that can no longer sever actin filaments (Weeds et al., 1993). The gelsolin fragments then break down further to produce an 8-kDa fragment that forms the amyloid fibrils giving rise to the disease. It is not yet clear what the consequence of a reduced circulating gelsolin activity actually is.

B. The Actin Depolymerizing Factor (ADF)/Cofilin Group

This second group comprises low molecular-weight actin filament severing proteins (Table 2), which in addition possess actin monomer binding activity (see below Actin Monomer Binding Proteins). The actin-binding functions of these proteins are not controlled by calcium but may be controlled by pH (Yonezawa et al., 1985; Hawkins et al., 1993; Hayden et al., 1993), phosphorylation (Baorto et al., 1992; Morgan et al., 1993) and perhaps phosphatidylinositol lipids (Yonezawa et al., 1990; Quirk et al., 1993).

Although the severing activity of this group is low in conventional assays when compared to that of the gelsolin group, two considerations make severing an important attribute of these proteins. ADF is present at high concentration in cells in relation to the concentration of the gelsolin group (Bamburg & Bray, 1987) and severs filaments in a strongly pH sensitive manner (Hayden et al., 1993; Hawkins et al., 1993). At low pH (within the physiological range), ADF binds to filaments in a strongly cooperative manner. This binding is stable and does not result in severing, but upon increasing the pH to above 7.3 severing is so rapid that it is very difficult to measure (Hawkins et al., 1993). Binding of this group to actin filaments

Table 2. The ADF/Cofilin Group

Protein	Source	Mol. Wt. kDa	pH sensitivity	Nuclear targeting
Actophorin	Acanthamoeba	15	–	?
ADF (Destrin)	Vertebrate	19	+	+
Cofilin	Vertebrate	21	+	+
Depactin	Sea urchin	20	–	?
Yeast cofilin	Saccharomyces	19	–+	–

seems to be the limiting step in severing, which is probably why low pH loaded filaments are severed so rapidly.

Actophorin, a member of the ADF/cofilin group from the soil amoeba *Acanthamoeba* also has severing activity, for which a mechanism has been proposed (Maciver et al., 1991a). According to this model, actophorin is able to sever the filament by binding an actin monomer within the filament, close to the C and N termini. This site constitutes an actin–actin contact (Holmes et al., 1990) and so only becomes exposed when the filament flexes and temporarily breaks this contact. Once bound, actophorin is able to prevent the filament from spontaneously "healing," that is, by reforming the actin–actin contact. The model proposes that actophorin severs filaments by increasing the likelihood that the other two actin–actin contacts that hold the filament together will spontaneously break. The ADF/cofilin group is likely to be represented in all eukaryotic organisms since cDNAs have been identified in several species of higher plants (Kim et al., 1993) and yeast. The precise role of the group in cells is not absolutely clear but the gene is essential for *Sacchromyces cerevisiae* (Moon et al., 1993; Iida et al., 1993). ADF and cofilin are targeted to the nucleus upon heat shock, which may indicate that thermal protection is at least one function (see below Nuclear Actin Binding Proteins). Present sequence data indicate that sea urchin depactin (Mabuchi, 1983) is the most divergent member of the group (Takagi et al., 1988), a finding that seems at odds with the suggestion that the echinoderms gave rise to the vertebrates. Perhaps echinoderms also express proteins more related to the ADF/cofilin type in addition to depactin. Other "outliers" of this group may include ABP1p from yeast and coactosin from *Dictyostelium* (Drubin et al., 1990). ABP1p, has an N-terminal domain particularly homologous to both yeast cofilin (Moon et al., 1993) and actophorin. Coactosin, a filament binding protein, has broad homology to members of the group, particularly vertebrate cofilin (de Hostos et al., 1993a).

III. FILAMENT CAPPING PROTEINS

Actin filaments grow by monomer addition exclusively at their ends, particularly their barbed ends. Capping proteins that bind to the barbed ends of filaments in cells, thereby preventing monomer addition or loss, are therefore placed at a "cellular bridgehead" for the control of the polymerization within cells or within local regions of individual cells. The capping protein group (Table 3) is composed of three main subgroups: (1) the Ca^{2+}-sensitive capping/severing group including severin and gelsolins, (2) the heterodimeric Ca^{2+}-insensitive capping group, and (3) a heterologous group including modified actin itself (Wegner & Aktories, 1988; Furuhashi & Hatano, 1990). The gelsolin group of actin filament severing proteins (see above) also cap filaments, but proteins (MCP, Mbh1 and gCap39) have been identified, that do not sever filaments despite being homologous to the gelsolin group (Weeds & Maciver, 1993). Other capping proteins exist that cannot at the moment be placed in any of these main groups. One of these, aginactin from *Dictyostelium* (Sauterer et al., 1991), is particularly exciting since its activity is modulated by external stimuli and is homologous to the heat shock cognate protein, Hsc70 (Eddy et al., 1993). More recent data indicate that Hsc70 may bind to *Dictyostelium* cap 32/34 protein, thus accounting for the capping activity of aginactin (Haus et al., 1993). The observation that aginactin-capping activity is stimulus-dependent may lead to an understanding of how the actin system interplays with the many systems in which Hsc70 is involved. Gelsolin itself remains the most widely studied of the capping proteins and it is anticipated that many of the structural questions will shortly be answered now that the capping domain of gelsolin (domain 1) has been cocrystallized with actin (McLaughlin et al., 1993).

A.   Heterodimeric Ca^{2+}-Independent Capping Proteins

Members of this group of heterodimeric proteins have been isolated from a variety of phyla, and both subunits share extensive homology with their counterparts from other species. The α subunits range between 32,000 and 36,000 in molecular mass, while the β subunits are generally smaller—ranging between 28,000 and 32,000. Although both subunits are required for capping activity, actin binding has only been ascribed to the β subunit. Puzzlingly, it is a region close to the carboxy-terminus that seems to possess actin-binding properties, despite the fact that at this particular part there is no detectable sequence homology between this and other species! (Hug et al., 1992). Homology between the β subunit and gCap39/gelsolin (other capping proteins) has been detected (Yu et al., 1990) but the significance of these observations is not at all certain since the sequences from the α and β subunits, which are allegedly akin to sequences in the well studied protein gelsolin, are known not to constitute an actin binding site (McLaughlin et al., 1993). Another possibility is that the similarities arise from the fact that both classes of protein bind phosphatidylinositol bisphosphate (see below), since the region of homology also corresponds to the lipid binding site (Yu et al., 1992; Janmey et al., 1992).

Several studies in a variety of cell types have revealed the intracellular localization of this group of capping proteins although such studies have not always

Table 3. Proteins Which Cap But Do Not Sever Filaments

	Source	Mol. wt. kDa	Nucleating Activity	Ca²⁺ Dependent	PIP₂ Regulated	References
Barbed end capping proteins						
Monomeric capping						
gCap39/Mbh1/MCP	Vertebrate	39	Yes (weak)	Yes	Yes	Southwick & DiNubile, 1986; Prendergast & Ziff, 1991.
Radixin	Vertebrate	82	No	No	?	Funayama et al., 1991.
Sea-Urchin capping protein	Echinoderm	20	?	No	?	Ishidate & Mabuchi, 1988.
Tensin	Vertebrate	165	Yes	No	Yes	Davis et al., 1991.
Cap100 (protovillin)	*Dictyostelium*	100	No	Yes	Yes	Hofmann et al., 1992; Hofmann et al., 1993.
Heterodimeric capping						
Cap Z	Vert. Muscle	36α/32β	indirectly	No	Yes	Casella et al., 1986; Heiss & Cooper, 1991.
	Dictyostelium	32α/34β	indirectly	No	Yes	Schleicher et al., 1984; Hartmann et al., 1989.
	Sacchromyces	32α/34β	indirectly	?	Yes	Amatruda & Cooper, 1992; Amatruda et al., 1990.

	Acanthamoeba	31/28	indirectly	No	?	
Miscellaneous						
Aginactin^	*Dictyostelium*	70(&32α/34β)	No	No	?	Isenberg et al., 1980; Cooper & Pollard, 1985.
ADP-ribosylated actin	ubiquitous?	43	No	No	?	Sauterer et al., 1991; Haus et al., 1993.
Cytochalsins	certain fungi	0.5	No	No	?	Wegner & Aktories, 1988.
Pointed end capping proteins						
Acumentin @	Vert. Macrophage	70	No	No	?	Southwick & Hartwig, 1982.
Tropomodulin #	Vert Erythrocyte	40.6	?	?	?	Fowler et al, 1993.
DNAse 1	Vertebrate	35	No	No	?	Podoslki & Steck, 1988.
β-actinin *	Vertebrate	35/37	No	No	Yes	Maruyama et al., 1990.
Phosphoryated actin	Slime mould	43	No	?	?	Howard et al., 1993; Maruta & Isenberg, 1984.

Notes: * - After a long controversy β-actinin is now known to be Cap Z and to cap the barbed end (Maruyama et al., 1990).

@ - Originally described as a pointed end capping protein but thought to be L-plastin (Hartwig & Kwiatkowski, 1991).

- The pointed end capping activity of tropomodulin is enhanced by the presence of tropomyosin.

^ Aginactin is now thought to be formed by HSP70 (Eddy et al., 1993), bound to Cap32/34 (Haus et al., 1993).

produced a consistent picture. Cap Z is localized to the Z-disc (hence the name) where it is thought to bind the barbed ends of the thin filaments. One of the proteins displays a nuclear distribution (Ankenbauer et al., 1989), while chicken Cap Z is concentrated in cell–cell junction complexes of epithelial cells (Schafer et al., 1992). Yeast capping proteins are found at the membrane, in regions rich in actin (Amatruda & Cooper, 1992). Capping protein null mutants and cells which over-express the protein, have less actin cables, and grow more slowly, yet remain viable (Amatruda et al., 1992).

B. Pointed End Capping Proteins

At the moment it seems that only DNase 1 (Podolski & Steck, 1988) and tropomodulin (Fowler et al., 1993) are actin binding proteins that bind the pointed ends: All the other actin filament capping proteins bind and cap the barbed end. The physiological significance of DNase 1 binding to actin is questionable since the two proteins would not normally be expected to meet, but pointed end capping of filaments by tropomodulin may play a role in maintaining (or limiting) the length of thin filaments in muscle, possibly in conjunction with nebulin, to date the largest ABP known (see below Cofilamentous Actin Binding Proteins). The possibility that actin itself may form a pointed end capping protein when modified by phosphorylation is discussed later (see below Actin as an Acting Binding Protein).

C. Nucleating Activity of Capping Proteins

Many studies have found that capping proteins alter the time course of polymerization of actin. This would of course be expected of proteins binding to the ends onto which actin monomer may bind, but some of these proteins are found to greatly increase the rate at which actin spontaneously polymerizes, especially under conditions that do not favor the formation of the rate limiting nuclei. It seems that some capping proteins behave as if they were able to substitute for the trimeric nuclei, thereby seeding polymerization. If such protein caps the barbed end, then the resulting filament will grow from the nonfavored pointed end. If, however, a pointed end capping protein was able to nucleate actin polymerization, the resulting polymerization would be expected to be very fast indeed, since polymerization would ensue at the heavily favored barbed end. At the time of writing, such pointed end capping/nucleating activity has not been described.

Many, but not all, capping proteins have nucleating activity (Table 3). A nucleating protein is able to directly seed actin polymerization by behaving as an actin trimer (the minimum actin nucleus) and therefore, no lag phase should be observed when a putative nucleating protein is present in spontaneous actin polymerization experiments (Gaertner et al., 1989) (Figure 2). Several proteins that do not quite fulfill this strict criteria are still classified as nucleating proteins. Members of the cap Z protein family are probably such proteins. *Acanthamoeba* capping protein

(Isenberg et al., 1980) has been found to accelerate actin polymerization but the data from these experiments fitted a model where some prenuclear actin–ABP complex was stabilized better than by simple nucleation (Cooper & Pollard, 1985). Many other members of the cap Z group (Heiss & Cooper, 1991; Kiliman & Isenberg, 1982) also accelerate actin polymerization with a detectable lag phase like *Acanthamoeba* capping protein; therefore, it is suspected that these also "nucleate" by stabilizing a prenuclear state.

Talin interacts with lipid membranes (Dietrich et al., 1993; Heise et al., 1991) and apparently nucleates actin polymerization (Kaufmann et al., 1992; Goldmann et al., 1992) *in vitro*. The protein is localized to the tip of actin bundles close to the leading edge of moving fibroblasts (Izzard, 1988), and so talin is thought to be a candidate for factors that mediate actin assembly at the leading edges of vertebrate cells (Isenberg & Goldmann, 1992).

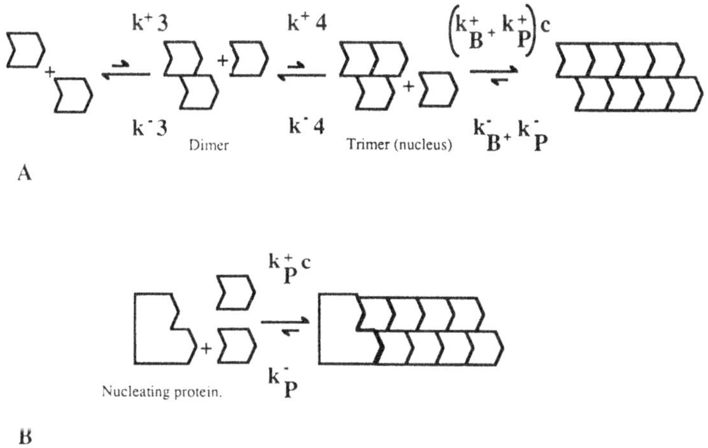

Figure 2. The nucleation of actin polymerization. **A.** Spontaneous polymerization of actin alone is a three step process. The first two steps are unfavorable but result in the formation of a "nucleus" consisting of an actin trimer. Elongation from this nucleus is favorable and is dominated by elongation from the barbed end. K_B rates for the barbed end. K_p rates for the pointed end, c = monomer concentration. (For values for K+3, K–3, K+4, and K–4 see Pollard & Cooper, 1986.) **B.** Nucleated actin polymerization proceeds rapidly as no unfavorable nucleus need be formed. The nucleating protein substitutes for an actin trimer, but elongation proceeds from the pointed end and so the rate of elongation from this nuclei is less than the rate of elongation from a spontaneous nucleus. K_P rates for the pointed end (see Gaertner et al., 1989 for further details of nucleated actin assembly).

IV. BUNDLING PROTEINS

The properties required for a protein to bundle filaments would seem to be few; any protein that binds actin at a site exposed on the filament, and also has an ability to self-associate, should bundle actin filaments. Alternatively, a bundling protein may be formed from a single protein that has two suitable actin-binding domains. Most actin-bundling proteins in nature adopt the first strategy, dimeric bundling proteins of the α-actinin type being particularly abundant. Examples of the second type of monomeric bundling proteins also exist (e.g., fimbrin, 30,000-D bundling protein and ABP-50, see Table 4). There also is the possibility that synapsin I, which is a monomeric bundling protein, may actually have three actin-binding sites (Südhof et al., 1989).

The tendency of filaments to form bundles is strongly dependent on the concentration of the bundling protein and actin. At low concentration many bundling proteins form isometric gels, connecting actin filaments where they cross instead of zipping filaments together laterally. It may be desirable to define a "critical bundling concentration" as being the minimal concentration of bundling protein that is necessary for bundle formation (Wachsstock et al., 1993). This critical bundling concentration is protein and condition specific (Hou et al., 1990; Wachsstock et al., 1993). α-Actinin from *Acanthamoeba* has a high critical bundling concentration (Pollard et al., 1982) while the 30-kD bundling protein from *Dictyostelium* has a very low critical bundling concentration (Brown, 1985). These differences in critical bundling concentration may be due to the length of the bundling protein; a very short protein (such as ABP-50 or Ef-1α from *Dictyostelium*) may perhaps be capable of only very close proximity lateral interaction with filaments, while a longer molecule such as α-actinin may be sufficiently flexible to be able to bind two filaments together in a number of orientations and not only laterally (Figure 3). Some evidence for the importance of molecular length in bundling capacity comes from a comparison of three types of α-actinin (Meyer & Aebi, 1990). However, others have pointed out that this generalization is at odds with some reported observations (Fechheimer & Furukawa, 1993) and that for the α-actinins affinity is a better predictor of bundling activity (Wachsstock et al., 1993).

Other factors are known to influence bundling. *Acanthamoeba* α-actinin forms optically isometric gels at high concentration, but if a needle is placed in the gel and pulled slightly, birefringence is seen to develop indicating that the disruption aligns the filaments allowing bundle formation (Pollard et al., 1982). The bundling activity of *Acanthamoeba* α-actinin is also stimulated by the presence of actophorin (Maciver et al., 1991b), a small filament severing protein also from *Acanthamoeba* (Cooper et al., 1986; Maciver et al., 1991a). Actophorin is thought to sever filaments as they form and the fragmented filaments become decorated with α-actinin, and then zipped together to form small bundles which are enlarged by additional polymerization. The result is that an heterogeneous "gel of bundles" is formed, whereas in the absence of actophorin, an isotropic gel of single cross–linked

Table 4. Actin Filament Bundling Proteins

Protein	Source	Mol. Wt kd	Bundle type	Bundles registered	Bundle orientation	Effect of Ca^{2+} on bunding	Other properties/comments	References
ABP-50/Ef-1α	Dictyostelium	50	very tight, 5 nm	yes	parallel?	none	bundling is pH and perhaps GTP dependent	Demma et al., 1990; Yang et al., 1990
α-actinin	Dictyostelium	95x2	loose	no	antiparallel	negative	comparatively weak bundling activity	Meyer & Aebi, 1990
	Acanthamoeba	95x2	loose	no	parallel	none	moderate bundling activity	Pollard et al., 1986
	Vert. Skel. Muscle	110x2	loose	no	?	none	concentrated at Z disk	Blanchard et al., 1989
	Vert. Smooth Muscle	100x2	loose	no	antiparallel	none	moderate bundling activity	Meyer & Aebi, 1990
	Vert. non muscle	103x2	loose	no	?	negative	cooperative binding saturates at 1:10 actins	Duhaiman & Bamburg, 1984
	Vert. Macrophage	103x2	loose	no	?	none	relationship to other non-muscle isotype unclear.	Pacaud & Harricane, 1993
Adducin	Vert. Erythrocyte	103&97	loose	yes	?	negative	protein kinase C substrate but function unclear	Mische et al., 1987
Dematin (Band 4.9)	Vert. Erythrocyte	48x3	tight	yes	?	none	phosphorylation abolishes bundling, similar to villin headpiece	Siegel & Branton, 1985; Rana et al., 1993
Fascin	Echinoderm	58	tight, 8.3 nm	yes	?	none	Echinoid fascin ia a homolog of Drosophila Singed gene	Kane, 1975; Bryan et al., 1993
Fimbrin (T-plastin)	Vertebrate	71x1	moderate, 11 nm	no	?	none	serine phosphorylated on head piece	Namba et al., 1992
L-plastin	Vert. Leukocytes	70x1	moderate	no	?	negative	serine phosphorylated on head piece	Namba et al., 1992
I-plastin	Vert. Brush border	71	?	?	?	negative	biochemical data lacking	de Arruda et al., 1990

Table 4. (Continued)

Protein	Source	Mol. Wt kd	Bundle type	Bundles registered	Bundle orientation	Effect of Ca^{2+} on bunding	Other properties/comments	References
Synapsin I	Vert. Neuron	74x1	moderate	no	?	none	CaM kinase II abolishes bundling	Bähler & Greengard, 1987
Scruin	Limmulus	102	very tight	yes	parallel	?	composed of six fold "Keltch-like" repeat which binds actin	Way et al., 1993
Villin	Vert. gut	95	moderate, 13 nm	no	?	negative	severs microfilaments in calcium	Bretscher & Weber, 1980
210kDa	Physarum	210x2?	not known	?	?	none	bundling inhibited by Ca^{2+} with calmodulin	Ishikawa et al., 1992
52 kDa	Physarum	52	tight	no	?	none	cross reacts to anti-EF1α, also bundles microtubules	Itano & Hatano, 1991
34-kd/p30a	Dictyostelium	34	tight	slight	?	negative	antibodies cross react widely, a Physarum cDNA is known	Fechheimer, 1987; St-Pierre et al., 1993
p30b	Dictyostelium	30x?	disordered	no	?	none	no sequence available yet	Brown, 1985

14

filaments is formed. It is assumed that α-actinin molecules become effectively trapped within the bundles, thereby greatly decreasing the apparent K_d. Evidence for such a mechanism has recently been provided in the case of gizzard α-actinin by Grazi et al. (1992) who found that α-actinin was enriched in actin filament bundles. Additionally, a computer simulation of actin and α-actinin gels recently predicted a shift from an isotropic to an amorphous phase upon filament shortening (Dufort & Lumsden, 1993).

A. The Geometry of the Filament Bundle

The geometry of packing may also be influenced by the rigidity of the bundling protein. If the protein is very rigid, connections between filaments may be constricted so that filaments are pulled in register; this is known to be the case for ABP-50 (EF-1α), fascin and scruin. The manner in which these bundling ABPs hold filaments in register determines the packing of filaments in bundles. Some ABPs are known to alter the twist in the actin filament (Stokes & DeRosier, 1987), making several packing arrangements possible. Fascin produces a hexagonal packing structure while ABP50 (EF-1α) produces a unique square packing structure (Owen et al., 1993). The physical size of the actin bundling protein also influences the packing of filaments in cells (Matsudaira et al., 1983), larger proteins being packed less closely than smaller ABPs. According to a model by Meyer and Aebi (1990), individual bundling proteins can produce bundles of differing tightness depending on the orientation of the ABP to the filament (Figure 3).

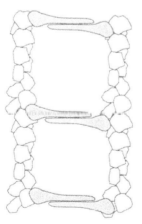 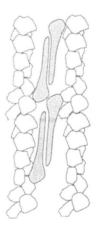

Figure 3. A model to explain the observed differences in packing of actin filaments bundled by α-actinin (after Meyer & Aebi, 1990). Many have observed that cross-sections of bundles with α-actinin show a range of filament-filament distance often within a single bundle (Maciver et al., 1991b). The figure shows the possible extremes, but it is thought that intermediate cases may exist.

Bundling proteins not only differ in the packing of the bundles that they produce, but also in the polarity of the actin filaments within the bundle. As actin filaments are often unipolar in the cell cortex immediately under the membrane, having their barbed ends closest to the plasma membrane, bundles at the perimeter of cells may be expected to have this orientation as well. However, bundles formed *in vitro* are also seen to be of uniform polarity in a number of bundling proteins. α-Actinins vary in their tendency to form polar bundles depending on their source (Meyer & Aebi, 1990) (Table 4).

B.   Some Consequences of Microfilament Bundling

The bundling of actin filaments produces some counterintuitive results in terms of the physical properties of actin solutions. Typically, it is found that with a low ratio of bundling protein to actin, viscosity increases due to cross-linking of filaments. At higher concentrations, where bundling becomes more prevalent, the viscosity diminishes as the actin is drawn into isolated bundles (Brown, 1985; Fechheimer & Furukawa, 1993). When actin is polymerized in the presence of α-actinin and actophorin (a filament severing protein), a meshwork of interconnecting bundles results that is very rigid (Maciver et al., 1991b). Recently it has been found that actin filaments in bundles depolymerize at a lower rate than would be the case for the same concentration of isolated filaments (Zigmond et al., 1992). This may be because the concentration of diffusing monomer may be sufficiently high to inhibit monomer loss. Similarly, the bundling protein dematin (band 4.9) is known to decrease the rate of polymerization (Siegel & Branton, 1985).

C.   Microfilament Bundles in Cells

Actin microfilaments exist in a variety of different structures within the same cell at the same time. The hair cell of the vertebrate inner ear is a particular example, where three actin filament containing structures exist, each with a distinct set of actin-binding proteins (Drenckhahn et al., 1991). This actin organization is greatest in the stress fibers found in most cultured cells, where the actin filaments form a miniature sarcomere. These bundles of actin filaments are held in register with myosin II filaments by a variety of bundling proteins. Large, tightly organized filament bundles are found around the nucleus where they form the "cellular geodome." Loose bundles known as arcs form at the base of the ruffling lamellipo-dium and move back toward the perinuclear region where they collapse and the filaments disassemble.

Bundles of actin filaments occur in cells, and several ABPs are known to bundle actin filaments *in vitro*. Are cellular actin bundles brought about by the actions of these same proteins? A strong argument in support of this hypothesis is that microvillar structures consisting of actin bundles are produced by the transfection of fibroblasts with villin cDNAs (Friedrich et al., 1989). Other evidence that suggests bundling ABPs are responsible for intracellular bundles comes from numerous studies in which these ABPs colocalize with bundles within cells. Examples include the 30-kDa protein in *Dictyostelium* (Johns et al., 1988; Fechheimer, 1987), scruin in the acrosomal process and fimbrin and villin in the microvillus.

V. PERPENDICULARLY CROSS-LINKING PROTEINS

Filamin (also known as ABP) has been described as a "molecular spring leaf" because of its elongated shape (Gorlin et al., 1990). It has been pointed out that, whereas rod-like molecules would tend to align in parallel, thus stabilizing actin bundles, filamin is "V" shaped and apparently flexible. The molecule is reluctant to form bundles even at very high concentration (Neiderman et al., 1986) as it cross-links filaments in a perpendicular manner (Figure 3), and this results in very rigid gels (Janmey et al., 1990); this is probably why cells lacking this protein have unstable cortexes (Cunningham et al., 1992). Despite immunological and hydrodynamic similarities within the filamin group, it is known that filamin from chicken gizzard does have bundling activity (Brotschi et al., 1978), and that this is stimulated by linear shear (Cortese & Freiden, 1988) in a manner similar to α-actinin (Pollard et al., 1982). The filamin dimer binds actin by an N-terminal domain homologous to other ABPs, but the ability of this protein to cross-link filaments perpendicularly may arise from weak actin filament interactions along the rod domain that are likely to stabilize both the rod domain and the microfilaments (Figure 4). In this configuration perpendicular interactions may be more favorable than parallel interactions. It has also been suggested (Barry et al., 1993) that the difference in bundling activity between isoforms of filamin may be due to two insertions in the sequence of the epithelial isoform that produce hinge regions in the rod domain. These hinge regions are postulated to facilitate orthogonal network formation rather than bundling. Filamin has recently been identified as "gyronemin," an intermediate filament associated protein (Brown & Binder, 1992).

It is possible that an ABP may only form nonparallel filament interactions by virtue of its inflexibility, having two actin binding domains presented in a perpendicular manner like a boss and clamp arrangement. Transgelin may be a candidate for this type of ABP (Figure 5). This protein is monomeric and small enough at 21 kDa to be rather inflexible. Transgelin aggregates actin filaments into dense meshworks, without any indication of bundle formation, with a maximum stoichiometry of 1 transgelin to 6 actins (Shapland et al., 1993).

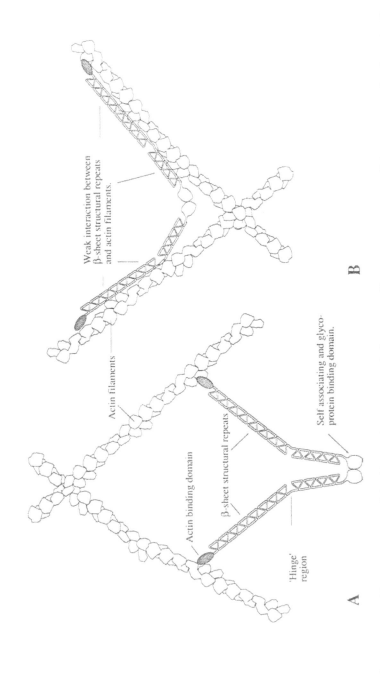

Figure 4. Possible arrangement of the filamin dimer in cross-linking microfilaments in a non-parallel manner. **A.** The dimer itself may be sufficiently inflexible for the two actin binding domains to be held at about 90° with respect to each other. **B.** Additional rigidity of the β-sheet rod like domain by weak interaction with the fulminate. In both cases it is expected that the hinge regions would influence the geometry of the filament–filamin interaction.

Actin filaments

Weak interaction between β-sheet structural repeats and actin filaments.

Actin binding domain

β-sheet structural repeats

Self associating and glyco-protein binding domain.

'Hinge' region

A

B

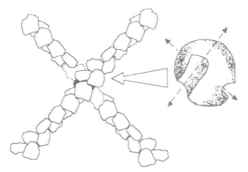

Figure 5. The exclusively non-parallel interaction between transgelin and actin filaments. Transgelin being comparatively small, may bind two filaments with F-actin sites on opposite sides and in an orientation about 90° different. This arrangement is comparable with the observations of Shapland et al. (1993).

VI. ACTIN MONOMER BINDING PROTEINS

There is a well-documented abundance of nonfilamentous actin in cells, despite the intracellular conditions greatly favoring the formation of F-actin. A number of low molecular weight actin binding proteins have been identified that are thought to bind G-actin and thus directly sequester monomers from the filamentous pool (Table 5). This pool of monomers is possibly drawn upon to support polymerization when cells are stimulated, for example, by chemoattractants. Profilin was considered to be the principle monomer sequestering agent, but this was questioned because of its relatively low affinity for actin. Indeed some work indicates that the actin released by profilin is insufficient to account for the additional polymerization of actin seen in stimulated neutrophils (Southwick & Young, 1990). More recent work suggests that monomer sequestration by profilin is not as simple as it had first appeared (Pantaloni & Carlier, 1993) and that profilin may actually promote actin polymerization in the presence of thymosin β4 (Theriot & Mitchison, 1993). Thymosinβ4 is a prime candidate for sequestering activity because it is an abundant, low molecular weight peptide/protein present in a variety of cell types (Safer et al., 1990; Weeds & Way, 1991; Sanders et al., 1992; Nachmias et al., 1993). The ADF/cofilin group also bind actin monomers, but their physiological role appears not to be sequester actin, and the bound actin may be a factor in controlling severing function since the bound form is unable to sever filaments.

In addition to filament capping and limitation of the G-actin pool by actin monomer sequestration, a number of other mechanisms may be (theoretically) employed to reduce the steady state level of actin polymerization. Actobindin from *Acanthamoeba castellanii* is a 9.8 kDa protein that has two actin binding sites (homologous to thymosin) and that is believed to catalyze the formation of an

Table 5. Actin Monomer Binding Proteins

Protein	Source	Mol. Wt (kDa)	K_d (μM)	Actin bound nucleotide preference	References
Profilin	Eukaryotes	11–14	1–5	ADP	Lal & Korn, 1985; Goldschmidt-Clermont et al., 1991.
Asp-56	Platelet	56	0.1	?	Gieselman and Mann, 1992.
Thymosin β4	Vertebrates	5	0.6	ATP	Safer, 1992; Carlier, 1993.
ADF/Cofilin	Vertebrates	19–21	0.1	ATP	Hayden et al., 1993.
Actophorin	Acanthamoeba	15	0.2	ADP	Cooper et al., 1986; Maciver et al., 1991a; Maciver & Weeds, 1994.
DNase1	Vertebrate	31	0.05	?	Mannhertz et al., 1980.

actin–actin dimer whose conformation is not compatible with the formation of a nucleus for actin polymerization (Bubb & Korn, 1993). Presently it is not clear if this "dimer" bears any relation to the "lower dimer" seen by cross-linking experiments (Millonig et al., 1988), but this "side reaction" mechanism may explain the large amounts of actin "dimer" found in polymerizing actin (Matsudaira et al., 1987). It also remains to be seen if this mechanism of actin polymerization inhibition is unique to actobindin and, indeed, if actobindin is found in cells other than *Acanthamoeba*. Hsp27, or 25-kDa IAP (Miron et al., 1991), is a heat shock inducible protein that inhibits the polymerization of actin by an uncertain mechanism. This protein rapidly decreases the viscosity of F-actin solutions (like the ADF/cofilin group), but is not an actin monomer sequestering protein since it does not cause an increase in the lag phase of spontaneous actin polymerization (Geiger, B. in Kreis & Vale, 1993). This makes the action of the protein distinct from that of either the ADF/cofilin group or the thymosins. Hsp27 apparently confers resistance to heat shock and stabilizes the microfilament system organization (Lavoie et al., 1993) and also is an ATP-independent chaperone (Jakob et al., 1993).

Several ABPs that reversibly sequester actin are also known to have other cellular functions. The ADF/cofilin group bind and sever filaments (as previously discussed) and in addition, may have a role in the heat shock response (see below Nuclear Actin Binding Proteins). Profilin is a small, loosely conserved protein, but most profilin variants so far examined have to greater or lesser degrees the ability to bind actin, poly–1–proline and polyphosphoinositides (PPI) (Machesky & Pollard, 1993). Profilin binding to actin has already been discussed but the binding to phosphatidylinositol 4,5-bisphosphate (PIP$_2$) and polyproline has implications in cell signal transduction. The PPIs are well known second messengers of diverse activating pathways. Generally, a primary event such as ligand occupancy on the cell surface leads to the cleavage of PIP$_2$ by phospholipase C γ (PLCγ), producing inositoltrisphosphate (IP_3) and diacylglycerol. IP_3 liberates Ca^{2+} from intracellular stores, and diacylglycerol activates isoforms of protein kinase C, both signals that in turn activate cells. The mechanism whereby PLCγ becomes activated has been illusive, the enzyme becomes phosphorylated on tyrosine in response to EGF, but this has no effect on the hydrolysis of PIP_2. Profilin protects PIP_2 against hydrolysis by the unphosphorylated PLCγ but tyrosine phosphorylated PLCγ is not inhibited by profilin, thus profilin may be a cellular off switch for such signals (Goldschmidt-Clermont et al., 1990). Details at this stage are still a little sketchy, but another hint as to a role for profilin in signaling is that overexpression of profilin compensates for a null mutant of yeast adenylate cyclase associated protein (CAP). Intriguingly, another actin sequestering protein, ASP56, is homologous to yeast CAP (Gieselmann & Mann, 1992), and the EGF receptor also binds actin via a region similar to the profilin actin binding motif (de Hartigh et al., 1992).

The significance of profilin binding to polyproline is uncertain, yet this interaction is perhaps similar to the Src Homology (SH3) system. The SH3 domain is present on many proteins involved in cell activation and some ABPs. This region recognizes a proline rich domain in other proteins (Cicchetti et al., 1992), and it is of course tempting to speculate that profilin binds polyproline via an SH3 domain. In fact a weak SH3 homology has been reported, but the polyproline binding domain maps outside this region (Björkegren et al., 1993). The fact that both profilin (Buss and Jockusch, 1989) and SH3 domain of spectrin become concentrated at the leading edge of fibroblast (Meriläinen et al., 1993) may not be coincidental!

VII. COFILAMENTOUS ACTIN-BINDING PROTEINS

This group of ABPs are largely responsible for controlling mechanisms in the interaction of microfilaments with other ABPs, most notably myosins. The most widely studied case is vertebrate muscle where tropomyosin and its associated proteins modulate interactions with myosin II in striated muscle, and caldesmon, in association with another protein set, do likewise in smooth muscle. Tropomyosin, caldesmon and some of their associated proteins are represented by homologous proteins in nonmuscle cells. Other proteins are included in the category of cofilamentous proteins because they bind the microfilament without other discernable actin-binding functions such as capping, severing bundling or gelating (Table 6).

Tropomyosins are elongated proteins with a high α-helical content, consisting of a variable number of sequence repeats. Two major forms exist, the larger form (~33 kDa) found predominantly in muscle spans 7 actins per dimer in the filament, and the smaller (~29 kDa) form from nonmuscle cells spans 6 actins in the filament. In humans, these isoforms are encoded by four genes and further diversity arises from alternative splicing. Tropomyosin exists as a parallel dimer of two identical or nonidentical molecules. Each dimer binds head to tail with other dimers along the

Table 6. Cofilamentous ABPs

ABP	Mol. Wt kD	Ca²⁺/calmodulin inhibited	Phosphorylation inhibited	pH sensitivity	Number of actin subunits spanned/ABP	References
ADF/Cofilin	~20	no	yes	yes	1	Hawkins et al., 1993; Hayden et al., 1993.
Caldesmon	~87	yes*	yes	?	14/28?	Matsumura & Yamashiro, 1993.
Calponin	32	yes*	yes	?	3	Winder & Walsh, 1993.
Coactosin	17	?	?	no	1	de Hostos et al., 1993a.
Coronin	55	?	?	?	?	de Hostos et al., 1991.
HSP 90	90	yes	?	?	?	Nishida et al., 1986.
HSP 100	100	yes	?	?	10	Koyasu et al., 1989.
Hisactophilin	17	?	?	yes	1	Scheel et al., 1989.
Nebulin	~750	possibly	?	?	100-200	Chen et al., 1993.
Tropomyosins						
nonmuscle	~33	no	no	no	6	Smillie in Kreis & Vale , 1993
muscle	~29	no	no	no	7	"

Note: *This may not be of physiological significance because of the relatively high concentrations required (Winder et al., 1993b).

filament length. The contraction of striated muscle, although far from fully understood, is thought to be controlled by calcium influx acting on a complex of troponins in cooperation with tropomyosin (for full details see the article by Thornell in Volume III of this series). In nonmuscle cells tropomyosins may have other functions in addition to modulating actin–myosin interactions. Tropomyosins inhibit severing of actin filaments by gelsolin (Ishikawa et al., 1989), perhaps by direct binding competition or conceivably by a direct interaction between gelsolin and tropomyosin (Koepf & Burtnick, 1992). The larger tropomyosin isotypes are more effective in gelsolin inhibition than the small isotypes. A sequence motif DAIKKKL is shared between tropomyosins and the ADF/cofilin group, which may explain why ADF competes with tropomyosin for filament binding and severing by ADF (Bernstein & Bamburg, 1982) and actophorin (Maciver et al., 1991a).

Caldesmon (Matsumara & Yamashiro, 1993), a smooth muscle protein that is also found in nonmuscle cells, shares many properties with tropomyosin and both have a generally high alpha-helical content and an elongated shape; however, there seems very little sequence similarity between the proteins (Leszyk et al., 1989). Like tropomyosin, caldesmon exists as a large form in muscle, and as a smaller form in nonmuscle cells. In the latter tissues it is thought that caldesmon may regulate acto–myosin interactions in a calcium sensitive manner (Walsh, 1993). Caldesmon enhances the binding of tropomyosin to actin filament, probably due to a specific interaction between the two. In nonmuscle cells, caldesmon may regulate the actin-binding function of a variety of other ABPs. Microfilaments decorated with caldesmon and tropomyosin are protected against the severing and capping activities of gelsolin (Ishikawa et al., 1989), and phosphorylation releases caldesmon from filaments at mitosis (Yamashiro et al., 1990) so it is supposed that caldesmon may regulate "rounding up" of cells at this time. In addition, it has been shown that caldesmon can to some extent overcome actin monomer sequestration by profilin to permit polymerization (Galazkiewicz et al., 1991). Caldesmon was thought to be a bundling protein, but it is now known that this activity is an artifact of disulfide cross-linking. Calponin (Winder & Walsh, 1993) is a 32 kDa monomeric protein restricted to smooth muscle tissues but related forms (e.g., transgelin, see perpendicular binding ABPs) of this protein also exist in nonmuscle cells. In addition to binding actin via a central segment of the molecule (Table 7), calponin also interacts with tropomyosin and Ca^{2+}/calmodulin. Calponin inhibits the cycling rate of the actin–myosin interaction and this is inhibited by calponin phosphorylation.

In addition to the above, a number of other ABPs have been characterized that bind F-actin but do not alter the viscosity of F-actin solutions. It is possible that these proteins also perform regulatory functions with respect to other ABP–actin interactions. Coronin is a 55-kDa ABP from *Dictyostelium* that is concentrated in actin-rich extensions and has homology to the β-subunit of the trimeric G-proteins (de Hostos et al., 1991); it is possible that coronin is a homologue of a 55-kDa ABP with similar properties previously isolated from *Acanthamoeba* (Ueno & Korn,

Table 7. Interactions of Proteins at the Focal Contact

	α-actinin	talin	tensin	vinculin	paxillin	β-integrin	zyxin	actin
α-actinin	●	O 8		● 1		● 5	● 4	●
talin	O 8	● 9		● 3, 7		● 2		● 11
tensin				● 12				● 12
vinculin	● 1	● 3	● 12	● 10	● 6			●
paxillin				● 6				
β-integrin	●	● 2						
zyxin	● 4						O 4	
actin	●	● 11	● 12	●				●

● Interaction known   O evidence against interaction.

Notes: References
[1]Wachsstock et al., 1987.
[2]Horwitz et al., 1986.
[3]Burridge & Mangeat, 1984.
[4]Crawford et al., 1992.
[5]Otey et al., 1990.
[6]Turner et al., 1990.
[7]Gilmore et al., 1993.
[8]Muguruma et al., 1992.
[9]Goldmann et al., 1992.
[10]Molony & Burridge, 1985.
[11]Kaufmann et al., 1991.
[12]Lin (in Kreis & Vale, 1993.)

1986). The significance of this observation is as yet unknown but a role in signal transduction to the cytoskeleton is an exciting possibility. Coactosin (also from *Dictyostelium*; de Hostos et al., 1993a) is a 17-kDa protein that has homology to the cofilin group but does not itself sever microfilaments. This similarity in the proteins may mean that they compete for the same binding site on the filament but a cofilin- like activity has not yet been reported in this organism.

VIII.   MEMBRANE-ASSOCIATED ACTIN-BINDING PROTEINS

A large amount of actin is seen tightly associated with the cell membrane by electron and fluorescence microscopy. The leading edge of tissue cells is also the site of actin polymerization (Wang, 1985). Much effort has been expended on the elucidation of the mechanisms of protein involvement leading to this membrane-associated, agonist-stimulated actin polymerization. A number of actin-binding proteins are known to be associated or controlled by components of the plasma membrane, which is especially enriched in actin filaments organized into the "cell cortex" (Luna & Hitt, 1992). ABPs are known to be associated with the plasma membrane by three distinct mechanisms: (1) Integral membrane proteins may span the membrane, be exposed at the cell surface and bind actin with the cytoplasmic face; (2) ABPs may specifically interact with other integral membrane proteins; (3) ABPs may bind lipid components of the cytoplasmic surface of the membrane. Additional information on membrane-associated ABP can be found in this volume in articles by Niggli and Isenberg.

The first group, integral membrane ABPs, is relatively small and includes ponticulin, actolinkin, the epidermal growth factor (EGF) receptor, and lymphocyte-specific protein (LSP1). Ponticulin is an abundant ABP (~1% total membrane protein or 1 million copies per cell) (Chia et al., 1993) from *Dictyostelium*, where it is responsible for the majority of the actin–membrane linkage (Hitt & Luna, 1993). Ponticulin (Wuestehube & Luna, 1987) has a reported nucleating activity (Chia et al., 1993), but because a lag phase was found to exist in these actin polymerization experiments, it is probable that the nucleus is formed only after low probability events involving the ponticulin molecule and actin monomers; a lateral filament has been suggested since monomers can exchange with both ends of filaments seeded by ponticulin. *Dictyostelium* mutants that lack the protein have much less actin associated with the membrane, yet are capable of cell locomotion (Hitt & Luna, 1993). Actolinkin is a similar protein from echinoderms but caps actin filaments (Mabuchi in Kreis & Vale, 1993). LSP1 is thought to mediate a physical connection between the actin cytoskeleton and mIgM on the surface of B lymphocytes. The cytoplasmic tail of LSP1 binds to F- but not G-actin, and does not decrease the lag time of spontaneous polymerization indicating that it does not nucleate or cap filaments. These properties are in accordance with the finding that this tail region is homologous to the actin-binding region of caldesmon (Jongstra-Bilen et al., 1992).

The second group, those ABPs that associate with other membranous proteins, includes the band 4.1 family, filamin, dystrophin, spectrin and catenin. A family of proteins related to band 4.1 has recently come to light. Band 4.1 itself binds glycophorin and band 3 binds to the spectrin–actin complex within the erythrocyte membranous cytoskeleton, and it is tempting to speculate that other members of the group perform similar functions, perhaps with distinct membrane protein specificity. Band 4.1 homologues, which have weak homology to talin (a capping protein) include ezrin (cytovillin) (Hanzel et al., 1991), radixin (a known filament capping protein), moesin and merlin. These proteins are found in the actin-rich cortex of many cell types in association with microvillar structures (Algrain et al., 1993; Lankes & Furthmayr, 1991). Some of the proteins show direct actin binding, but others are thought to interact with other ABPs. Tyrosine phosphorylation of these proteins has been correlated with malignant transformation (Faziola et al., 1993). Filamin, dystrophin and utrophin possess glycoprotein binding domains at the C terminus. It is suggested that filamin binds GPIb on platelets (Fox, 1985), the Fc receptor on B-lymphocyte (Ohta & Hartwig, 1991), and a component of an osmotically sensitive ion channel in melanoma cells (Cantiello et al., 1993). A specific isoform of filamin is localized to the focal contact region (Pavalko et al., 1989), possibly as a result of interaction with a focal contact specific protein. Several melanoma cell lines that completely lack filamin are unable to volume regulate when exposed to hypotonic stimulus. Thus, an additional role for some isoforms of filamin may be to form a self-regulating "stretch receptor" in concert with the cell cortex. It is very tempting to speculate that dystrophin together with the dystroglycan complex to which it binds, performs a similar ion transport control function in muscle, since this would fit with the pathology of some aspects of the disease (see Ozawa Volume IV this series). Catenins are thought to mediate interactions between the adhesion molecules E-cadherin/uvomorulin and the actin cytoskeleton (Ozawa et al., 1990).

The last group, those that interact with lipids noncovalently (Isenberg, 1991; see article in this volume by Isenberg), is probably the largest. In addition to the following, all those proteins that bind polyphosphoinositides (see below) may also be considered members of this wide group. The annexins are a group of proteins that bind anionic lipids in a calcium dependent manner. Many functions have been ascribed to this group, for example, an extracellular function such as an anticoagulant activity, but the weight of evidence suggests that this group forms a connection between the plasma membrane and the actin based cytoskeleton. There are eight distinct annexin types in mammals, and possibly several more. An actin binding motif has been identified (Table 8), which is within an annexin repeat and thus present in a recognizable form in many of the group members. Annexin VI has two of these actin binding repeats and this may explain its unique ability (within the annexins) to bundle actin and localize to bundles within the cell. The apparent conserved actin binding motif probably means that actin binding is a general property of the annexins. An annexin-like protein "comitin" from *Dictyostelium* (and a mammalian immuno-analog) are enriched at the golgi apparatus (Weiner et al., 1993). All of these ABPs associate with membranes by noncovalent interactions with lipids, but at least one ABP, MARCKS, is directly and covalently bound to lipid components. MARCKS is an mnenomic for "myristoylated alanine rich C-kinase substrate." The protein is covalently linked to myristic acid close to the N-terminus and thus is membrane attached. A short central alpha-helical portion (which is similar to a region of adducin) contains the four protein kinase C phosphorylated residues, the calmodulin binding site, and the actin binding site (Hartwig et al., 1992). MARCKS colocalizes with talin and vinculin in macrophages, although the mechanism for this association is not yet clear.

Table 8. Actin Binding Motifs in ABPs

ABP	Actin binding motif	Reference
Actobindin/Thymosin β4	LKHAET & LKKTET	Vancompernolle et al., 1991
ABP120	LVGIGAEELVDKNLKMTLGMIWTIILR	Bresnick et al., 1990
Annexin	VLRIMVSR	Jones et al., 1992
Aldolase	ADESTGSIAKRLQSIGTENTE	O'Reilly & Clarke, 1993
CapZ *Dictyostelium* β subunit	NQKYKQLQRELSQVL	Hug et al., 1992
Caldesmon	NLKGAANAEAGSEKLKEKQQEAAVE	Wang et al., 1991
Calponin	AEKQQRRFQPEKLREGRNIIGLQMGTNKFASQQGMTAY	Mezgueldi et al., 1992
Cofilin	DAIKKK	Yonezawa et al., 1989
Cofilin	WAPECAPLKSKM	Yonezawa et al., 1991
Cortactin	GFGGKFGVQxDRVDKSAVGFEYQGKTEKHESQKDYSK	Wu & Parsons, 1993
EGF receptor	DDVVDADEYLIPQ	den Hartigh et al., 1992
Gelsolin	DESGAAAIFTVQLDDY	McLaughlin et al., 1993
HSP27	LLPSESALLPAPGSPYGR	Vandekerckhove & Vancompernolle, 1992
MARCKS	KRFSFKKSKLSGFSFKKN	Hartwig et al., 1992
Myosin S1	IRICRKG	Eto et al., 1991
	YRGKKRQ	Vandekerckhove & Vancompernolle, 1992
	EGGGGKKGGKKKGSSF	"
	LHQLRCNGVLEGIRICRKGEPNRILYGSFLQRYRVL	"
Profilin	GᵂAANVVEKLADYLIGQGF	Vandekerckhove et al., 1989
Villin	PAAFSALPRWKQQNLKKEKGLF	Friederich et al., 1992

27

A. The Focal Contact

The focal contact is a specialized and discrete region of the plasma membrane that forms the closest contact between a cell and the underlying substrate (Burridge et al., 1988). A large number of proteins, some of which also bind actin, is concentrated at the focal contact of cultured vertebrate cells, especially fibroblasts (Table 7), and this complex forms the terminus for stress fibers (see article by Burridge, Volume II in this series). It has recently become apparent that the focal contact contains its own signal transduction machinery based on the phosphorylation of tyrosine (Burridge et al., 1992). Clustering of integrins, which occurs at the focal contact naturally or by "capping" cells with antibodies directed against integrins, produces tyrosine phosphorylation of p125[fak], itself a constituent of the focal contact and a tyrosine kinase (Kornberg et al., 1992). As the integrins themselves are not kinases and no direct connection between them and p125[fak] has been found, it is thought that other molecules may mediate phosphorylation. It has been suggested that tyrosine phosphorylation may be a signal controlling the formation of new focal contacts and thus may be involved in cell spreading (Kornberg et al., 1992).

IX. ACTIN AS AN ACTIN BINDING PROTEIN

In the search for "the actin binding motif," many have assumed that some actin-binding proteins will themselves contain domains that are actinlike (Tellam et al., 1989). Since actin binds itself it was argued that some ABPs, such as capping proteins, may have arisen from duplicate actin genes. Apparent proof for such hypotheses was furnished by Maruta et al. (1984) who reported that cap42(a) was a nonpolymerizable variant of actin. Later studies found that cap42(a) is fragmin and that cap42(b) is actin (Ampe & Vandererckhove, 1987). However, the situation was clarified when it was found that the EGTA-resistant fragmin–actin complex (cap42(a+b)) was phosphorylated (Maruta & Isenberg, 1983) at the pointed end of the actin subunit in the complex (Furuhashi et al., 1992a). Phosphorylation occurs at residues Thr203 and Thr202 (Gettemans et al., 1992) and the resulting complex blocks nucleation and capping (Furuhashi & Hatano, 1990). The actin kinase itself binds tightly enough to copurify with the complex, and so may have pointed end capping activity alone (Furuhashi & Hatano, 1992). *Dictyostelium* also phosphorylates actin but on tyrosine in a stage specific manner (Schweiger et al., 1992). *Dictyostelium* actin phosphorylation has also been associated with shape changes (Howard et al., 1993). It is possible that this phosphorylation of a minor fraction results in a capping activity accounting for the overall actin depolymerization seen in these cells (Carlier, 1993). A number of bacterial toxins are known to ADP-ribosylate actin (Aktories & Wegner, 1989; Ohishi et al., 1990) and to convert them into barbed end capping proteins (Wegner & Aktories, 1988). Only nonskeletal muscle actins are ribosylated on Arg177 by botulinum C2 toxin. In summary, actin becomes modified in a number of situations, which alters its actin-binding function, but no actin binding protein binds actin via actinlike domains. Ironically, a number of actin related proteins have now been discovered, yet none of these are known to bind actin (Clark & Meyer, 1993).

X. CONTROL OF ACTIN BINDING PROTEINS BY THE PHOSPHATIDYLINOSITIDES

In 1985, Lassing and Lindberg reported that PIP_2 bound to the actin binding protein profilin. Soon after, it was reported that the same was true for gelsolin (Janmey & Stossel, 1987). These rather surprising observations were quickly extended to many other ABPs including α-actinin (Fukami et al., 1992), filamin (Furuhashi et al., 1992b) and a number of capping proteins (Table 3).

The fact that ABPs bind polyphosphatidylinositol lipids is now certain. However, there are two difficulties in establishing the physiological significance of the binding. First, PIP_2 is an extremely highly-charged molecule and in solution forms small micelles producing a very high charge density. It is hardly remarkable, therefore, that some proteins display some attraction for these micelles. Related to this is the difficulty in mimicking natural yet biochemically characterized lipid surfaces. Techniques have only recently begun to be developed to form definable membranes. This is not to say that all these interactions are explained totally by nonspecific charge interactions, since several lines of evidence point to the specificity of such interaction with at least some actin binding proteins. The interaction is especially convincing in the case of profilin, members of the gelsolin group, and α-actinin. PIP_2 has been incorporated with other lipids in LUVETS (large unilammelar vesicles made by the extrusion technique), thus mimicking as closely as is currently possible an actual membrane with a definable lipid composition (Machesky et al., 1990).

XI. THE QUESTION OF REDUNDANCY IN ACTIN BINDING PROTEINS

The huge number of ABPs in cells, and the high concentration at which some are present, may mean that an overprescription for actin-binding sites exists in cells, that is, there are more ABPs present than actin to bind them. The identity and quantity of ABPs bound to an actin filament at any given moment is likely to be very subtly controlled by a huge variety of factors including the presence of calcium, pH, the concentration of the protein in question, and other proteins competing for the same site on actin (Pope et al., 1994). Some proteins that bind very weakly or not at all to actin *in vitro* (e.g., transgelin and aldolase) in "physiological" salt concentrations are known to bind actin in cells (Pagliaro &

Taylor, 1988), possibly through interactions with other ABPs. Another possible complication is that some ABPs are known to modulate the angular twist between subunits in the actin filament (Stokes & DeRosier, 1987), thereby altering (albeit subtly) the exposed surface of the actin filament, and concomitantly the type of ABP bound. The fact that there seem to be more ABPs than actin-binding sites has led to the suggestion that an element of redundancy exists, that is, one ABP performs an overlapping function with another.

The slime mould *Dictyostelium discoideum* has proven to be a remarkably fruitful organism for the study of the actin based eukaryotic cytoskeleton. Making use of the fact that this organism is haploid, it is possible to disrupt specific gene expression and (it is hoped) to illuminate the function of the gene product in question. This approach was first used to abolish myosin II expression (DeLozanne & Spudich, 1987; Knecht & Loomis, 1987), which showed (rather unexpectedly) that myosin II was not essential for the locomotion of the cell (cell locomotion was slightly slower than in wild type cells). What was abundantly clear was that myosin II is absolutely required for classical cytokinesis, indicating that myosin II was providing the contractile force for division of the cytoplasm by a "purse string" like mechanism. However surprising these findings were, more was to follow. It was soon evident that not only could cells move without α-actinin, a major gelation protein in *Dictyostelium* (Brier et al., 1983), but such cells could go through the entire cycle of slug and fruiting body formation, apparently normally. A double mutant lacking both α-actinin and ABP-120 was also able to move (Witke et al., 1992)! Not all genes can be individually removed without affecting cell motility. Deletion of coronin (see above Cofilamentous Proteins) from *Dictyostelium* impairs both cytokinesis and cell motility (de Hostos et al., 1993b).

The finding that cells may "do without" certain functional genes is certainly not limited to genes encoding actin binding proteins. Gene "knockout" experiments in transgenic animals lead to the astonishing conclusion that an organism as complex as a mouse may survive apparently asyptomatically without functioning genes for *c-src, TGF β1* and tenascin! (Erickson, 1993). However, taken together, these finding have led to the suggestion that the microfilament system is redundant; that is, the function of one actin-binding protein is duplicated, or at least is partially fulfilled, by another (Bray & Vasiliev, 1989). It is supposed that because the cytoskeleton performs such a vital function, redundancy would avert mutational catastrophe! However, systems such as the duplication of the genome, which is even more vital to the existence of cells, are not redundant. In addition, another fact that seems at odds with the "redundancy" hypothesis is that actin-binding proteins from the amoebae to man show high homology. α-Actinin from *Dictyostelium* (an organism that apparently may safely dispense with the gene) shows ~60% identity to vertebrate α-actinins (Beggs et al., 1992). If the requirements for a bundling protein are as simple as those outlined earlier, it is very surprising that such a degree of homology is necessary. The finding that a cell lacking a particular actin-binding protein retains a locomotory ability or even goes through the normal life cycle may not constitute a realistic test for cellular normality. It may be expected that actin-binding proteins produce a cortex with a flexible rigidity needed in the field but not on the petri dish! Perhaps the middle ground is closer to the truth: Actin-binding proteins may be able to partially fulfill roles that are the normal domain of another due to similar actin-binding function. Natural selection is expected to greatly exaggerate minuscule differences between similar organisms (e.g., *Dictyostelium* amoebae with and without a particular ABP), differences at present perhaps below the resolution of available research tools.

XII. ACTIN BINDING DOMAINS

Many actin-binding proteins share similar amino acid motifs with other actin-binding proteins (Tellam et al., 1989). The similarity in these motifs may have come about coincidently by parallel evolution or, much more likely, because they have arisen from a common ancestor. Not surprisingly, motifs are shared between proteins with similar function; for example, proteins such as bundlers and crosslinkers, which must bind to the sides of filaments, are more likely to have similar domains than actin monomer sequestering proteins. A particular motif in the *Dictyostelium* protein ABP-120 (Condeelis et al., 1982) was first identified as being the actin-binding domain (Bresnick et al., 1990), and it is now known to exist in many other ABPs, including β-spectrin, α-actinin, fimbrin (de Arruda et al., 1990), filamin, adducin, dystrophin and utrophin (Winder et al., 1994). Similar though these actin-binding domains are, it remains likely that the actual proteins bind in physically distinct ways (Lebart et al., 1993) to overlapping but not identical sites on the actin molecule (Figure 6).

Table 8 shows amino acid sequences that have been determined by peptide competition, crystallography and other techniques to constitute actin-binding motifs. In some cases, these, or similar sequences also exist in other ABPs and are thus suspected to perform similar actin-binding functions. The actin-binding region in caldesmon seems to be an especially common motif, found in a variety of other ABPs such as dystrophin, tropomyosin (Lezyk et al., 1989), LSP1 (Jongstra-Bilen et al., 1992), *Dictyostelium* 30 kDa protein (Fechheimer et al., 1991) and myosin light chain kinase (Kanoh et al., 1993).

A. Nuclear Actin Binding Proteins

A number of actin-binding proteins are known to exist in the nucleus, some of which normally reside there and others that become targeted to the nucleus by a stimulation such as heatshock. Actin also is thought to be a normal constituent of nuclei, but fractionation studies are not often conclusive because of the large amount of contaminating cytoplasmic actin. High concentrations of actin have been found in nuclei mechanically isolated from oocytes. It is reported that actin is

Figure 6. The contact sites on the actin molecule for (A). DNase 1 (Kabsch et al., 1990), (B). α-actinin type proteins, (C). Gelsolin segment 1 (McLaughlin et al., 1993) and (D). Profilin (Schutt et al., 1993). The actin molecule is shown as a tube outlining the peptide chain, in the same "standard" orientation (Kabsch et al., 1991). Hydrogen bonding sites on actin are shown as black spheres centered on the Cα positions for all except the α-actinin group, where the black segments highlight regions found by crosslinking studies to be contact sites. The actin model is derived from the gelsolin segment 1 complex is used in B, C and D, and is similar to the DNase 1 complexed molecule except that the DNase 1 binding loop in subdomain 2 is ill-defined and is therefore not shown. (Figure generously provided by Dr. Paul J. McLaughlin, LMB Cambridge.)

required for error-free transcription by RNA polymerase II (Egly et al., 1984) and that the endonuclease activity of poly(A)polymerase/endoribonuclease IV is regulated by F-actin binding (Schröder et al., 1982). However, it is not exactly clear what the function of nuclear actin is or what function nuclear ABPs perform. ADF and cofilin (see above Severing Proteins) possess a putative nuclear localization sequence (NLS), defined by homology with SV40 large T antigen (Matsuzaki et al., 1988). Both proteins associate with actin in the nucleus upon heat shock or treatment with DMSO (Nishida et al., 1987; Ono et al., 1993) and they are found in tight bundles that are perhaps helical. Neither ADF nor cofilin are expected to produce bundles of filaments themselves, however, but the presence of other ABPs in these nuclear bundles has not yet been found. It is known that the bundles are not recognized by antibodies to α-actinin in immunofluorescence studies. The NLS motif (KKRKK) has been shown to be necessary for nuclear targeting by site directed mutagenesis (Iida et al., 1992), and furthermore, by microinjecting BSA conjugated to the motif the sequence has been shown to be sufficient to target the BSA to the nucleus on heat shock (Abe et al., 1993).

Mbh1 is a member of the gelsolin group with a postulated NLS and exhibits nuclear localization and has a basic/helix-loop-helix DNA binding motif (Prendergast & Ziff, 1991). Mbh1 is equivalent to a previously characterized capping protein from mouse gCap39. Mbh1/gCap39 is known to bind F-actin in a calcium sensitive manner and is homologous to gelsolin segments 1 to 3; it is a capping protein but does not sever filaments. gCap39 is phosphorylated and phospho-gCap39 is enriched in the nucleus (Onoda & Yin, 1993a,b). A heterodimeric capping protein (see above) has been identified in the nucleus of *Xenopus laevis* oocytes at concentrations about 12 times that of the cytosol (Ankenbauer et al., 1989). Antibodies to this protein also demonstrate the presence of similar proteins in vertebrate cells that are presumably different from previously identified heterodimeric capping proteins since these are not nuclei specific. A similar protein has been isolated from the nuclei of *Acanthamoeba* that binds both actin and DNA but the relationship between these proteins is presently unknown (Rimm & Pollard, 1989). Cap50, a calcyclin binding annexin, is located in the nucleus, and although this protein has not been shown to bind actin (Mizutani et al., 1992), it possesses a motif very similar to that shown in other annexins to be responsible for actin binding (Jones et al., 1992).

XIII. CONCLUSIONS AND FUTURE RESEARCH PROSPECTS

There is now not much doubt that the cytoskeleton, largely the actin based cytoskeleton, is responsible for the generation and control of forces driving cell locomotion. The above discussion and other chapters in the series should indicate the complexity of the total actin–myosin system in cells. Further details of individual ABPs are of course necessary, but a major challenge is to establish computer-aided models to predict and explain the often complex behavior and properties of microfilaments and multi-ABP complexes *in vitro* and *in vivo*. These models are likely to be both proposed and tested by the results of genetic manipulation studies. ABPs can already be studied by "knock out" technology in *Dictyostelium*, but "gene-targeting technology" is progressing and this will make it possible to disrupt the function of both alleles of a particular ABP in vertebrate cells and in whole vertebrate animals using similar manipulations of germ cells.

Acknowledgments

I thank Drs. Steven J. Winder and Paul J. McLaughlin and Alan G. Weeds for proof reading and many suggestions. This work was supported by the MRC.

REFERENCES

Abe, H., Nagaoka, R., & Obinata, T. (1993). Cytoplasmic localization and nuclear transport of cofilin in cultured myotubes. Exp. Cell Res. 206, 1–10.

Aktories, K. & Wegner, A. (1989). ADP-ribosylation of actin by clostridial toxins. J. Cell Biol. 109, 1385–1387.

Algrain, M., Turunen, O., Vaheri, A., Louvard, D., & Arpin, M. (1993). Ezrin contains cytoskeletal and membrane binding domains accounting for its proposed role as a membrane-cytoskeletal linker. J. Cell Biol. 120, 129–139.

Ampe, C. & Vandekerckhove, J. (1987). The F-actin capping proteins of *Physarum polycephalum*: cap42(a) is very similar, if not identical, to fragmin and is structurally and functionally very homologous to gelsolin; cap42(b) is *Physarum* actin. EMBO J. 6, 4149–4157.

Amatruda, J.F., Cannon, J.F., Tatchell, K., Hug, C., & Cooper, J.A. (1990). Disruption of the actin cytoskeleton in yeast capping protein mutants. Nature 344, 352–354.

Amatruda, J.F. & Cooper, J.A. (1992). Purification, characterization, and immunofluorescence localization of *Saccharomyces cerevisiae* capping protein. J. Cell Biol. 117, 1067–1076.

Amatruda, J.F., Gattermeir, D.J., Karpova T.S., & Cooper, J.A. (1992). Effects of null mutations and overexpression of capping protein on morphogenesis, actin distribution and polarized secretion in yeast. J. Cell Biol. 119, 1151–1162.

André, E., Brink, M., Gerisch, G., Isenberg, G., Noegel, A., Schleicher, M., Segall, J.E., & Wallraf, E. (1989). A *Dictyostelium* mutant deficient in severin, an F-actin fragmenting protein, shows normal motility and chemotaxis. J. Cell Biol. 108, 985–995.

Ankenbauer, T., Kleinschmidt, J.A., Walsh, M.J., Weiner, O.H., & Franke, W.W. (1989). Identification of a widespread nuclear actin binding protein. Nature 342, 822–825.

Arpin, M., Pringault, E., Finidori, J., Garcia, A., Jeltsch, J., Vandekerckhove, J., & Louvard, D. (1988). Sequence of human villin: A large duplicated domain homologous with other actin severing proteins and a unique small carboxy-terminal domain related to villin specificity. J. Cell Biol. 107, 1759–1766.

Bader, M.-F., Trifaró, J.-M., Langley, O.K., Thriersé, D., & Aunis, D. (1986). Secretion cell actin-binding proteins: Identification of a gelsolin-like protein in chromaffin cells. J. Cell Biol. 102, 636–646.

Bähler, M. & Greengard, P. (1987). Synapsin I bundles F-actin in a phosphorylation-dependent manner. Nature, 326, 704–707.

Bamburg, J.R. & Bray, D. (1987). Distribution and cellular localization of actin depolymerizing factor. J. Cell Biol. 105, 2817–2825.

Bamburg, J.R., Harris, H.E., & Weeds, A.G. (1980). Partial purification and characterization of an actin depolymerization factor from brain. FEBS Lett. 121, 178–182.

Baorto, D.M., Mellado, W., & Shelanski, M.L. (1992). Astrocyte process growth induction by actin breakdown. J. Cell Biol. 117, 357–367.

Barry, C., Xie, J., Lemmon, V., & Young, A.P. (1993). Molecular characterization of a multipromotor gene encoding a chicken filamin protein. J. Biol. Chem. 268, 25577–25586.

Beggs, A.H., Byers, T.J., Knoll, J.H.M., Boyce, F.M., Bruns, G.A., & Kunkel, L.M. (1992). Cloning & characterization of two human skeletal muscle α-actinin genes located on chromosomes 1 and 11. J. Biol. Chem. 267, 9281–9288.

Bernstein, B.W. & Bamburg, J.R. (1982). Tropomyosin binding to F-actin protects the F-actin from disassembly by brain actin–depolymerizing factor (ADF). Cell Motility 2, 1–8.

Björkegren, C., Rozycki, M., Schutt, C.E., Lindberg, U., & Karlsson, R. (1993). Mutagenesis of human profilin locates its poly (L) proline binding site to a hydrophobic patch of aromatic amino-acids. FEBS Lett. 333, 123–126.

Blanchard, A., Ohanian, V., & Critchley, D. (1989). The structure and function of α-actinin. J. Muscle Res. Cell Mot. 10, 280–289.

Bray, D. & Vasiliev, J. (1989). Networks from mutants. Nature, 338, 203.

Bresnick, A., Warren, V., & Condeelis, J. (1990). Identification of a short sequence essential for actin binding by *Dictyostelium* ABP-120. J. Biol. Chem. 265, 9236–9240.

Bretscher, A. & Weber, K. (1980). Villin is a major protein of the microvillus cytoskeleton which binds both G and F actin in a calcium-dependent manner. Cell 20, 839–847.

Brier, J., Fechheimer, M., Swanson, J., & Lansing Taylor, D. (1983). Abundance, relative gelation activity, and distribution of the 95,000-dalton actin-binding protein from *Dictyostelium discoideum*. J. Cell Biol. 97, 178–185.

Brotschi, E.A., Hartwig, J.H., & Stossel, T.P. (1978). The gelation of actin by actin-binding protein. J. Biol. Chem. 253, 8988.

Brown, S.S. (1985). A Ca^{2+} insensitive actin-crosslinking protein from *Dictyostelium discoideum*. Cell Motility 5, 529–543.

Brown, K.D. & Binder, L.I. (1992). Identification of the intermediate filament-associated protein gyronemin as filamin. J. Cell Sci. 102, 19–30.

Bryan, J., Edwards, R., Matsudaira, P., Otto, J., & Wulfkuhle, J. (1993). Fascin, an echinoid actin-bundling protein, is a homolog of the Drosophila singed gene product. Proc. Natl. Acad. Sci. USA 90, 9115–9119.

Bubb, M.R. & Korn, E.D. (1993). A catalytic model for the inhibition of actin polymerization by *Acanthamoeba* actobindin. Mol. Biol. Cell 4, 152a.

Burridge, K. & Mangeat, P. (1984). An interaction between vinculin and talin. Nature 308, 744–746.

Burridge, K., Fath, K., Kelly, T., Nuckolls, G., & Turner, C. (1988). Focal adhesions: Transmembrane junctions between the extracellular matrix and cytoskeleton. Ann. Rev. Cell Biol. 4, 487–525.

Burridge, K., Turner, C.E., & Romer, L.H. (1992). Tyrosine phosphorylation of paxillin and pp125[fak] accompanies cell adhesion to extracellular matrix: A role in cytoskeletal assembly. J. Cell Biol. 119, 893–903.

Buss, F., Temm-Grove, C., Henning, S., & Jockusch, B.M. (1989). Distribution of profilin in fibroblasts correlates with the presence of highly dynamic actin filaments. Cell Mot. Cytoskel. 22, 51–61.

Cantiello, H.F., Prat, A.G., Bonventre, J.V., Cunningham, C.C., Hartwig, J.H., & Ausiello, D.A. (1993). Actin-binding protein contributes to cell volume regulatory ion channel activation in melanoma cells. J. Biol. Chem. 268, 4596–4599.

Carlier, M.F. (1993). Dynamic actin. Current Biol. 3, 321–324.

Casella, J.F., Maack, D.J., & Lin, S. (1986). Purification and initial characterization of a protein from skeletal muscle that caps the barbed ends of actin filaments. J. Biol. Chem. 261, 10915–10921.

Chen, M-J.G., Shih, C.L., & Wang, K. (1993). Nebulin as an actin zipper. Biol. Chem. 268, 20327–20334.

Chia, C.P., Shariff, A., Savage, S.A., & Luna, E.J. (1993). The integral membrane protein, ponticulin, acts as a monomer in nucleating actin assembly. J. Cell Biol. 120, 909–922.

Cicchetti, P., Mayer, B.J., Thiel, G., & Baltimore, D. (1992). Identification of a protein that binds to the SH3 region of abl and is similar to bcr and GAP-rho. Science 257, 803–806.

Clark, S.W. & Meyer, D.I. (1993). Long lost cousins of actin. Curr. Biol. 3, 54–55.

Condeelis, J., Geosits, S., & Vahey, M. (1982). Isolation of a new actin-binding protein from Dictyostelium discoideum. Cell Motility 2, 273–285.

Cooper, J.A. & Pollard, T.D. (1985). Effect of capping protein on the kinetics of actin polymerization. Biochemistry 24, 793–799.

Cooper, J.A., Blum, J.D., Williams, Jr, R.C., & Pollard, T.D. (1986). Purification and characterization of actophorin, a new 15,000-dalton actin binding protein from Acanthamoeba castellanii. J. Biol. Chem. 261, 477–485.

Cortese, J.D. & Freiden, C. (1988). Microheterogeneity of actin gels formed under controlled linear shear. J. Cell Biol. 107, 1477–1487.

Crawford, A.W., Michelsen, J.W., & Beckerle, M.C. (1992). An interaction between zyxin and α-actinin. J. Cell Biol. 116, 1381–1393.

Cunningham, C.C., Stossel, T.P., & Kwiatkowski, D.J. (1991). Enhanced motility in NIH 3T3 fibroblasts that overexpress gelsolin. Science 251, 1233–1236.

Cunningham, C.C., Gorlin, J.B., Kwiatkowski, D.J., Hartwig, J.H., Janmey, P.A., Byers, H.R., & Stossel, T.P. (1992). Actin-binding protein requirement for cortical stability and efficient locomotion. Science 256, 325–327.

Davis, S., Lu, M.L., Lo, S.H., Lin, S., Butler, J.A., Drucker, B.J., Roberts, T.M., An, Q., & Chen, L.B. (1991). Presence of an SH2 domain in the actin-binding protein tensin. Science 252, 712–715.

de Arruda, M.V., Watson, S., Lin, C.-S., Leavitt, J., & Matsudaira, P. (1990). Fimbrin is a homologue of the cytoplasmic phosphoprotein plastin and has domains homologous with calmodulin and actin gelation proteins. J. Cell Biol. 111, 1069–1079.

den Hartigh, J.C., van Bergen en Henegouwen, P.M.P., Verkleij, A.J., & Boonstra, J. (1992). The EGF receptor is an actin-binding protein. J. Cell Biol. 119, 349–355.

de Hostos, E.L., Bradtke, B., Lottspeich, F., & Gerisch, G. (1993a). Coactosin, a 17kDa F-actin binding protein from Dictyostelium discoideum. Cell Mot. Cytoskel. 26, 181–191.

de Hostos, E.L., Bradtke, B., Lottspeich, F., Guggenheim, R., & Gerisch, G. (1991). Coronin, an actin binding protein from Dictyostelium discoideum localized to cell surface projections, has sequence similarities to G protein β subunits. EMBO J. 10, 4097–4104.

de Hostos, E.L., Rehfuess, C., Bradtke, B., Waddell, D.R., Albrecht, R., Murphy, J., & Gerisch, G. (1993b). Dictyostelium mutants lacking the cytoskeletal protein coronin are defective in cytokinesis and cell motility. J. Cell Biol. 120, 163–173.

de Lozanne, A. & Spudich, J.A. (1987). Disruption of the Dictyostelium myosin heavy chain gene by homologous recombination. Science 236, 1086–1091.

Demma, M., Warren, V., Hock, R., Dharmawardhane, S., & Condeelis, Y. (1990). Isolation of an abundant 50,000-Dalton actin filament bundling protein from Dictyostelium amoebae. J. Biol. Chem. 265, 2286–2291.

Dietrich, C., Goldmann, W.H., Sackmann, E., & Isenberg, G. (1993). Interaction of NBD-talin with lipid monolayers. FEBS Lett. 324, 37–40.

Drenckhahn, D., Engel, K., Höfer, Merte, C., Tilney, L., & Tilney, M. (1991). Three different actin filament assemblies occur in every hair cell: Each contains a specific actin crosslinking protein. J. Cell Biol. 112, 641–651.

Drubin, D.G., Mulholland, J., Zhu, Z., & Botstein, D. (1990). Homology of a yeast actin-binding protein to signal transduction proteins and myosin-1. Nature 343, 288–290.

Dufort, P.A. & Lumsden, C.J. (1993). Cellular automation model of the actin cytoskeleton. Cell Mot. Cytoskeleton 25, 87–104.

Duhaiman, A.S. & Bamburg, J.R. (1984). Isolation of brain α-actinin. Its characterization and a comparison of its properties with those of muscle α-actinins. Biochemistry 23, 1600–1608.

Eddy, R.J., Sauterer, R.A., & Condeelis, J.S. (1993). Aginactin, an agonist-regulated F-actin capping activity is associated with an Hsc70 in *Dictyostelium*. J. Biol. Chem. 268, 23267–23274.

Egly, J.M., Miyamoto, N., Moncollin, V., & Chambon, P. (1984). Is actin a transcription initiation factor for RNA polymerase B? EMBO J. 3, 2363–2371.

Erickson, H.P. (1993). Gene knockouts of *c-src*, transforming growth factor b1, and tenascin suggest suferflous, nonfunctional expression of proteins. J. Cell Biol. 120, 1079–1081.

Eto, M., Morita, F., Nishi, N., Tokura, S., Ito, T., & Takahashi, K. (1991). Actin polymerisation promoted by a heptapeptide, an analog of the actin-binding S site on myosin head. J. Biol. Chem. 266, 18233–18236.

Faziola, F., Wong, W.T., Ullrich, S.J., Sakaguchi, K., Appella, E., & Paolo di Fiore, P. (1993). The ezrin-like family of tyrosine kinase substrates: Receptor-specific pattern of tyrosine phosphorylation and relationship to malignant transformation. Oncogene 8, 1335–1345.

Fechheimer, M. (1987). The *Dictyostelium discoideum* 30,000-Dalton protein is an actin filament-bundling protein that is selectively present in filopodia. J. Cell Biol. 104, 1539–1551.

Fechheimer, M., Murdock, D., Carney, M., & Glover, C.V.C. (1991). Isolation and sequencing of cDNA clones encoding the *Dictyostelium discoideum* 30,000-dalton actin-bundling protein. J. Biol. Chem. 266, 2883–2889.

Fechheimer, M. & Furukawa, R. (1993). A 27,000-D core of the *Dictyostelium* 34,000-D protein retains Ca^{2+}-regulated actin cross-linking but lacks bundling activity. J. Cell Biol. 120, 1169–1176.

Fowler, V.M., Sussman, M.A., Miller, P.G., Flucher, B.E., & Daniels, M.P. (1993). Tropomodulin is associated with the free (pointed) ends of the thin filaments in rat skeletal muscle. J. Cell Biol. 120, 411–420.

Fox, J.E.B. (1985). Identification of actin binding protein as the protein linking the membrane skeleton to glycoproteins on platelet plasma membranes. J. Biol. Chem. 260, 11970–11977.

Friederich, E., Huet, C., Arpin, M., & Louvard, D. (1989). Villin induces microvilli growth and actin distribution in transfected fibroblasts. Cell 59, 461–475.

Friederich, E., Vancampernolle, K., Huet, C., Goethals, M., Finidori, Y., Vandekerckhove, J., & Louvard, D. (1992). An actin-binding site containing a conserved motif of charged amino acid residues is essential for the morphogenic effect of villin. Cell. 70, 81–92.

Fukami, K., Furuhashi, K., Inagaki, M., Endo, T., Hatano, S., & Takenawa, T. (1992). Requirement of phosphatidylinsoitol 4,5-bisphosphate for α-actinin function. Nature 359, 150–152.

Funayama, N., Nagafuchi, A., Sato, N., Tsukita, S., & Tsukita, S. (1991). Radixin is a novel member of the band 4.1 family. J. Cell Biol. 115, 1039–1048.

Furuhashi, K., Hatano, S., Ando, S., Nishizawa, K., & Inagaki, M. (1992a). Phosphorylation by actin kinase of the pointed end domain on the actin molecule. J. Biol. Chem. 267, 9326-9330.

Furuhasi, K. & Hatano, S. (1990). Control of actin filament length by phosphorylation of fragmin-actin complex. J. Cell Biol. 111, 1081–1087.

Furuhashi, K. & Hatano, S.(1992) Identification of actin kinase activity in purified fragmin-actin complex. FEBS Lett. 310, 34–36.

Furuhashi, K., Inagaki, M., Hatano, S., Fukami, K., & Takenawa, T. (1992b). Inositol phospholipid-induced suppression of F-Actin-gelating activity of smooth muscle filamin. Biochem. Biophys. Res. Comm. 184, 1261–1265.

Gaertner, A., Ruhnau, K., Schröer, E., Selve, N., Wanger, M., & Wegner, A. (1989). Probing nucleation, cutting and capping of actin filaments. J. Muscle Res. Cell Mot. 10, 1–9.

Galazkiewicz, B., Buss, F., Jochusch, B.M., & Dabrowska, R. (1991). Caldesmon-induced polymerization of actin from profilactin. Eur. J. Biochem. 195, 543–547.

Gettemans, J., De Ville, Y., Vandekerckhove, J., & Waelkens, E. (1992). *Physarum* actin is phosphorylated as the actin-fragmin complex at residues Thr203 and Thr202 by a specific 80 kDa kinase. EMBO J. 11, 3185–3191.

Gieselmann, R. & Mann, K. (1992). Asp-56, a new actin sequestering protein from pig platelets with homology to CAP, an adenylate cuclase-associated protein from yeast. FEBS Lett. 298, 149–153.

Gilmore, A.P., Wood, C., Ohanian, V., Jackson, P., Patel, B., Rees, D.J.G., Hynes, R.O., & Critchley, D.R. (1993). The cytoskeletal protein talin contains at least two distinct vinculin binding domains. J. Cell Biol. 122, 337–347.

Goldmann, W.H., Niggli, V., Kaufman, S., & Isenberg, G. (1992). Probing actin and liposome interactions of talin and talin-vinculin complexes: A kinetic, thermodynamic and lipid labeling study. Biochemistry 31, 7665–7671.

Goldschmidt-Clermont, P.J., Machesky, L., Baldassare, J., & Pollard, T.D. (1990). The actin binding protein profilin binds to PIP$_2$ and inhibits its hydrolysis by phospholipase C. Science 247, 1575–1578.

Goldschmidt-Clermont, P.J., Machesky, L., Doberstein, S.K., & Pollard, T.D. (1991). Mechanism of the interaction of human platelet profilin with actin. J. Cell Biol. 113, 1081–1089.

Gorlin, J., Yamin, R., Egan, S., Stewart, M., Stossel, T., Kwiatokowski, D., & Hartwig, J. (1990). Human endothial actin-binding protein (ABP-280, non-muscle filamin): A molecular spring leaf. J. Cell Biol. 111, 1089–1105.

Grazi, E., Cuneo, P., Magri, E., & Schwienbacher, C. (1992). Preferential binding of α-actinin to actin bundles. FEBS Lett. 314, 348–350.

Hartmann, H., Noegel, A.A., Eckerkorn, C., Rapp, S., & Schleicher, M. (1989). Ca^{2+}-independent F–actin capping proteins. J. Biol. Chem. 264, 12639–12647.

Hartwig, J.H. & Kwiatkowski, D.J. (1991). Actin-binding proteins. Curr. Opp. Cell Biol. 3, 87–97.

Hartwig, J.H., Thelen, M., Rosen, A., Jamney, P.A., Nairn, A.C., & Aderem, A.A. (1992). MARCKS is an actin filament crosslinking protein regulated by protein kinase C and calcium-calmodulin. Nature 356, 618–622.

Haus, U., Trommler, P., Fisher, P.R., Hartmann, H., Lottspeich, F., Noegel, A.A., & Schleicher, M. (1993). The heat shock cognate protein from Dictyostelium affects actin polymerization through interaction with the actin-binding protein cap32/34. EMBO J. 12, 3763–3771.

Hanzel, D., Reggio, H., Bretscher, A., Forte, J.G., & Mangeat, P. (1991). The secretion-stimulated 80K phosphoprotein of parietal cells is ezrin, and has properties of a membrane cytoskeletal linker in the induced apical microvilli. EMBO J. 10, 2363–2373.

Hawkins, M., Pope, B., Maciver, S.K., & Weeds, A.G. (1993). Human actin depolymerizating factor mediates a pH sensitive destruction of actin filaments. Biochemistry 32, 9985–9993.

Hayden, S.M., Miller, P.S., Brauweiler, A., & Bamburg, J.R. (1993). Analysis of the interaction of actin depolymerizing factor with G- and F-actin. Biochemistry 32, 9994–10004.

Heintzelman, M.B., Frankel, S.A., Artavanis-Tsakonas, S., & Mooseker, M.S. (1993). Cloning of a secretory gelsolin from Drosophila melanogaster. J. Mol. Biol. 230, 709–716.

Heise, H., Bayerl, T., Isenberg, G., & Sackmann, E. (1991). Human platelet P-235, a talin like actin binding protein, binds selectively to mixed lipid bilayers. Biochem. Biophys. Acta 1061, 121–131.

Heiss, S.G. & Cooper, J.A. (1991). Regulation of capZ, an actin capping protein of chicken muscle, by anionic phospholipids. Biochemistry 30, 8753–8758.

Hitt, A.L. & Luna, E.J. (1993). Characterization of ponticulin null cells: Evidence that ponticulin is the major actin-plasma membrane link in Dictyostelium. Mol. Biol. Cell 4, 168a.

Hofmann, A., Eichinger, L., Andre, E., Reiger, D., & Schleicher, M. (1992). Cap100, a novel phosphatidylinositol 4,5-bisphosphate-regulated protein that caps actin filaments but does not nucleate actin assembly. Cell Motil. Cytoskel. 23, 133–144.

Hofmann, A., Noegel, A.A., Bomblies, L., Lottspeich, F., & Schleicher, M. (1993). The 100kDa F-actin capping protein of Dictyostelium amoebae is a villin prototype ('protovillin'). FEBS Lett. 328, 71–76.

Holmes, K.C., Popp, D., Gebhard, W., & Kabsch, W. (1990). Atomic model of the actin filament. Nature 347, 44–49.

Horwitz, A., Duggan, K., Buck, C., Beckerle, M., & Burridge, K. (1985). Interaction of plasma membrane fibronectin receptor with talin—a transmembrane linkage. Nature 320, 531–533.

Hou, L., Luby-Phelps, L., & Lanni, F. (1990). Brownian motion of inert tracer macromolecules in polymerized and spontaneously bundled mixtures of actin and filamin. J. Cell Biol. 110, 1645–1654.

Howard, P.K., Sefton, B.M., & Firtel, R.A. (1993). Tyrosine phosphorylation of actin in *Dictyostelium* associated with cell-shape changes. Science 259, 241–244.

Hug, C., Miller, T.M., Torres, M.A., Casella, J.F., & Cooper, J.A. (1992). Identification and characterization of an actin binding site of CapZ. J. Cell Biol. 116, 923–931.

Iida, K., Matsumoto, S., & Yahara, I. (1992). The KKRKK sequence is involved in heat shocked-induced nuclear translocation of the 18-kDa actin binding protein, cofilin. Cell Struct. Funct. 17, 39–46.

Iida, K., Moriyama, K., Matsumoto, S., Kawasaki, H., Nishida, E., & Yahara, I. (1993). Isolation of a yeast essential gene, COF1, that encodes a homologue of mammalian cofilin, a low-Mr actin-binding and depolymerizing protein. Gene 124, 115–120.

Isenberg, G., Aebi, U., & Pollard, T.D. (1980). An actin-binding protein from *Acanthamoeba* regulates actin filament polymerization and interactions. Nature 288, 455–459.

Isenberg, G. (1991). Actin binding proteins–lipid interactions. J. Muscle Res. Cell Mot. 12, 136–144.

Isenberg, G. & Goldmann, W.H. (1992). Actin-membrane coupling: A role for talin. J. Muscle Res. Cell Mot. 13, 587–589.

Ishidate, S. & Mabuchi, I. (1988). A novel actin filament-capping protein from sea urchin eggs: A 20,00-molecular weight protein-actin complex. J. Biochem. 104, 72–80.

Ishikawa, R., Yamashiro, S., & Matsumura, F. (1989). Differential modulation of actin -severing activity of gelsolin by multiple isoforms of cultured rat cell tropomyosin. Potentiation of protective ability of tropomyosins by 83-kDa non-muscle caldesmon. J. Biol. Chem. 264, 7490–7497.

Ishikawa, R., Okagaki, T., & Kohama, K. (1992). Regulation by Ca^{2+}-calmodulin of the actin-bundling activity of *Physarum* 210-kDa protein. J. Muscle Res. Cell Mot. 13, 321–328.

Itano, N. & Hatano, S. (1991). F-actin bundling protein from *Physarum polycephalum*: Purification and its capacity for co-bundling of actin filaments and microtubules. Cell Mot. Cytoskel. 19, 244–254.

Izzard, C.S. (1988). A precursor of the focal contact in cultured fibroblasts. Cell Mot. Cytoskel. 10, 137–142.

Jakob, U., Gaestel, M., Engel, K., & Buchner, J. (1993). Small heat shock proteins are molecular chaperones. J. Biol. Chem. 268, 1517–1520.

Janmey, P.A. & Stossel, T.P. (1987). Modulation of gelslin function by phosphatidylinositol 4,5-bisphosphate. Nature 325, 362–364.

Janmey, P.A., Hvidt, S., Lamb, J., & Stossel, T.P. (1990). Resemblance of actin binding protein/actin gels to covalently crosslinked networks. Nature 345, 89–92.

Janmey, P.A., Lamb, J., Allen, P.G., & Matsudaira, P.T. (1992). Phosphoinositide-binding peptides derived from the sequences of gelsolin and villin. J. Biol. Chem. 267, 11818–11823.

Johns, J.A., Brock, A.M., & Pardee, J.D. (1988). Colocalization of F-Actin and 34-kilodalton actin bundling protein in *Dictyostelium* amoebae and cultured fibroblasts. Cell Motility Cytoskeleton 9, 205–218.

Jones, P.G., Moore, G.J., & Waisman, D.M. (1992). A nonapeptide to the putative F-actin binding site of annexin-II tetramer inhibits its calcium-dependent activation of actin filament bundling. J. Biol. Chem. 267, 13993–13997.

Jongstra-Bilen, J., Janmey, P.A., Hartwig, J.H., Galea, S., & Jongstra, J. (1992). The lymphocyte-specific protein LSP1 binds to F-actin and the cytoskeleton through its COOH-terminal basic domain. J. Cell Biol. 118, 1443–1453.

Kabsch, W., Mannhertz, H.G., Suck, D., Pai, E.F., & Holmes, K.C. (1990). Atomic structure of the actin:DNase 1 complex. Nature 347, 37–49.

Kane, R.E. (1975). Preparation and purification of polymerized actin from sea urchin egg extracts. J. Cell Biol. 66, 305–315.

Kanoh, S., Ito, M., Niwa, E., Kawano, Y., & Hartshorne, D.J. (1993). Actin-binding from smooth muscle light chain kinase. Biochemistry 32, 8902–8907.

Kaufmann, S., Pickenbrock, T., Goldmann, W.H., Barmann, M., & Isenberg, G. (1991). Talin binds to actin and promotes filament nucleation. FEBS Lett. 284, 187–191.

Kilimann, M.W. & Isenberg, G. (1982). Actin filament capping protein from bovine brain. EMBO J. 1, 889–894.

Kim, S.-R., Kim,Y., & An, G. (1993). Molecular cloning and characterization of pollen-preferential genes encoding putative actin depolymerizing factor. Plant Mol. Biol. 21, 39–45.

Knecht, D.A. & Loomis, W.F. (1987). Antisense RNA inactivation of myosin heavy chain gene expression in *Dictyostelium discoideum*. Science 236, 1081–1085.

Koepf, E.K. & Burtnick, L.D. (1992). Interaction of plasma gelsolin with tropomyosin. FEBS Lett. 309, 56–58.

Koffer, A., Edgar, A.J., & Bamburg, J.R. (1988). Identification of two species of actin depolymerizing factor in cultulres of BHK cells. J. Mus. Res. Cell Mot. 9, 320–328.

Kornberg, L., Earp, H.S., Parsons, J.T., Schaller, M., & Juliano, R.L. (1992). Cell adhesion or integrin clustering increases phosphorylation of a focal adhesion-associated tyrosine kinase. J. Biol. Chem. 267, 23439–23442.

Koyasu, S., Nishida, E., Miyata, Y., Sakai, H., & Yahara, I. (1989). HSP100, a 100-kDa heat shock protein, is a Ca^{2+}-calmodulin regulated actin binding protein. J. Biol. Chem. 264, 15083–15087.

Kries, T. & Vale, R. (1993). Guidebook to the cytoskeletal and motor proteins. Sambrook and Tooze, Oxford University Press.

Kwiatkowski, D.J. (1988). Predominant induction of gelsolin and actin-binding protein during myeloid differentiation. J. Biol. Chem. 263, 13852–13862.

Kwiatkowski, D.J., Mehl, R.A., Izumo, S., Nadal-Ginard, B., & Yin, H.L. (1988). Muscle is the major source of plasma gelsolin. J. Biol. Chem. 263, 8239–8243.

Kwiatkowski, D.J., Stossel, T.P., Orkin, S.H., Mole, J.E., Colten, H.R., & Yin, H.L. (1986). Plasma and cytoplasmic gelsolins are encoded by a single gene and contain a duplicated actin-binding domain. Nature 323, 455–458.

Lal, A.A. & Korn, E.D. (1985). Reinvestigation of the inhibition of actin polymerization by profilin. J. Biol. Chem. 260, 10132–10138.

Lamb, J.A., Allen, P.G., Tuan, B.Y., & Janmey, P.A. (1993). Modulation of gelsolin function. Activation at low pH overrides Ca^{2+} requirement. J. Biol. Chem. 268, 8999–9004.

Lankes, W.T. & Furthmayr, H. (1991). Moesin: A member of the protein 4.1-talin-ezrin family of proteins. Proc. Natl. Acad. Sci. USA 88, 8297–8301.

Lassing, I. & Lindberg, U. (1985). Specific interaction between phosphatidyl-inositol 4,5–bisphosphate and profiliactin. Nature 314, 472–474.

Lavoie, J., Hickey, E., Weber, L.A., & Landry, J. (1993). Modulation of actin microfilament dynamics and fluid phase pinocytosis by phosphorylation of heat shock protein 27. J. Biol. Chem. 268, 24210–24214.

Lebart, M.C., Mejean, C., Roustan, C., & Benyamin, Y. (1993). Further characterization of the α-actinin–actin interface and comparison with filamin-binding sites on actin. J. Biol. Chem. 268, 5642–5648.

Luna, E.J. & Hitt, A.L. (1992). Cytoskeleton-plasma membrane interactions. Science 258, 955–964.

Leszyk, J., Mornet, D., Audemard, E., & Collins, J. (1989). Amino acid sequence of a 15 kilodalton actin-binding fragment of turkey gizzard caldesmon: Similarity with dystrophin, tropomyosin and the tropomyosin-binding region of troponin T. Biochem. Biophys. Res. Comm. 160, 210–216.

Mabuchi, I. (1983). An actin-depolymerizing protein (Depactin) from starfish oocytes: Properties and interaction with actin. J. Cell Biol. 97, 1612–1621.

Machesky, L., Goldschmidt-Cleremont, P., & Pollard, T.D. (1990). The affinities of human platelet and *Acanthamoeba* profilin isoforms for polyphosphoinositides account for their relative abilities to inhibit phospholipase. C. Cell Regulation 1, 937–950.

Machesky, L. & Pollard, T.D. (1993). Profilin as a potential mediator of membrane-cytoskeleton communication. Trends Cell Biol. 3, 381–385.

Maciver, S.K., Zot, H.G., & Pollard, T.D. (1991a). Characterization of actin filament severing by actophorin from *Acanthamoeba castellanii*. J. Cell Biol. 115, 1611–1620.

Maciver, S.K., Wachsstock, D., Schwarz, W.H., & Pollard, T.D. (1991b). The actin filament severing protein actophorin promotes the formation of rigid bundles of actin filaments crosslinked with α-actinin. J. Cell Biol. 115, 1621–1628.

Maciver, S.K. (1994). Actin-binding proteins. In: *The Encyclopaedia of Molecular Biology* (Kendrew, J., Ed.) Blackwell Scientific, Cambridge, MA.

Maciver, S.K. & Weeds, A.G. (1993). The interaction between actophorin and actin: actophorin preferentially binds monomeric ADP-actin over ATP-bound actin. Mol. Biol. Cell 4, 152a.

Maekawa, S., Toriyama, M., Hisanaga, S.-I., Yonezawa, N., Endo, S., Hirokawa, N., & Sakai, H. (1989). Purification and characterization of a Ca^{2+}-dependent actin filament severing protein from bovine adrenal medulla. J. Biol. Chem. 264, 7458–7465.

Mannhertz, H.G., Goody, R.S., Konrad, M., & Nowak, E. (1980). The interaction of pancreatic deoxyribonuclease 1 and skeletal muscle actin. Eur. J. Biochem. 104, 367–379.

Maruta, H., Knoerzer, W., Hinssen, H., & Isenberg, G. (1984). Regulation of actin polymerization by non-polymerizable actin-like proteins. Nature 312, 424–427.

Maruta, H. & Isenberg, G. (1984). Ca^{2+}-dependent actin-binding phosphoprotein in *Physarum polycephalum*. J. Biol. Chem. 259, 5208–5213.

Maruta, H. & Isenberg, G. (1983). Ca^{2+}-dependent actin-binding phosphoprotein in physarum polycephalum II. Ca^{2+}-dependent F-actin capping activity of subunit a and its regulation by phosphorylation of subunit b. J. Biol. Chem. 258, 10151–10158.

Maruyama, K., Kurokawa, H., Oosawa, M., Shimaoka, S., Yamamoto, H., Ito, M., & Maruyama, K. (1990). β-actinin is equivalent to cap Z protein. J. Biol. Chem. 265, 8712–8715.

Matsudaira, P., Mandelkow, E., Renner, W., Hesterberg, L.K., & Weber, K. (1983). Role of fimbrin and villin in determining the interfilament distances of actin bundles. Nature 301, 209–214.

Matsudaira, P., Bordas, J., & Koch, M. (1987). Synchrotron x-ray diffraction studies of actin structure during polymerization. Proc. Natl. Acad. Sci. USA 84, 3151–3155.

Matsumura, F. & Yamashiro, S. (1993). Caldesmon. Curr. Opp. Cell Biol. 5, 70–76.

Matsuzaki, F., Matsumoto, S., Yahara, I., Yonezawa, N., Nishida, E., & Sakai, H. (1988). Cloning and characterization of porcine brain cofilin cDNA. Cofilin contains the nuclear transport signal sequence. J. Biol. Chem. 263, 11564–11568.

McLaughlin, P.J., Gooch, J.T., Mannhertz, H.-G., & Weeds, A.G. (1993). Structure of gelsolin segment 1-actin complex and the mechanism of filament severing. Nature 364, 685–692.

Meriläinen, J., Palovuori, R., Sormunen, R., Wasenius, V.-M., & Lehto, V.-P. (1993). Binding of the α-fodrin SH3 domain to the leading lamellae of locomoting chicken fibroblasts. J. Cell Sci. 105, 647–654.

Meyer, R.K. & Aebi, U. (1990). Bundling of actin filaments by α-actinin depends on its molecular length. J. Cell Biol. 110, 2013–2024.

Mezgueldi, M., Fattoum, A., Derancourt, J., & Kassab, R. (1992). Mapping of the functional domains in the amino-terminal region of calponin. J. Biol. Chem. 267, 15943–15951.

Millonig, R., Salvo, H., & Aebi, U. (1988). Probing actin polymerization by intermolecular crosslinking. J. Cell Biol. 106, 785-796.

Miron, T., Vancompernolle, K., Vandekerckhove, J., Wilchek, M., & Geiger, B. (1991). A 25-kD inhibitor of actin polymerization is a low molecular mass heat shock protein. J. Cell Biol. 114, 255–261.

Mische, S.M., Mooseker, M.S., & Morrow, J.S. (1987). Erythrocyte adducin: A calmodulin-regulated actin-bundling protein that stimulates spectrin-actin binding. J. Cell Biol. 105, 2837–2845.

Mizutani, A., Usuda, N., Tokumitsu, H., Minami, H., Yasui, K., Kobayashi, R., & Hidaka, H. (1992). CAP-50, a newly identified annexin, localizes in nuclei of cultured fibroblast 3Y1 cells. J. Biol. Chem. 267, 13498–13504.

Molony, L. & Burridge, K. (1985). Molecular shape and self-association of vinculin and metavinculin. J. Cell Biochem. 29, 31–36.

Moon, A.L., Janmey, P.A., Andrea Louie, K., & Drubin, D.G. (1993). Cofilin is an essential component of the yeast cortical cytoskeleton. J. Cell Biol. 120, 421–435.

Morgan, T.E., Lockerbie, R.O., Minamide, L.S., Browning, M.D., & Bamburg, J.R. (1993). Isolation and characterization of a regulated form of actin depolymerizing factor. J. Cell Biol. 122, 623–633.

Muguruma, M., Matsumura, S., & Fukazawa, T. (1992). Augmentation of α-actinin induced gelation of actin by talin. J. Biol. Chem. 267, 5621–5624.

Nachmias, V.T., Cassimeris, L., Golla, R., & Safer, D. (1993). Thymosin beta 4 (Tβ4) in activated platelets. Eur. J. Cell Biol. 61, 314–320.

Namba, Y., Ito, M., Zu, Y., Shigesada, K., & Maruyama, K. (1992). Human T cell l-plastin bundles actin filaments in a calcium-dependent manner. J. Biochem. 112, 503–507.

Neiderman, R., Amrein, P.C., & Hartwig, J. (1986). Three-dimentional structure of actin filaments and of an actin gel made with actin-binding protein. J. Cell Biol. 96, 1400–1413.

Nishida, E., Iida, K., Yonezawa, N., Koyasu, S., Yahara, L., & Sakai, H. (1987). Cofilin is a component of intranuclear and cytoplasmic rods induced in cultured cells. Proc. Natl. Acad. Sci. USA 84, 5262–5266.

Nishida, E., Koyasu, S., Sakai, H., & Yahara, I. (1986). Calmodulin-regulated inding of the 90-kDa heat shock protein to actin filaments. J. Biol. Chem. 261, 16033–16036.

Nodes, B.R., Shackelford, J.E., & Lebhertz, H.G. (1987). Synthesis and secretion of serum gelsolin by smooth muscle tissue. J. Biol. Chem. 262, 5422–5427.

Ohta, Y., Stossel, T.P., & Hartwig, J.H. (1991). Ligand-sensitive binding of actin binding protein to immunoglobulin G Fc receptor I (FcγRI). Cell 67, 275–282.

Ohishi, I., Morikawa, Y., & Baba, T. (1990). ADP-Ribosylation of nonmuscle actin by component I of Botulinum C_2 toxin inactivates the ability to interact with unmodified actin. J. Biochem. 107, 420–425.

Ono, S., Abe, H., Nagaoka, R., & Obinata, T. (1993). Colocalization of ADF and cofilin in intranuclear actin rods of cultured muscle cells. J. Muscle Res. Cell Mot. 14, 195–204.

Onoda, K, Yu, F.-X., & Yin, H.L. (1993a). gCap39 is a nuclear and cytoplasmic protein. Cell Mot. Cytoskel. 26, 227–238.

Onoda, K. & Yin, H.L. (1993b). gCap39 is phosphorylated. Stimulation by okadaic acid and preferential association with nuclei. J. Biol. Chem. 268, 4106–4112.

O'Reilly, G. & Clarke, F. (1993). Identification of an actin binding region in aldolase. FEBS Lett. 321, 69–72.

Otey, C.A., Pavalko, F.M., & Burridge, K. (1990). An interaction between α-actinin and β1-integrin subunit in vitro. J. Cell Biol. 111, 721–729.

Owen, C.H., DeRosier, D.J., & Condeelis, J. (1993). Actin crosslinking protein EF-1a of Dictyostelium discoideum has a unique bonding rule that allows square-packed bundles. J. Struct. Biol. 109, 248–254.

Ozawa, M., Ringwald, M., & Kemler, R. (1990). Uvomorulin-catenin complex formation is regulated by a specific domain in the cytoplasmic region of the cell adhesion molecule. Proc. Natl. Acad. Sci. USA 87, 4246–4250.

Pacaud, M. & Harricane, M.C. (1993). Macrophage α-actinin is not a calcium-modulated actin-binding protein. Biochemistry 32, 363–374.

Pagliaro, L. & Taylor, D.L. (1988). Aldolase exists in both the fluid and solid phases of cytoplasm. J. Cell Biol. 107, 981–991.

Pantaloni, D. & Carlier, M.-F. (1993). How profilin promotes actin filament assembly in the presence of thymosin β4. Cell 75, 1007–1014.

Pavalko, F.M., Otey, C.A., & Burridge, K. (1989). Identification of a filamin isoform enriched at the ends of stress fibers in chick embryo fibroblasts. J. Cell Sci. 94, 109–118.

Podolski, J.L. & Steck, T.L. (1988). Association of deoxyribonuclease 1 with the pointed ends of actin filaments in human red blood cell membrane skeletons. J. Biol. Chem. 263, 638–645.

Pollard, T.D., Aebi, U., Cooper, J.A., Elzinga, M., Fowler, W.E., Griffith, L.M., Herman, I.M., Heuser, J., Isenberg, G., Keihart, D.P., Levy, J., MacLean-Fletcher, S., Maupin, P., Mooseker, M.S., Runge, M., Smith, P.R., & Tseng, P. (1982). The mechanism of actin-filament assembly and crosslinking. In "Cell and Muscle Motility" 2, Dowben, R.M., & Shay, J.W., Eds. Plenum Press, New York.

Pollard, T.D. & Cooper, J.A. (1986). Actin and actin binding proteins. A critical evaluation of mechanisms and functions. Ann. Rev. Biochem. 55, 987–1035.

Pollard, T.D., Tseng, P.C.-H., Rimm, D.L., Bichell, D.P., Williams, R.C., Sinard, J., & Sato, M. (1986). Characterization of alpha-actinin from *Acanthamoeba*. Cell Mot. Cytoskel. 6, 649–661.

Pope, B., Way, M., Matsudaira, P.T., & Weeds, A.G. (1994). Characterisation of the F-actin domains of villin: Classification of F-actin binding into two groups according to their binding sites on actin. FEBS Lett. (in press).

Prendergast, G.C. & Ziff, E.B. (1991). Mbh1: A novel gelsolin/severin-related protein which binds actin in vitro and exhibits nuclear localization in vivo. EMBO J. 10, 757–766.

Quirk, S., Maciver, S.K., Ampe, C., Doberstein, S.K., Kaiser, D.A., VanDamme, J., Vanderckhove, J.S., & Pollard, T.D. (1993). Primary structure of *Acanthamoeba* actophorin. Biochemistry 32, 8525–8533.

Rana, A.P., Ruff, P., Maalouf, G.J., Speicher, D.W., & Chishti, A.H. (1993). Cloning of human erythroid dematin reveals another member of the villin family. Proc. Natl. Acad. Sci. USA 90, 6651–6655.

Rimm, D.L. & Pollard, T.D. (1989). Purification and characterization of an *Acanthamoeba* nuclear actin binding protein. J. Cell Biol. 109, 585–591.

Rodríguez Del Castillo, A., Vitale, M.L., & Trifaró, J.-M. (1992). Ca^{2+} and pH determine the interaction of chromaffin cell scinderin with phosphatidylserine and phosphatidylinositol 4,5,–bisphosphate and its cellular distribution during nicotinic–receptor stimulation and protein kinase C activation. J. Cell Biol. 119, 797–810.

Sabbatini, A., Petrini, M., Arnaud, P., & Galbraith, R.M. (1993). Vitamin D binding protein is produced by human monocytes. FEBS Lett. 323, 89–92.

Safer, D. (1992). The interaction of actin with thymosin β4. J. Muscle Res. Cell Mot. 13, 269–271.

Safer, D., Golla, R., & Nachmais, V. (1990). Isolation of a 5-Kilodalton actin sequestering peptide from human blood platelets. Proc. Natl. Acad. Sci. USA 87, 2536–2539.

Sanders, M.C., Goldstein, A.L., & Wag, Y.L. (1992). Thyomosin β4 (Fx peptide) is a potent regulator of actin polymerization in living cells. Proc. Natl. Acad. Sci. USA 89, 4678–4682.

Sauterer, R.A., Eddy, R.J., Hall, A.L., & Condeelis, J.S. (1991). Purification and characterization of aginactin, a newly identified agonist-regulated actin capping protein from *Dictyostelium* amoebae. J. Biol. Chem. 266, 24533–24539.

Schafer, D.A., Mooseker, M.S., & Cooper, J.A. (1992). Localization of capping protein in chicken epithelial cells by immunofluorescence and biochemical fractionation. J. Cell Biol. 118, 335–346.

Scheel, J., Zeigelbauer, K., Kupke, T., Humbel, B.M., Noegel, A.A., Gerisch, G., & Scleicher, M. (1989). Hisactophilin, a histidine-rich actin binding protein from *Dictyostelium discoideum*. J. Biol. Chem. 264, 2832–2839.

Schleicher, M., Gerisch, G., & Isenberg, G. (1984). New actin-binding proteins from *Dictyostelium discoideum*. EMBO J. 3, 2095–2100.

Schröder, H.C., Zahn, R.K., & Müller, W.E.G. (1982). Role of actin and tubulin in the regulation of poly(A)polymerase- endonuclease IV complex from calf thymus. J. Biol. Chem. 257, 2305–2309.

Schutt, C.L., Myslik, J.C., Rozycki, M.D., Goonesekere, N.C.W., & Lindberg, U. (1993). The structure of crystalline profilin-β-actin. Nature 365, 810–816.

Schweiger, A., Mihalache, O., Ecke, M., & Gerisch, G. (1992). Stage-specific tyrosine phosphorylation of actin in *Dictyostelium discoideum* cells. J. Cell Sci. 102, 601–609.

Shapland, C., Hsuan, J.J., Totty, N.F., & Lawson, D. (1993). Purification and properties of transgelin: A transformation and shape change sensitive actin-gelling protein. J. Cell Biol. 121, 1065–1073.

Siegel, D.L. & Brandon, D. (1985). Partial purification and characterization of an actin-bundling protein, band 4.1, from human erythrocytes. J. Cell Biol. 100, 775–785.

Southwick, F.S. & DiNubile, M.J. (1986). Rabbit alveolar macrophages contain a Ca^{2+}-sensitive, 41,000-dalton protein which reversibly blocks the "barbed" ends of actin filaments but does not sever them. J. Biol. Chem. 261, 14191–14195.

Southwick, F.S. & Hartwig, J.H. (1982). Acumentin, a protein in macrophages which caps the 'pointed' end of actin filaments. Nature 297, 303–307.

Southwick, F.S. & Young, C.L. (1990). The actin released from profilin-actin complexes is insufficient to account for the increase in F-actin in chemoattractant-stimulated polymorphonuclear leukocytes. J. Cell Biol. 110, 1965–1973.

St-Pierre, B., Couture, C., Laroche, A., & Pallotta, D. (1993). Two developmentally regulated mRNAs encoding acting-binding proteins in *Physarum polycephalum*. Biochim. Biophys. Acta 1173, 107–110.

Stokes, D.L. & DeRosier, D.J. (1987). The variable twist of actin and its modulation by actin-binding proteins. J. Cell Biol. 104, 1005–1017.

Südhof, T.C., Czernik, A.J., Kao, H.T., Takei, K., Johnston, P.A., Horiuchi, A., Kanazir, S.D., Wagner, M.A., Perin, M.S., De Camilli, P., & Greengard, P. (1989). Synapsins: Mosaics of shared and individual domains in a family of synaptic vesicle phosphoproteins. Science 245, 1474–1480.

Takagi, T., Konishi, K., & Mabuchi, I. (1988). Amino-acid sequence of starfish oocyte depactin. J. Biol. Chem. 263, 3097–3102.

Tellam, R.L., Morton, D.J., & Clarke, F.M. (1989). A common theme in the amino-acid sequences of actin and many many actin-binding proteins? Trends Biochem. Sci. 14, 130–133.

Theriot, J.A. & Mitchison, T.J. (1993). The three faces of profilin. Cell 75, 835–838.

Turner, C.E., Glenney, Jr., J.R., & Burridge, K. (1990). Paxillin: A new vinculin binding protein present in focal adhesions. J. Cell Biol. 111, 1059–1068.

Ueno, T. & Korn, E.D. (1986). Isolation and partial characterization of a 110-kD dimer actin-binding protein. J. Cell Biol. 103, 621–630.

Vancompernolle, K., Vandekerckhove, J., Bubb, M.R., & Kord, E.D. (1991). The interfaces of actin and *Acanthamoeba* actobindin. J. Biol. Chem. 266, 15427–15431.

Vandekerckhove, J., Kaiser, D.A., & Pollard, T.D. (1989). *Acanthamoeba* actin and profilin can be cross-linked between glutamic acid 364 of actin and lysine 115 of proficin. J. Cell Biol. 109, 619–626.

Vandekerckhove, J. & Vancompernolle, K. (1992). Structural relationships of actin–binding proteins. Curr. Op. Cell Biol. 4, 36–42.

Vitale, M.L., Rodríguez Del Castillo, A., Tchakarov, L., & Trifaró, J.-M. (1991). Cortical filamentous actin disassembly and scinderin redistribution during chromaffin cell stimulation precede exocytosis, a phenomenon not exhibited by gelsolin. J. Cell Biol. 113, 1057–1067.

Walsh, M. (1993). Calcium-dependent mechanisms of regulation of smooth muscle contraction. Biochem. Cell Biol. 69, 771–800.

Wachsstock, D.H., Schwartz, W.H., & Pollard, T.D. (1993). Affinity of α-actinin for actin determines the structure and mechanical properties of actin filament gels. Biophys. J. 65, 205–214.

Wachsstock, D.H., Wilkins, J.A., & Lin, S. (1987). Specific interaction of vinculin with α-actinin. Biochem. Biophys. Res. Comm. 146, 554–560.

Wang, Y.-L. (1985). Exchange of actin subunits at the leading edge of living fibroblasts: Possible role of treadmilling. J. Cell Biol. 101, 597–602.

Wang, C.L.A., Wang, L.W.C., Xu, S., Lu, R.C., Saavedra-Alanis, V., & Bryan, J. (1991). Localization of the calmodulin and the actin-binding sites of caldesmon. J. Biol. Chem. 266, 9166–9172.

Way, M. & Mastudaira, P.T. (1993). Kelch like repeats in scruin define a new family of F-actin binding proteins found in poxviruses, *Drosophila* and humans. Mol. Biol. Cell 4, 4a.

Weeds, A.G. (1982). Actin-binding proteins, regulators of cell architecture and motility. Nature 296, 811–816.

Weeds, A.G., Gooch, J., McLaughlin, P., & Maury, C.P.J. (1993). Variant plasma gelsolin responsible for familial amyloidosis (Finnish type) has defective actin severing activity. FEBS Lett. 335, 119–123.

Weeds, A.G. & Maciver, S.K. (1993). F-actin capping proteins. Curr. Opp. Cell Biol. 5, 63–69.

Weeds, A.G. & Way, M. (1991). Is thymosin-b4 the missing link? Curr. Biol. 1 (5), 307–308.

Wegner, A. & Aktories, K. (1988). ADP-ribosylated actin caps the barbed ends of actin filaments. J. Biol. Chem. 263, 13739–13742.

Weiner, O.H., Murphy, J., Griffiths, G., Schleicher, M., & Noegel, A.A. (1993). The actin-binding protein comitin (p24) is a component of the golgi apparatus. J. Cell Biol. 123, 23–34.

Wills, F.L., McCubbin, W.D., & Kay, C.M. (1993). Characterization of smooth muscle calponin and calmodulin complex. Biochemistry 32, 2321–2328.

Winder, S.J., Hemmings, L., Bolton, S.J., Maciver, S.K., Tinsley, J.M., Davies, K.E., Critchley, D.R., & Kendrick-Jones, J. (1995). Utrophin actin binding domain: Analysis of actin binding and cellular targeting. J. Cell Sci. 108, 63–71.

Winder, S.J. & Walsh, M.P. (1993). Calponin: Thin filament-linked regulation of smooth muscle contraction. Cellular Signalling, 5 , 677–686.

Witke, W., Schleicher, M., & Noegel, A. (1992). Redundancy in the microfilament system: Abnormal development of *Dictyostelium* cells lacking two F-actin crosslinking proteins. Cell 68, 53–62.

Wu, H. & Parsons, J.T. (1993). Cortactin, an 80/85-Kilodalton pp60 [src] substrate, is a filamentous actin binding protein enriched in the cell cortex. J. Cell Biol. 120, 1417–1426.

Wuestehube, L.J. & Luna, E.J. (1987). F-Actin binds to the cytoplasmic surface of ponticulin, a 17-kD integral glycoprotein from *Dictyostelium discoideum* plasma membranes. J. Cell Biol. 105, 1741–1751.

Yamashiro, S., Yamakita, Y., Ishikawa, R., & Matsumura, F. (1990). Mitosis-specific phosphorylation causes 83kD a nonmuscle caldesmon to dissociate from microfilaments. Nature 344, 675–678.

Yang, F., Demma, M., Warren, V., Dhamawardhane, S., & Condeelis, J. (1990). Identification of an actin-binding protein from *Dictyostelium* as elongation factor 1a. Nature 347, 494–496.

Yin, H.L. (1988). Gelsolin: Calcium- and polyphosphoinositide-regulated actin modulating protein. Bioessays 7, 176–179.

Yonezawa, N., Nishida, E., & Sakai, H. (1985). pH control of actin polymerization by cofilin. J. Biol. Chem. 260, 14410–14412.

Yonezawa, N., Nishida, E., Ohba, M., Seki, M., Kumagai, H., & Sakai, H. (1989). An actin-interacting heptpeptide in the cofilin sequence. Eur. J. Biochem. 183, 235–238.

Yonezawa, N., Nishida, E., Iida, K., Yahara, I., & Sakai, H. (1990). Inhibition of the interaction of cofilin, destrin and deoxyribonuclease 1 with actin by phosphoinositides. J. Biol. Chem. 265, 8382–8386.

Yonezawa, N., Nishida, E., Iida, K., Kumagai, H., Yahara, I., & Sakai, H. (1991). Inhibition of actin polymerization by a synthetic dodecapeptide patterned on the sequence around the actin-binding site of cofilin. J. Biol. Chem. 266, 10485–10489.

Yonezawa, N., Nishida, E., & Sakai, H. (1985). pH control of actin polymerization by cofilin. J. Biol. Chem. 260, 14410–14412.

Yu, F.-X., Johnston, P.A., Südhof, T.C., & Yin, H.L. (1990). gCap39, a calcium ion- and polyphospho-inositide-regulated actin capping protein. Science 250, 1413–1415.

Yu, F.-X., Sun, H.-Q., Janmey, P.A., & Yin, H.L. (1992). Identification of a polyphosphoinositde-binding sequence in an actin monomer-binding domain of gelsolin. J. Biol. Chem. 267, 14616–14621.

Zigmond, S.H., Furukawa, R., & Fechheimer, M. (1992). Inhibition of actin filament depolymerization by the *Dictyostelium* 30,000-D actin bundling protein. J. Cell Biol. 119, 559–567.

CONTROL OF MICROTUBULE POLYMERIZATION AND STABILITY

Jesús Avila and Javier Díaz Nido

The Cytoskeleton, Volume 1
Structure and Assembly, pages 47–85.
Copyright © 1995 by JAI Press Inc.
All rights of reproduction in any form reserved.
ISBN: 1-55938-687-8

I. INTRODUCTION

Microtubules are protein polymers that, together with microfilaments and interme-diate filaments, constitute the cytoskeleton. They are involved in essential cell functions including the regulation of cell shape, cell movement, cell division and intracellular organelle transport and organization. Microtubules are present in almost every eukaryotic cell, although they are more abundant in neurons than in any other cell type.

A. Microtubule Structure

Microtubules are long, helical, hollow fibers with a diameter of 24 nm and variable length. The helical polymer usually contains 13 subunits per turn. Thus, a microtubule can be considered to be composed of 13 longitudinal strands, which are known as protofilaments. This helical structure has an asymmetric form in which both ends can be distinguished (Amos, 1979).

B. Microtubule Components

Microtubule protein composition has been determined after the isolation of microtubules by a procedure that consists of cycles of *in vitro* temperature-depend-ent assembly and disassembly of the polymers (Shelanski et al., 1973). Brain tissue has been used as the preferred source for microtubule isolation because of its abundance. The major component of isolated microtubules is a protein called tubulin, which can itself polymerize into microtubules in the presence of GTP and Mg^{2+}. However, other minor protein constituents are also found (See Figure 1). These tubulin-binding proteins were originally referred to as microtubule-associ-ated proteins (MAPs) and have the ability to promote the polymerization of tubulin into microtubules (see section below on MAPs). Other minor protein constituents are known as microtubule-binding proteins (MTBPs) or microtubule-interacting proteins (MTIPs), as their binding to tubulin is transient and may be influenced by the presence of other molecules, including nucleotides such as ATP or GTP. Proteins having such requirements include dynein, kinesin and dynamin ATPases, and they are known as microtubule-based motor proteins.

MAP-1A
MAP-1B
MAP-2

TAU
$-\alpha$-TUBULIN

$-\beta$-TUBULIN

Figure 1. Microtubule protein composition. Microtubules were prepared from brain extracts through temperature-dependent cycles of assembly-disassembly. Analysis by SDS-gel electrophoresis of this brain microtubule preparation reveals the presence of α and β tubulin subunits as the major component. Other proteins present in the preparation are referred to as microtubule-associated proteins (MAPs). These include MAP 1A, MAP 1B, MAP 2 and tau proteins.

C. Tubulin Subunits and Isoforms

Tubulin is a globular heterodimeric protein composed of two subunits, α and β. A new subunit, γ tubulin, has been recently discovered (Oakley, 1992), but the nature of its association with the other subunits is not yet clear.

In vertebrates, there are different genes coding for different isoforms of α- and β-tubulin subunits. For example, in mammals, 12 different genes have been described that code for 6 different α- and 6 different β-tubulin isoforms (Lewis and Cowan, 1990). In some cases, these isoforms show tissue specificity and their structure is highly conserved from organism to organism (Lewis and Cowan, 1990). In brain, there is expression of almost all these isoforms, a fact that has been correlated with the noteworthy morphological complexity exhibited by neurons (Gozes and Littauer, 1978). Whether or not the diversity of tubulin isoforms may be correlated with the remarkable diversity of functions performed by microtubules (mentioned above) has also been the subject of discussion (Joshi and Cleveland, 1990), although no clear answer is available at the moment. The determination of the amino acid sequences of the different tubulin isoforms has shown that the main differences among them are localized at the carboxy-terminal region of α- and β-tubulin isoforms, a region corresponding to a regulatory domain (see below).

In addition to these tubulin isoforms, also referred to as isotypes, that are encoded by distinct genes, novel tubulin isoforms arise as the consequence of posttranslational modifications. Four major types of tubulin modifications have been described as taking place on microtubules: acetylation of the α-tubulin subunit (Le Dizet and Piperno, 1987), phosphorylation of the β-tubulin subunit at its carboxy-terminal region (Gard and Kirschner, 1985; Serrano et al., 1987; Díaz-Nido et al., 1990a), detyrosination and tyrosination of the carboxy terminus of the α-tubulin subunit (Barra et al., 1988) and polyglutamination at the carboxy-terminal region of both tubulin subunits (Eddé et al., 1990). No modifications of the γ tubulin subunit are known to date.

D. Tubulin Structure

The study of the structural and functional domains of tubulin has been performed using several approaches. One of these takes advantage of the structural homology between certain proteins with similar functional features and of the existence of several empirical methods to predict the secondary structure of a protein from its primary sequence. In this way, structural similarities between some regions of tubulin (which is a GTP-binding protein, see below) and those of other nucleotide-binding proteins including actin or cAMP-dependent protein kinase have been suggested (Ponstingl et al., 1982). In addition, microtubule structure has been studied by electron microscopy and X-ray diffraction (Amos, 1979; Beese et al., 1987). Tubulin subunit shape appears to be elongated with a maximum diameter of 4 nm, and it seems to consist of three or four major structural domains. A detailed structure of tubulin is not available, since all attempts to obtain tubulin crystals for analysis by X-ray diffraction have been unsuccessful to date.

Using limited proteolysis, four main structural domains in each tubulin subunit have been identified (Serrano et al., 1986a), and the interactions of these tubulin domains with other molecules such as GTP, MAPs or calcium (which may regulate the polymerization–depolymerization of tubulin) have been analyzed (Linse and Mandelkow, 1988; Serrano et al., 1984a, b, 1985, 1986b; Cross et al., 1991).

Functionally, tubulin subunits can be considered as consisting of two parts, which can be separated by limited proteolysis with subtilisin, a protease that cleaves tubulin subunits near their carboxy-terminal ends (Serrano et al., 1984a, b; Sackett et al., 1985). After subtilisin cleavage, the larger tubulin fragment contains the amino-terminal end, the GTP-binding site and the regions involved in the interaction of tubulin subunits, whereas the smaller tubulin fragment contains the carboxy-terminal end and the binding sites for MAPs and calcium (Serrano et al., 1984a, b, 1986a, b). Interestingly, the larger fragment retains the capacity to polymerize (Serrano & Avila, 1985), a function that may be controlled by the presence of the smaller (regulatory) fragment in the intact molecule, possibly through the interaction of this regulatory domain with factors that facilitate (MAPs) or prevent (calcium) microtubule assembly (Serrano et al., 1984a, b).

E. Tubulin Folding

Newly synthesized tubulin subunits in a reticulocyte cell extract require the help of other proteins (molecular chaperones) to be folded in the proper way in order to have the capacity for self-assembly (Yaffe et al., 1992; Zabala and Cowan, 1992). A multisubunit toroid particle, which contains a protein known as TCP-1, hsp70 heat shock protein, and four to six unidentified proteins, is required for the proper folding of tubulin subunits (Yaffe et al., 1992). A two-step mechanism, involving ATP hydrolysis in the first step and GTP hydrolysis in the second step, has been suggested for the formation of the α-β tubulin heterodimer from newly synthesized tubulin subunits (Fontalba et al., unpublished). In this mechanism, TCP-1 protein, an ATPase, appears to be directly involved (Yaffe et al., 1992). The functional role of the heat shock protein hps 70, which binds to the carboxy-terminal region of the tubulin molecule as MAPs (Sánchez et al., unpublished), is not yet clear.

More recently, the existence of multiprotein complexes containing α, β and γ tubulin, heat shock protein hsp70 and elongation factor 1α have been described (Marchesi and Ngo, 1993). It is not known whether such complexes are related to the maintainance of a specifically folded conformation of tubulin.

II. MICROTUBULE DYNAMICS

A. Models of Microtubule Dynamics

The different cell functions in which microtubules are involved depend on their ability to polymerize and depolymerize. This ability does not correlate with the turnover of tubulin, which is a protein with a long life ($t_{1/2}$ of about 1 day in a mammalian cell), compared with the average half-life of a cytoplasmic microtubule (in a mammalian nonneuronal cell) of about 5 minutes.

Microtubule dynamics have been extensively analyzed in recent years (Kirschner and Mitchison, 1986; Caplow, 1992; Erikson and O'Brien, 1992; Avila, 1990). From the simpler point of view, the cycle of a microtubule involves its polymerization, which can be divided into two steps, nucleation and elongation, and its depolymerization. Actually, following several *in vitro* and *in vivo* studies, it can be concluded that a description of microtubule polymerization–depolymerization is far more complex.

The theoretical analysis of *in vitro* microtubule polymerization has been based on the hypothesis that it involves a nucleation step, followed by an elongation step.

In vitro, the nucleation step depends on the tubulin concentration, since a minimum (or critical) concentration of tubulin is required to form the nucleation structure (seed) of the microtubule that may then be elongated. However, the nucleation step differs in *in vitro* conditions from the *in vivo* situation, since, in the latter case, the initiation of microtubule polymerization usually occurs at the centrosome, or related structures known as microtubule organizing centers (MTOCs) (Solty and Borisy, 1985; Karsenti and Maro, 1986). An important feature of microtubules organized at these MTOCs is their orientation. Microtubule polymers have two distinct ends, and all microtubules growing from an MTOC have the same orientation, with their fast-growing (fast-elongating) ends located away from the organizing center. These ends are known as "plus" ends, whereas those located close to the MTOCs are the "minus" ends. A role for γ-tubulin as a "minus" end-capping protein in the MTOCs has been proposed (Oakley, 1992; Joshi et al., 1992). However, some questions remain about the *in vivo* nucleation of microtubules in several cell types, including differentiated epithelial cells (Bré et al., 1987, 1990), and neurons (Baas and Ahmad, 1992; Baas and Joshi, 1992; Brown et al., 1992). Particularly in neurons, all axonal microtubules have a uniform plus-end-distal polarity, whereas dendrite microtubules are nonuniform in their polarity orientations, with approximately equal numbers of microtubules having plus-end and minus-end-distal orientations (Burton, 1988; Baas et al., 1988, 1989).

Concerning microtubule elongation, a model was proposed on the basis of a simple mathematical equation (Oosawa and Askura, 1975) that describes the incorporation of tubulin into the microtubule polymer until a concentration of unpolymerized tubulin equal to that of the critical concentration required for the initiation of the process is reached. At that point the net incorporation of tubulin into microtubules is zero, and there is an equilibrium between unpolymerized and polymerized tubulin. This model implies that no chemical changes occur during the assembly process.

However, when this theoretical model is compared with experimental observations performed on living fibroblasts, it does not fit. Microtubules within the cell are more dynamic (i.e., have a shorter half-life) than that predicted by the model. This may be due to chemical changes in tubulin during the polymerization process. Indeed, there is GTP hydrolysis during the assembly process (Carlier, 1982).

Tubulin is a GTP-binding protein (Weisenberg and Deery, 1976; Linse and Mandelkow, 1988; Shivanna et al., 1993) containing one binding site in each α- and β-tubulin subunit. It has been suggested that GTP binds in an exchangeable manner to the β subunit, whereas the binding of the other GTP molecule to the α subunit is nonexchangeable (Carlier, 1982). Unpolymerized tubulin has low GTPase activity (Carlier and Pantaloni, 1981), and exchangeable GTP is hydrolyzed only after tubulin is incorporated into the polymer (Carlier and Pantaloni, 1981). For other GTP-binding proteins with a low GTPase activity, it has been described that GTP hydrolysis is accelerated when the protein binds to another protein (GAP), which activates the GTPase activity (Pai et al., 1989). Thus, tubulin may act as its own GAP upon interaction of different heterodimers in the polymer (Erikson and O'Brien, 1992).

The possible correlation between tubulin polymerization and GTP hydrolysis constitutes the basis for another theoretical model to explain microtubule dynamics, which has been referred to as the "treadmilling" model (Margolis and Wilson, 1978). This model predicts that, at a monomer–polymer equilibrium in which the net quantity of assembled microtubules is maintained constant, the polymerized microtubules will show a similar average length (Margolis and Wilson, 1978).

However, two experimental observations argue against this treadmilling model: (a) microtubule polymerization and GTP hydrolysis are indeed uncoupled, as there is a lag between these two events (Carlier and Pantaloni, 1981) and (b) the analysis of a large population of *in vitro*-assembled microtubules shows a great divergence in polymer length, which increases when the incubation time of the polymerization assay is prolonged (Mitchison and Kirschner, 1984). The latter observation suggests the presence of two microtubule populations in conditions under which the net polymer amount is maintained: a growing microtubule population and a shrinking microtubule population (Mitchison and Kirschner, 1984).

Considering the previous observations, Mitchison and Kirschner (1984) proposed the current theoretical model to explain microtubule dynamics. This "dynamic-instability" model (Figure 2) indicates that there is slow polymerization (elongation or growth) for some microtubules and fast depolymerization (shortening) for others. These two processes could be regulated by two different (polymerization, or growth, and depolymerization, or shortening) rate constants. In addition, as a transition in a microtubule from slow elongation to rapid shortening (an event called catastrophe) is possible and the reverse transition from shortening to elongation (termed rescue) is also possible, two transition frequencies (the frequencies of catastrophe and rescue events) should also be considered. Thus, four mathematical parameters control microtubule dynamics (Kirschner and Mitchison, 1986).

Briefly, the Mitchison and Kirschner, or dynamic-instability model, indicates that unpolymerized tubulin binds GTP, and GTP-bound tubulin is then incorporated into the microtubule; subsequently, GTP bound to tubulin may be hydrolyzed to GDP. Microtubules containing GDP-bound tubulin are less stable than those containing

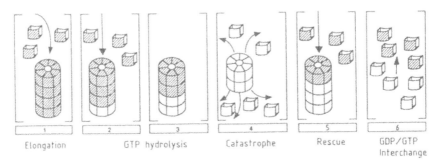

Figure 2. Dynamic instability of microtubules. Soluble GTP-bound tubulin (filled blocks) has the ability to polymerize into microtubules which thus become elongated (1). Once tubulin is bound to the polymer, the GTP on tubulin is hydrolyzed to GDP. In this way, GTP-bound tubulin is converted into GDP-bound tubulin (white blocks) (2 and 3). When the proportion of GTP-tubulin at a microtubule end decreases below a threshold level, the microtubule polymer starts to rapidly depolymerize, an event which is referred to as "catastrophe" (4). If a microtubule end incorporates GTP-bound tubulin back, it stops depolymerizing, which is "rescue" (5). Additionally, soluble GDP-tubulin can interchange GDP for GTP, yielding GTP-tubulin that may polymerize into microtubules (6).

GTP-bound tubulin. When the GDP-tubulin/GTP-tubulin ratio reaches a threshold, the microtubule depolymerizes rapidly, and catastrophe occurs. After this, some microtubules (only very few) may again incorporate GTP-bound tubulin and consequently be rescued. In addition, depolymerized GDP-tubulin can interchange GDP and GTP-bound tubulin is able to polymerize into new microtubules, and a new cycle can begin.

B. Factors that May Modify Microtubule Dynamics

Distinct microtubule populations may show different dynamic characteristics. In other words, there may be polymers with different stabilities, and these stability differences may be correlated with the various functions in which microtubules are involved (Avila, 1991). Of great relevance in this regard is the observation and quantification of the behavior of individual microtubules *in vivo*. Although the dynamic instability of microtubules has been directly observed in living cells after microinjection of fluorescently-labeled tubulin (Cassimeris et al., 1988; Sammak and Borisy, 1988; Schulze and Kirschner, 1988; Shelden and Wadsworth, 1993), some differences between microtubule dynamic behavior *in vitro* and *in vivo* have been shown. Thus, microtubules can remain for many minutes in a metastable state *in vivo*, exhibiting neither growth nor shrinkage (Samumack and Borisy, 1988; Schulze and Kirschner, 1988), whereas this is rarely seen *in vitro* (Walker et al., 1988). The average microtubule growth rate is higher in living cells than it is *in vitro* (Cassimeris et al., 1988); the frequency of catastrophes and rescues is also higher *in vivo* than *in vitro* (Walker et al., 1988). Finally, there are striking differences in the dynamic properties of microtubules from different cell types. As a case in point, the average rates of growth and shortening are significantly lower in epithelial cells than they are in fibroblasts; the frequency of rescue events is significantly higher in epithelial cells than in fibroblasts, and the frequency of catastrophe events is similar for both cell types (Shelden and Wadsworth, 1993). This means that microtubules in epithelial cells are less dynamic than microtubules in fibroblasts (Pepperkok et al., 1990; Wadsworth and McGrail, 1990; Shelden and Wadsworth, 1993). In neurons, microtubule dynamics are progressively decreased during the development and maturation of axons and dendrites (Lim et al., 1989; Baas et al., 1991). The microtubule dynamic instability behavior seems thus to be specifically regulated *in vivo*, suggesting the existence of factors able to control microtubule dynamics.

In some cases, the different dynamic characteristics have been correlated to the presence of distinct tubulin isoforms. For example, microtubules from the marginal band of chicken erythrocytes contain an unique and specific β-tubulin isoform and are much more stable than those containing other β-tubulin isoforms (Murphy, 1991). The main structural differences between this erythrocyte β-tubulin isoform and the others are located in the carboxy-terminal region (Murphy, 1991; Wang et al., 1986), which contains the binding site for MAPs (Serrano et al., 1985; Littauer et al., 1988; Cross et al., 1991).

Indeed, the presence of MAPs seems to be the major factor controlling the dynamic characteristics of microtubules in most cell types. MAPs bound to individual microtubules *in vitro* increase the microtubule growth rate and the frequency of rescue events, and decrease the frequency of catastrophe events and the rate of microtubule depolymerization (Bré and Karsenti, 1990; Horio and Hotani, 1986; Hotani and Horio, 1988). The significance of MAPs in the regulation of the assembly and dynamics of microtubules *in vivo* is reviewed below (see section below on MAPs).

The stabilization of microtubules is facilitated not only by their lateral interaction with MAPs, but also by the association of microtubule ends with other subcellular structures (Kirschner, 1980). The stabilization of some mitotic microtubules by kinetochores constitutes a clear case in point (see below, Microtubule Dynamics in Interphase and Mitosis). Thus, the fast dynamic characteristics of some microtubules may favor the rapid recycling of those microtubules that have missed their proper targets within the cell (Erikson and O'Brien, 1992).

C. Microtubule Dynamics in Interphase and Mitosis

Perhaps the most striking example of modulation of microtubule dynamics *in vivo* occurs during the transition from interphase to mitosis in proliferating cells. The few long and relatively stable cytoplasmic microtubules nucleated by the centrosome disappear at the beginning of mitosis and are converted into the numerous short and highly dynamic microtubules nucleated by the mitotic poles (which result from the division of the centrosome) (Karsenti, 1991; Mitchison, 1989). All of these microtubules have a uniform polarity orientiation with their "plus" ends distal to the mitotic poles.

The plus ends of some of these mitotic microtubules are captured by the kinetochores, which are specialized nucleoprotein structures present in the centromeric regions of mitotic chromosomes (Hayden et al., 1990). The attachment of some microtubules nucleated by the mitotic poles to the kinetochores results in their selective stabilization and the subsequent formation of the mitotic spindle in prometaphase (Hayden et al., 1990; Karsenti et al., 1984; Karsenti, 1991; Mitchison, 1989; Sawin and Mitchison, 1991).

The molecular mechanisms responsible for the increased microtubule–nucleating ability of mitotic poles (Centonze and Borisy, 1990; Kuriyama and Borisy, 1981; Buendía et al., 1992; Ohta et al., 1993) and for increased microtubule dynamic instability (Belmont et al., 1990; Verde et al., 1990; Gotoh et al., 1991a) occurring at the beginning of mitosis are not yet entirely known, although their dependence on protein phosphorylation is clear (Buendía et al., 1992; Gotoh et al., 1991a; Ohta et al., 1993; Verde et al., 1990; Centonze and Borisy, 1990). Indeed, certain protein kinases are activated and protein phosphorylation becomes prominent at the interphase–mitosis transition. In particular, activation of cdc2 kinase is the key control point at the onset of mitosis (Dunphy et al., 1988; Gautier et al., 1988; Labbé et al., 1988; Nurse, 1990). Addition of purified mitotic cdc2 kinase to *Xenopus* egg extracts results in the activation of the mitotic regime of microtubule dynamics (Verde et al., 1990) and of the nucleating activity of centrosomes (Buendía et al., 1992; Ohta et al., 1993). However, the *in vivo* substrates of this kinase involved in the regulation of microtubule dynamics and nucleation have not been identified. A protein kinase cascade is possibly implicated in the stimulation of microtubule dynamics, as this effect is also observed after addition of a purified MAP kinase to *Xenopus* egg extract (Gotoh et al., 1991a). MAP kinase, originally described as a mitogen-activated protein kinase (Ray and Sturgill, 1988), is also activated downstream of cdc2 kinase in *Xenopus* eggs (Gotoh et al., 1991b). Possibly, the phosphorylation of some MAPs during mitosis facilitates changes in microtubule dynamics (see section below on the MAP2/MAP4/Tau protein family).

D. Markers for Stable Microtubules

Certain posttranslational modifications, including detyrosination, phosphorylation or acetylation, occur mainly on assembled tubulin (Barra et al., 1988; Bulinski et al., 1988; Bulinski and Gundersen, 1991; Gundersen et al., 1987; Serrano et al., 1987; Díaz-Nido et al., 1990a). The probability of modification may thus be higher for stably assembled tubulin. Experimental observations indicate that this is the case, since stable microtubules appear to be modified to a greater degree than dynamic polymers (Bulinski and Gundersen, 1991) and the presence of posttranslationally modified tubulin may be an indication of the stability of the polymers. In particular, highly dynamic microtubules are rich in tyrosinated α tubulin, while the stable polymers are generally deficient in this α-tubulin variant as the consequence of extensive detyrosination (Bulinski et al., 1988; Gundersen et al., 1987). As an example of the use of tyrosinated α tubulin as a marker for microtubules, the proportion and distribution of stable and dynamic microtubules within axons and dendrites of cultured neurons has been studied (Baas et al., 1991; Ahmad et al., 1993). These studies emphasize the existence of regional differences in microtubule dynamics in distinct cellular compartments.

III. ROLE OF MICROTUBULE-ASSOCIATED PROTEINS IN THE REGULATION OF MICROTUBULE ASSEMBLY AND DYNAMICS

A. Microtubule-Associated Proteins and Microtubule-Binding Proteins

A heterogenous group of brain proteins that co-purified with tubulin through several cycles of temperature-dependent *in vitro* polymerization–depolymerization were originally identified as microtubule-associated proteins (MAPs) (Murphy and Borisy, 1975; Weingarten et al., 1975; Sloboda and Rosenbaum, 1979). A large and rapidly increasing number of tubulin-binding proteins has been described (Vallee, 1990; Bloom, 1992; Lee, 1993) and it has become clear that the term "microtubule-associated protein" is no longer adequate for all these proteins, because some of them interact only transiently with tubulin. Alternative designations such as microtubule-binding proteins or microtubule-interacting proteins should be used instead. Here, we will use the term microtubule-associated proteins (MAPs) only for those tubulin-binding proteins able to stimulate the assembly of tubulin and remain bound to the resulting microtubules. All other proteins that interact with microtubules will be referred to as microtubule-binding proteins (MTBPs) (Rickard and Kreis, 1991).

MAPs isolated from mammalian brain and other tissues were conventionally classified into families based on their relative resistance to thermal denaturation and their apparent molecular mass (as measured by sodium dodecyl sulfate (SDS-PAGE) polyacrylamide gel electrophoresis). A group of relatively heat-sensitive polypeptides known as MAP1 proteins was distinguished from another group of MAPs that showed greater resistance to thermal denaturation (Vallee, 1985, Vallee and Bloom, 1984). This latter group of heat-stable MAPs comprised high-molecular-weight polypeptides referred to as MAP2, MAP3 and MAP4, as well as lower molecular weight MAPs known as tau proteins (Cleveland et al., 1977; Olmsted, 1991; Wiche et al., 1991). The recent determination of the primary structures of most of these MAPs by molecular cloning techniques (Lee et al., 1988; Lewis et al., 1988; Noble et al., 1989; Himmler, 1989; Kindler et al., 1990; Goedert et al.,

Table 1. Major Brain Microtubule-Associated Proteins

MAPs	Mr(10^3)	Distribution
MAP 1A	350	Neurons (dendrites, axons). Glia.
MAP 1B	320	Neurons (axons, dendrites). Glia.
MAP 2A, B	270	Neurons (dendrites)
MAP 2C, D	70	Neurons (axons, dendrites). Glia.
MAP 3, 4	180–220	Glia. Non-neuronal cells.
TAU (several isoforms)	50–70	Neurons (mainly axons)

1991; 1992; Chapin and Bulinski, 1991; Aizawa et al., 1991a, West et al., 1991; Hammarback et al., 1991; Langkopf et al., 1992; Couchie et al., 1992) has revealed that mammalian MAPs can actually be classified in two major groups according to the amino acid sequences of their tubulin-binding domains. One group corresponds to the MAP1 protein family, whereas another includes the MAP2/MAP4/MAP3/tau protein superfamily). See Table 1.

The study of MAPs from nonmammalian sources is leading to the discovery of novel MAP families. A protein identical to histone H1 has unexpectedly been found associated with sea urchin sperm flagellar microtubules and is thought to be responsible for the extreme stabilization of ciliary and flagellar microtubules (Multigner et al., 1992). Novel MAPs have also been identified in microorganisms more amenable to genetic analyses than mammalian cells, particularly in yeast (Barnes et al., 1992) and *Trypanosoma* (Hemphill et al., 1992). Interestingly, the MAP identified in *Trypanosoma* has a tubulin-binding domain that is not related to that of any other MAP and does not appear to interact with the carboxy terminus of tubulin, the binding site for all mammalian brain MAPs (Serrano et al., 1984b; 1985; Cross et al., 1991). Nevertheless, here we will focus only on mammalian MAPs, as they are better characterized than MAPs from any other source.

These MAPs bind to tubulin dimers (or oligomers), stimulating their assembly into microtubules, and remain bound to the microtubule surface, thus decreasing the dynamics of microtubules (Pryer et al., 1992). The molecular basis for microtubule stabilization by MAPs is still unknown, although a clue is provided by the fact that the affinity of GTP for tubulin is increased in the presence of MAPs (Hamel et al., 1983). As an intramolecular interaction of the GTP-binding site and the carboxy-terminal region of β tubulin is plausible (Padilla et al., unpublished), the binding of MAPs to the carboxyterminus of tubulin may interfere with such an intramolecular interaction, thus removing any restraint to GTP binding. Given the importance of GTP binding to tubulin, according to the model of Mitchison and Kirschner (1984), this would explain the effect of MAPs on the dynamic instability of microtubules.

MAPs may also influence the protofilament number of microtubules (Bohm et al., 1984). As MAPs are long, fibrous molecules that project outward from the microtubule surface (Vallee and Borisy, 1977; Sloboda and Rosenbaum, 1979), a role for MAPs as "spacer" molecules controlling the distance between microtubules has also been proposed (Black, 1987; Lee and Brandt, 1992; Chen et al., 1992). Curiously, MAPs seem unable to form bridges or cross-links between microtubules, despite the fact that some MAP molecules may form dimers (Wille et al., 1992; Lee and Brandt, 1992; García de Ancos et al., 1993). However, there is some controversy on this point (Lee and Brandt, 1992). MAPs may also serve as anchors for a variety of cytosolic proteins, including several protein kinases (Theurkauf and Vallee, 1983; Serrano et al., 1989; Obar et al., 1989). Finally, putative roles for MAPs in linking microtubules with other cytoskeletal polymers, particularly intermediate filaments (Heimann et al., 1985; Hirokawa et al., 1988a; Leterrier et al., 1982) and actin filaments (Griffith and Pollard, 1982; Selden and Pollard, 1983), as well as in connecting microtubules with other cell organelles such as mitochondria (Linden et al., 1989; Rendon et al., 1990), have been suggested.

As for the other MTBPs, their possible functional roles are even more diverse than those of MAPs. Some MTBPs are enzymes responsible for certain posttranslational modifications of tubulin. These include tubulin-tyrosine carboxypeptidase, the enzyme that detyrosinates tubulin (Argaraña et al., 1980), and pp39 (c-mos), the protein kinase thought to be responsible for the meiosis-specific phosphorylation of tubulin in oocyte spindles (Zhou et al., 1991). Other MTBPs can serve to anchor certain organelles or plasma membrane proteins to microtubules. These include gephyrin, a protein that presumably links membrane glycine receptors to subsynaptic microtubules in neurons (Kirsch et al., 1991; Prior et al., 1992), and a novel myelin/oligodendrocyte-specific protein (Dyer et al., 1991). The larger group of MTBPs undoubtedly corresponds to motor proteins (Bloom, 1992). These are nucleotide-sensitive microtubule-binding proteins with ATPase (or GTPase) activity that allow them either to carry specific organelles on microtubules or to catalyze the sliding of one microtubule on another. These mechano-chemical (energy-transducing) activities may explain the existence of microtubule-based organelle transport and may also contribute to mitotic forces in anaphase. The variety of motor MTBPs, which are grouped in several families (kinesin, dynein, dynamin), is beyond the scope of this review. Finally, a novel group of MTBPs may have a regulatory role on microtubule dynamics. This group includes a microtubule-severing homo-oligomeric protein complex composed of 56-kDa polypeptide subunits (Shiina et al., 1992) that seems to be activated during mitosis in *Xenopus* eggs (Vale, 1991; Shiina et al., 1992).

Even this nonexhaustive review, which only pinpoints certain relevant examples, makes it apparent that the variety of MAPs and MTBPs, together with their high degree of cell type specificity, may be considered responsible for the variety of functions performed by microtubules in eukaryotic cells as well as for the differential regulation of these functions in distinct cell types. In this respect, a great deal of attention has been paid recently to the putative role of MAPs in modulating microtubule assembly and dynamics in mammalian cells.

B. Regulation of Microtubule Assembly and Dynamics by MAPs

We have already mentioned that the extent of tubulin polymerization and the parameters of dynamic instability determined for microtubules assembled *in vitro* from purified tubulin do not represent microtubule assembly or dynamics *in vivo*, presumably because cellular microtubules interact with cell-specific MAPs *in vivo*. The variations in these parameters existing among distinct cell types may therefore be ascribed to the presence of specific MAPs (Shelden and Wadsworth, 1993).

As a case in point, neuronal microtubules seem to be far more stable than microtubules in non-neuronal cells (Baas et al., 1991; Lim et al., 1989; Okabe and Hirokawa, 1988; Seitz-Tutter et al., 1988), a fact that can be correlated with the extremely high concentration of MAPs within neurons as compared with non-neuronal cells (Matus, 1988). This has been corroborated either by microinjection of certain neuron-specific MAPs into non-neuronal cells (Drubin and Kirschner, 1986) or by transfection of these cells with cloned cDNA coding for certain neuronal MAPs (Kanai et al., 1989; 1992; Knops et al., 1991; Takemura et al., 1992; Chen et al., 1992; Weisshaar et al., 1992; Lee and Rook, 1992; Umeyama et al., 1993), which induces stabilization of microtubules *in situ*.

Attenuation of neuron-specific MAP expression in neuronal cells by treatment with antisense oligodeoxynucleotides inhibits the assembly and stabilization of microtubules that accompany neurite extension (Lim et al., 1989), thus blocking neurite outgrowth at distinct stages (Caceres and Kosik, 1990; Caceres et al., 1991; Dinsmore and Solomon, 1991; Brugg et al., 1993). The functional role of neuronal MAPs in controlling *in vivo* microtubule assembly and dynamics during neuronal morphogenesis is therefore well established; however, little is known about the detailed molecular mechanisms responsible for these effects.

Early *in vitro* studies demonstrated that the addition of brain MAPs to pure tubulin solutions enhances the nucleation of microtubule assembly, increases the rate of microtubule elongation and decreases the rate of microtubule disassembly (Sloboda et al., 1976; Murphy et al., 1977a,b; Job et al., 1985). The addition of brain MAPs to isolated centrosomes *in vitro* causes an increase in the number of microtubules nucleated per centrosome and a noteworthy reduction in their instability (Bré and Karsenti, 1990). This is mainly due to a large decrease in the frequency of catastrophe and an increase in the frequency of rescue events (Pryer et al., 1992). However, most of these studies have been performed with complex unfractionated MAPs obtained from brain extracts. Likewise, *in vivo* transfection or microinjection studies cannot rule out the interaction of exogenous MAPs with endogenous cellular proteins, thus giving rise to a complex effect. A recent study showed that unfractionated heat-stable MAPs practically abolish the dynamic instability of tubulin polymers, which is, however, observed in the presence of either purified MAP2 or purified tau proteins (Pryer et al., 1992). This study also indicates that MAP2 is more effective than tau in increasing the frequency of rescue events (Pryer et al., 1992). There is evidence that also favors the possibility that some MAPs (for instance, MAP1) are more effective than others (tau proteins) in promoting microtubule nucleation (Díez et al., 1985). Thus, it has become clear that a thorough analysis of the ability of each MAP to modulate microtubule assembly and dynamics is required. In particular, the structural and functional domains of individual MAPs must be characterized in further detail. The use of pure MAP proteins and protein fragments obtained from surrogate prokaryotic cell expression systems may be useful in this respect (Brandt and Lee, 1993).

C. Regulation of the Functionality of MAPs

The effects of MAPs on microtubule assembly and stabilization can be modulated either by the binding of other regulatory molecules to MAPs or by posttranslational modifications (largely phosphorylation and dephosphorylation) of the MAP molecules.

A case in point is the association of MAPs with calcium/calmodulin (Kumagai et al., 1986; Vera et al., 1988), as it competitively inhibits the binding of MAPs to tubulin *in vitro*, presumably because both calcium/calmodulin and tubulin compete for a common site on MAPs (Padilla et al., 1990). This may constitute a physiologically important regulator mechanism, since MAPs, alternately, might interact with calmodulin and tubulin, depending on the intracellular level of calcium. This would favor or suppress microtubule dynamics in response to those extracellular signals (growth factors, hormones, neurotransmitters, neuromodulators, adhesion molecules) that control intracellular calcium variations.

Reversible posttranslational modifications of MAPs, mainly phosphorylation and dephosphorylation, can also play a major role in the modulation of cellular microtubule assembly and dynamics in response to those extracellular signals (growth factors, hormones, neurotransmitters, neuromodulators, proteases, adhesion molecules) that control the activity of different protein kinases and phosphatases. Reversible phosphorylation of proteins is the best established mechanism for the rapid and efficient regulation of intracellular events (Westheimer, 1987; Fischer and Krebs, 1989), especially in neuronal cells (Walaas and Greengard, 1991). Indeed, several identified MAPs are known to be good substrates for phosphorylation and dephosphorylation, both *in vitro* and *in vivo* (Avila and Díaz-Nido, 1991; Brugg and Matus, 1991; Riederer, 1992). Phosphorylation may either promote or inhibit the binding of MAPs to microtubules, thus influencing microtubule assembly and dynamics (Díaz-Nido et al., 1988, 1990b; Avila and Díaz-Nido, 1991; Brugg and Matus, 1991; Murthy and Flavin, 1983; Jameson et al., 1980; Jameson and Caplow, 1981; Lindwall and Cole, 1984a, b; Rickard and Kreis, 1991; Cheley et al., 1992; Ulloa et al., 1993a; Drechsel et al., 1992; Correas et al., 1992; García de Ancos et al., 1993; Aizawa et al., 1991b; Gustke et al., 1992;

Johnson, 1992; Scott et al., 1993; Yamamoto et al., 1983; Hoshi et al., 1987, 1988, 1992; Baudier et al., 1987). Phosphorylation of MAPs may also control their function as "spacers" between microtubules, as the introduction of many negatively charged phosphate moieties on MAP molecules may lead to an increase in the "stiffness" of these fibrous molecules and to an enhanced electrostatic repulsion between MAP-coated microtubules (Hagestedt et al., 1989; Wille et al., 1992; Lee and Brandt, 1992).

However, the variety of sites on MAP molecules that are modified by different protein kinases and separately or interactively influence the conformation, the affinity for tubulin and the effects of MAP molecules on microtubules are still unknown in most cases. A growing research effort in the identification of phosphorylation sites on MAPs and in the elucidation of the functional consequences of these phosphorylation events is now underway in a number of different laboratories.

Likewise, little is known about the regulation of changes in MAP phosphorylation and in microtubule organization in living cells. In proliferating cells, the mitosis-specific phosphorylation of some MAPs (Vandre et al., 1991; Tombes et al., 1991; Aizawa et al., 1991b) may contribute to the building of the mitotic spindle (see above). In neurons, some neurotransmitter-induced reorganizations in the microtubule cystoskeleton (Bigot et al., 1991) may be due to changes in the phosphorylation state of certain MAPs (Halpain and Greengard, 1990; Montoro et al., 1993; Díaz-Nido et al., 1993). The correlation between the phosphorylation state of some MAPs and the growth of neurites is also well established (Aletta et al., 1988; Díaz-Nido et al., 1988, 1990b; Ulloa et al., 1993a, b).

Further research is clearly required in order that we may understand the signal transduction pathways leading to changes in the phosphorylation state of MAPs in response to distinct extracellular signals, as well as to comprehend the molecular mechanisms underlying the functional effects of these changes in MAP phosphorylation on the assembly and dynamics of cellular microtubules.

In summary, we can hypothesize that, *in vivo*, the state of microtubule assembly and dynamics in definite compartments of cells will depend on

(i)   the presence of specific sets of MAPs;
(ii)   the presence of regulatory molecules able to modify the interaction of MAPs with microtubules; and
(iii) the phosphorylation state of these MAPs.

The regulation of these factors may be largely accomplished by:

(i)   cell specific gene expression, RNA splicing pattern and subcellular-specific compartmentalization of MAPs and regulatory molecules;
(ii)   changes in second messenger levels, for instance in the intracellular concentration of calcium, which may affect regulatory molecules as calmodulin; and

(iii)   modulation of the activities of the protein kinases and phosphatases able to modify specific sites on MAP molecules in response to extracellular signals.

There is currently a great deal of evidence favoring the importance of factors (i) and (iii). MAPs are actually a very heterogeneous group of proteins, individual members showing cell-specific expression as well as a subcellular-specific compartmentalization; and MAP molecules can be phosphorylated on distinct sites with different functional consequences. While the first factor may be instrumental in generating differences in microtubule assembly and dynamics among distinct cell types, and even in different compartments of the same cells, the latter factor (phosphorylation) may allow modulation of the assembly and dynamics of microtubules in particular cell compartments in immediate response to certain extracellular stimuli. We will refer to these factors in the following section, where we will briefly review the MAPs that to date have been better characterized in mammalian tissues (mainly in brain) at a molecular level.

D.   The MAP1 Protein Family

General Characteristics

MAP1 is a protein family consisting of two distinct but related proteins, MAP1A and MAP1B (Schoenfeld et al., 1989; Garner et al., 1990; Langkopf et al., 1992). MAP1B (Bloom et al., 1985) has also been referred to as MAP1.2 (Greene et al., 1983; Aletta et al., 1988), MAP1X (Binder et al., 1984; Calvert and Anderton, 1985) and MAP5 (Riederer et al., 1986). MAP1A has an apparent molecular mass of 350,000, as determined by SDS-PAGE, or 299,000 as calculated from its amino acid sequence (Langkopf et al. 1992). MAP1B has an apparent molecular mass of 320,000 as determined in SDS-PAGE or 255,000 as calculated from its amino acid sequence (Noble et al., 1989). These two MAPs are encoded by two distinct genes (Garner et al., 1990), which show, however, extensive regional amino acid similarities including a positively charged segment close to the amino terminus that contains multiple repeats of a (Lys/Arg) (Lys/Arg) (Glu/Asp) motif (Noble et al., 1989; Langkopf et al., 1992). This amino-terminal protein fragment seems to be the microtubule-binding domain, as inferred from transfection studies (Noble et al., 1989). The functional roles played by other regions highly conserved in both MAP1A and MAP1B molecules are still unknown (Langkopf et al., 1992). A series of 12 imperfect 15 amino acid repeats of unknown function is present close to the carboxy terminus of MAP1B and is absent from the MAP1A molecule (Noble et al., 1989; Langkopf et al., 1992). See Figure 3.

There are three low molecular weight proteins referred to as light chains, LC1 (34 kDa), LC2 (30 kDa) and LC3 (19 kDa), associated with the microtubule-binding domains of both MAP1A and MAP1B (Vallee and Davis, 1983; Schoenfeld et

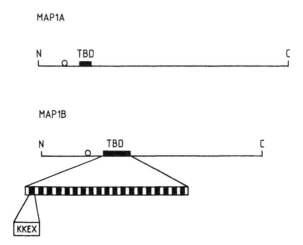

Figure 3. Molecular structure of MAP1 protein. Microtubule-associated proteins MAP 1A and MAP 1B show extensive homology, including a very similar positively-charged segment close to the amino terminus which contains multiple repeats of a (K/R)-(K/R)-(E/D) motif. These repeated sequences seem to be the tubulin-binding domain (TBD) of the molecules.

al., 1989). Curiously, some of these proteins are derived from polyprotein precursors. Thus, MAP1A and LC2 amino acid sequences are translated as a pre-MAP1A/LC2-polyprotein precursor (Langkopf et al., 1992). Likewise, MAP1B and LC1 are coded by a single mRNA that gives rise to a pre-MAP1B/LC1-polyprotein precursor (Hammarback et al., 1991). Although light chains are not strictly needed for the binding of either MAP1A or MAP1B to microtubules (Vallee and Davis, 1983; Noble et al., 1989), microtubule binding may be modulated by their presence. Electron microscopy examination of purified native MAP1A (Shiomura and Hirokawa, 1987) and MAP1B (Sato-Yoshitake et al., 1989) by the rotary shadowing technique has shown that both proteins are long, thin, filamentous and flexible molecules with a small spherical structure at one end. It is plausible that the globular appearance of one end of these molecules is due to the complex between the microtubule-binding domains and the light chains of these MAPs.

Although these MAPs, or immunologically related proteins, are widely distributed in different cultured cell types, they are notably more abundant in neurons than in non-neuronal cells (Vallee et al., 1986; Díaz-Nido and Avila, 1989; Tucker et al., 1989). The expression of these proteins in the mammalian brain is under strong developmental control. MAP1B has been shown to be the first MAP that is expressed in neurons *in situ* (Tucker et al., 1988; 1989; Tucker, 1990). The expression of MAP1B is down-regulated during brain development (Binder et al., 1984; Bloom et al., 1985; Riederer et al., 1986; Tucker et al., 1989; Garner et al., 1990), whereas the expression of MAP1A is up-regulated (Binder et al.,1984;

Tucker et al., 1984; Garner et al., 1990). *In situ* immunohistochemical studies have shown that MAP1B is highly concentrated in developing neurons, particularly within their growing axons, and exhibits a more moderate and widespread expression in all compartments of mature neurons (Riederer et al., 1986; Schoenfeld et al., 1989). MAP1A is mainly abundant in dendrites of mature neurons (Shiomura and Hirokawa, 1987; Huber and Matus, 1984; Bloom et al., 1984; Schoenfeld et al., 1989). MAP1B is practically the only MAP present in the axons of olfactory neurons *in situ* (Viereck et al., 1989) and is also the major MAP in neurites of cultured PC12 (Brugg and Matus, 1988) and neuroblastoma cells (Díaz-Nido et al., 1991). In view of these data, it is tempting to speculate that MAP1B is important in neurite growth in the developing brain and in neurite plasticity in the adult brain, whereas MAP1A may be required for dendrite maintenance in the adult brain.

The essential role of MAP1B in the initiation of neurite growth has recently been demonstrated using antisense oligodeoxynucleotides that block MAP1B expression (Brugg et al., 1993). However, the molecular mechanism responsible for this effect is not yet clear. When MAP1B is expressed in transfected non-neuronal cells, no process extension is observed, although microtubules become partially stabilized (Takemura et al., 1992). In transfected cells, microtubule stabilization by MAP1B is less efficient than that produced by other MAPs (Takemura et al., 1992). However, these results should be interpreted cautiously, as the proteolytic processing of pre-MAP1B/LC1-polyprotein or other posttranslational modifications may not operate properly in transfected cells. In addition, MAP1B may be more efficient in promoting microtubule nucleation than in stabilizing microtubules (Díez et al., 1985). A detailed understanding of the effect of MAP1B (and MAP1A) on microtubule assembly and dynamics is clearly required before advancing any molecular interpretation of their *in vivo* functions.

Phosphorylation

An important role for the phosphorylation of MAP1A and MAP1B can be anticipated: Both MAP1A and MAP1B are extensively phosphorylated in the living brain (Díaz-Nido, 1990b). Phosphorylation of MAP1B has been more thoroughly studied, as it occurs during neurite growth in a variety of cell lines of neuronal origin (Aletta et al., 1988; Díaz-Nido et al., 1988).

The existence of at least two major modes of MAP1B phosphorylation has been recently described. Mode I induces a marked upward shift in the electrophoretic mobility of the protein and might be catalyzed by proline-directed protein kinases, while mode II barely modifies MAP1B electrophoretic mobility and is presumably catalyzed by casein kinase II (Ulloa et al., 1993a,b). This latter mode of phosphorylation may favor the binding of MAP1B to microtubules (Díaz-Nido et al., 1988, 1990b; Ulloa et al., 1993a). The functional consequences of mode I MAP1B phosphorylation are still unknown.

Interestingly, the two modes of MAP1B phosphorylation are independently regulated during brain development. Mode I phosphorylation strongly diminishes during development in most adult brain regions, with the exception of the olfactory bulb where axonal growth from the olfactory epithelium persists into adulthood (Fischer and Romano-Clarke, 1990; Viereck et al., 1989; Ulloa et al., 1993b). Mode I-phosphorylated MAP1B is localized to growing axons and axonal maturation is accompanied by dephosphorylation (Mansfield et al., 1992; Ulloa et al., unpublished results). Mode II MAP1B phosphorylation is maintained in the adult brain (Ulloa et al., 1993b) and is present both in axons and dendrites (Ulloa et al., unpublished results). These results are compatible with a specific role for mode I phosphorylation in regulating the function of this protein during axonal growth and a more general role for mode II phosphorylation in controlling MAP1B binding to microtubules in all neuronal compartments at all developmental stages.

E.   The MAP2/MAP4/Tau Protein Superfamily

General Characteristics

MAP2, MAP4 and tau proteins are protein families belonging to a superfamily of MAPs clearly different from the MAP1 family. The presence of a carboxy-terminal, microtubule-binding domain consisting of a proline-rich, cationic region followed by three or four imperfect tandem repeats containing homologous 18 amino acid motifs is the conserved superfamily feature not found in other MAPs (Goedert et al., 1991; West et al., 1991; Aizawa et al., 1991a). These proteins are also highly resistant to thermal denaturation (Vallee and Bloom, 1984; Vallee, 1985; Hernández et al., 1986). MAP2, MAP4 and tau proteins are filamentous molecules, with their carboxy-terminal portions bound to the microtubule surface and their larger amino-terminal fragments projecting outward from the microtubule wall (Voter and Erickson, 1992; Hirokawa et al., 1988b; Wille et al., 1992).

The molecular basis of microtubule binding in these MAPs has been studied thoroughly. A synthetic peptide containing a single 18 amino acid repeat is sufficient to bind to tubulin, initiate microtubule assembly and stabilize microtubules in vitro, with relatively low efficiency (Joly et al., 1989; Joly and Purrich, 1990; Melki et al., 1991; Yamauchi et al., 1993). However, microtubule binding is increased with a higher number of repeats (Butner and Kirschner, 1991; Goedert and Jakes, 1990). The spacing between repeated sequences on the molecule may be sufficient to allow each of these repeats to bind to neighboring tubulin molecules on the microtubule surface (Lewis et al., 1988). Analogously, repeated sequences may bind tubulin dimers, forcing them to form oligomers that may easily be incorporated into microtubules, thus stimulating microtubule assembly.

However, transfection studies of tau protein fragments (Lee and Rook, 1992; Kanai et al., 1992) and in vitro experiments using recombinant tau and MAP4 protein fragments (Aizawa et al., 1991a; Butner and Kirschner, 1991; Brandt and

Lee, 1993) show that the flanking segments on both sides of the tubulin-binding repeats, particularly the proline-rich cationic segment preceding the repeats, also contribute to the binding of these MAPs to microtubules. In fact, the fragment containing the proline-rich region of the MAP4 molecule has even greater micro-tubule-binding and microtubule-assembly-stimulating abilities *in vitro* than the fragment containing the three repeats (Aizawa et al., 1991a). However, the proline-rich region of the tau molecule is notably less efficient in the binding to tubulin *in vitro* than the fragment containing the three repeats (Butner and Kirschner, 1991) emphasizing the existence of molecular differences between members of the same MAP superfamily. Indeed, tau and MAP2 exhibit far more sequence similarity in the regions flanking the repeats than the corresponding regions of MAP4. This may constitute an important property of tau and MAP2, which are neuronal MAPs as compared with MAP4, which is abundant in nonneuronal cells.

The proline-rich region, together with the three repeats of the tau molecule, are essential for *in vitro* microtubule nucleation. The proline-rich region is not necessary for the growth of centrosome-nucleated microtubules *in vitro*, which requires only the presence of the three repeats (Brandt and Lee, 1993). This opens the possibility of a differential regulation of microtubule nucleation and elongation *in vivo* through the modification of specific regions of the MAP molecules. Also of great relevance is the influence on microtubule organization of the flanking regions on both sides of the repeats. Thus, a recombinant truncated tau fragment containing the repeats and lacking both the amino and carboxy termini induces the bundling of microtubules *in vitro*, whereas either full-length tau protein or a fragment containing the amino-terminal region and the repeats and lacking the carboxy terminus do not bundle microtubules *in vitro* (Brandt and Lee, 1993). It has been suggested that bundling results from charge attraction between microtubules following the neutralization of the acidic carboxy-terminal region of tubulin by MAP binding (Melki et al., 1991) or subtilisin-mediated proteolysis (Serrano et al., 1984a). Accordingly, the large number of prolines in the amino-terminal region and the negatively charged amino acids at the carboxy terminus of tau may hinder microtubules from coming close to one another, thus inhibiting bundling *in vitro* (Brandt and Lee, 1993). However, microtubule bundling has been observed *in vivo* in cells transfected with cDNA for MAP2 and tau proteins (Kanai et al., 1989, 1992; Chen et al., 1992; Weisshaar et al., 1992; Lee and Rook, 1992; Knops et al., 1992; Umeyama et al., 1993). Moreover, the intermicrotubule distance is different between MAP2 and tau-bundled microtubules *in vivo* (Chen et al., 1992). This suggests the existence of specific *in vivo* modifications at certain sites on MAP molecules that may control the formation of microtubule bundles. This again emphasizes the importance of differences between distinct proteins of the same superfamily, as well as the contribution of posttranslational modifications to the regulation of the conformation, and therefore function, of MAP molecules. Indeed, there is ample evidence showing a large heterogeneity of these MAPs in different cell types, thus supporting a role for phosphorylation in controlling their functions.

Heterogeneity

MAP2, MAP4 and tau are each encoded by a single-copy gene, but considerable heterogeneity is generated by alternative splicing of primary transcripts.

There are several forms of MAP2 that arise from an alternative splicing that is under developmental control in neurons. A 9-Kb mRNA codes for a high molecular weight protein called MAP2B (1828 amino acids, apparent molecular mass of 270,000 as determined in SDS-PAGE), while a 6-Kb mRNA codes for a smaller form referred to as MAP2C (467 amino acids, apparent molecular mass of 70,000 as determined in SDS-PAGE) (Kindler et al., 1990). MAP2C consists of amino-terminal and carboxy-terminal domains of high molecular weight MAP2B joined together but lacking the 1372 amino acid intervening sequence. The carboxy-terminal segment contains the microtubule-binding domain, whereas the amino-terminal domain contains a binding site for the regulatory subunit of cAMP-dependent protein kinase (Obar et al., 1989). There are two forms of high molecular weight MAP2, MAP2A and MAP2B, of which only MAP2B has been fully characterized.

MAP2C is expressed in the developing brain and is strongly down-regulated during brain maturation, MAP2B is expressed in both developing and adult brain, and MAP2A appears only after brain maturation (Nunez, 1988; Matus, 1988; Tucker, 1990). MAP2C appears in postmitotic neuroblasts before the appearance of high molecular weight MAP2 and has a widespread distribution; it is present in neuronal cell bodies, dendrites and axons as well as in glial cells (Tucker, 1990). In contrast, high molecular weight MAP2 is a neuron-specific protein selectively localized in dendrites and neuronal cell bodies (Cáceres et al., 1984; De Camilli et al., 1984; Huber and Matus, 1984; Tucker, 1990). The specific compartmentalization of MAP2 into dendrites may be due to the selective sorting of the corresponding 9-Kb mRNA into dendrites (Garner et al., 1988). It is plausible that distinct MAP2 molecules contribute to microtubule stabilization and organization in different neuronal compartments. Thus, transfection of fibroblasts with cDNA of MAP2C results in the formation of centrosome-independent microtubule bundles (Weisshaar et al., 1992), which are extremely stable (Umeyama et al., 1993). Actually, MAP2C has a greater ability to form more stable microtubule bundles than does tau in transfected cells (Takemura et al., 1992). The spacing between microtubules depends on the MAP expressed in transfected cells. In cells expressing high molecular weight MAP2, the distance is similar to that found in dendrites, whereas the spacing between microtubules in cells expressing MAP2C or tau is shorter and similar to that found in axons (Chen et al., 1992). Practically all microtubules in MAP-induced bundles of transfected cells exhibit a uniform polarity, with the plus ends pointing distally, independently of the MAP expressed (Chen et al., 1992). This indicates that the mixed-polarity orientation of microtubules within dendrites must be determined by factors in addition to high molecular weight MAP2 expression (Chen et al., 1992).

Similarly to MAP2, there are at least five forms of MAP4 generated by alternative splicing (West et al., 1991). These include proteins previously identified as MAP3 (Huber and Matus, 1990). In contrast to other MAPs, MAP4 proteins are predominantly expressed in non-neuronal tissues. In the brain, MAP4 is only expressed in glial cells and in immature neuroblasts (Bulinski and Borisy, 1980; Olmsted et al., 1986; Huber and Matus, 1990).

Finally, there are a large number of tau protein isoforms, also generated by alternative splicing (Himmler, 1989; Goedert et al., 1991, 1992; Couchie et al., 1992; Kosik et al., 1989; Lee, 1990). Several tau proteins with apparent molecular masses ranging from 55,000 to 68,000 (SDS-PAGE) are found in the central nervous system. Isoforms containing three repeated tubulin-binding motifs are predominantly expressed in developing brain, while isoforms containing four tubulin-binding repeats are expressed in the adult brain (Goedert et al., 1991; Kosik et al., 1989; Lee, 1990). These latter tau isoforms are more efficient in *in vitro* microtubule binding (Goedert and Jakes, 1990; Butner and Kirschner, 1991). Additional tau isoforms with an apparent molecular mass of 110,000 (SDS-PAGE) have been identified in the peripheral nervous system (Georgieff et al., 1991). This high molecular weight tau contains four repeats in its microtubule-binding domain, as for adult brain tau proteins, but it has an additional 254 amino acid insertion in the amino-terminal region of the molecule (Goedert et al., 1992; Couchie et al., 1992).

In the brain, tau proteins are mainly associated with axonal microtubules (Binder et al., 1985; Brion et al., 1988); although the presence of some tau proteins within neuronal cell bodies and dendrites has also been reported (Papasozomenos and Binder, 1987). It is generally accepted that the main function for tau *in vivo* could be the stabilization of axonal microtubules. In fact, inhibition of tau expression by antisense oligodeoxynucleotides does not affect the initiation of neurite outgrowth, but blocks axonal elongation and maturation (Cáceres and Kosik, 1990; Cáceres et al., 1991). In support of this view is the stabilization of microtubules observed after microinjection or transfection experiments, in which tau is introduced into non-neuronal cells (Drubin and Kirschner, 1986; Kanai et al., 1989, 1992; Knops et al., 1992; Lee and Rook, 1992; Chen et al., 1992; Montejo et al., unpublished). However, transfection studies have indicated that microtubule stabilization by tau may be less efficient than by MAP2C (Takemura et al., 1992). These results agree with *in vitro* studies showing that purified MAP2 is more effective in augmenting rescue events than tau (Pryer et al., 1992). Detailed analyses suggest that tau generates a partially stable, but still dynamic, state in microtubules (Drechsel et al., 1992). Thus, different tau isoforms may control microtubule stabilization during axonal growth and maturation.

Further comparisons between individual MAP2, MAP4 and tau isoforms in *in vitro* assays using recombinant proteins and protein fragments as well as in transfection studies are clearly needed in order to understand how the structural differences among these MAPs contribute to distinct states of microtubule assembly and stabilization in different cell types and cell compartments.

Phosphorylation

The role played by reversible and multiple phosphorylations and dephosphorylations of these MAPs in the temporal and spatial regulation of microtubule assembly and dynamics in living cells, is another topic of interest (Avila and Díaz-Nido, 1991; Brugg and Matus, 1991; Riederer, 1992). Research in progress is directed towards:

(i)   the identification of the sites on MAP molecules undergoing modification;
(ii)   the determination of the functional consequences of these modifications;
(iii)   the identification of the protein kinases and phosphatases responsible for these modifications; and
(iv)   the characterization of the physiological regulation of these phosphorylation and dephosphorylation events, as well as their alterations in some pathological conditions.

A large number of protein kinases that phosphorylate MAPs have been identified. Some phosphorylation sites, as well as the functional consequences of the modifications, are now being determined. However, much less information about the *in vivo* regulation of these modifications is available.

MAP2 has been identified as one of the preferred *in vitro* substrates for cAMP-dependent protein kinase (Sloboda et al., 1975; Theurkauf and Vallee, 1983), calcium/calmodulin-dependent protein kinase (Yamamoto et al., 1983), protein kinase C (Hoshi et al., 1988) and proline-directed kinases such as MAP2 kinase (Hoshi et al., 1992) and cdc2 kinase (Faruki et al., 1992). See Figure 4. Current evidence suggests that these phosphorylation events may also occur *in vivo* (Tsuyama et al., 1987; Díaz-Nido et al., 1990b; Brugg and Matus, 1991; Riederer, 1992; Montoro et al., 1993; Díaz-Nido et al., 1993). Highly phosphorylated MAP2,

MAP₂

Figure 4. MAP2 phosphorylation by proline-directed protein kinases. The diagram shows the positions on the MAP2B molecule of motifs containing (S/T)-P sequences which are consensus targets for proline-directed protein kinases. White circles represent S-P motifs, whereas (black) filled circles represent T-P motifs. A large cluster of T-P motifs is found immediately before the tubulin-binding domain (which is represented as a white square close to the carboxy terminus of the molecule).

containing up to 46 phosphates per molecule, binds less efficiently to tubulin than underphosphorylated MAP2 containing up to 16 phosphates per molecule (Tsuyama et al., 1987). However, completely dephosphorylated MAP2 seems to be the least efficient in tubulin binding (Brugg and Matus, 1991). *In vitro* studies have shown that extensive phosphorylation of MAP2 with purified protein kinases decreases its binding to tubulin, thus diminishing its ability to promote microtubule assembly (Jameson et al., 1980; Jameson and Caplow, 1981; Murthy and Flavin, 1983; Hoshi et al., 1988, 1992). At least in the cases of protein kinases C and MAP2 kinase, this has been correlated with the phosphorylation of sites on the microtubule binding domain of the MAP2 molecule (Hoshi et al., 1988, 1992). However, nothing is known about the protein kinases responsible for phosphorylation at other sites that can stimulate the binding of MAP2 to microtubules (Brugg and Matus, 1991). The correlation between the phosphorylation state of MAP2 and synaptic plasticity in the developing visual cortex of the cat (Aoki and Siekevitz, 1985) supports a physiological role for MAP2 phosphorylation and dephosphorylation in triggering cytoskeletal changes in response to certain neurotransmitters. In fact, a rapid and selective MAP2 dephosphorylation after activation of NMDA-type glutamate receptors has been described in rat hippocampus (Halpain and Greengard, 1990). MAP2 phosphorylation may be catalyzed by the calcium/calmodulin-dependent phosphatase calcineurin (Montoro et al., 1993) and might lead to a stabilization of the microtubule cytoskeleton (Bigot et al., 1991). On the other hand, high external potassium induces a depression in excitatory synaptic transmission in hippocampus and an increase in the phosphorylation of MAP2 (Díaz-Nido et al., 1993).

MAP4 can also be phosphorylated by different protein kinases, including proline-directed protein kinases such as MAP2 kinase (Hoshi et al., 1992) and cdc2 kinase (Aizawa et al., 1991b). These phosphorylation events seem to take place on the proline-rich region of the microtubule binding domain of MAP4, causing a decrease in the microtubule assembly-promoting activity of the protein (Aizawa et al., 1991b; Hoshi et al., 1992). Both cdc2 kinase (Dunphy et al., 1988; Gautier et al., 1988; Labbé et al., 1988) and MAP2 kinase (Gotoh et al., 1991b) are activated during mitosis and may therefore be responsible for the mitotic-specific increase in the phosphorylation of MAP4 that has been observed in cultured cells (Vandre et al., 1991; Tombes et al., 1991). The phosphorylation of MAP4 during mitosis might contribute to the enhanced dynamics of mitotic microtubules (Belmont et al., 1990; Gotoh et al., 1991a; Verde et al., 1990). However, factors other than MAP4 phosphorylation may be required to explain the dynamics of mitotic microtubules (Faruki et al., 1992).

Finally, tau proteins (Figure 5) are also modified by several protein kinases including cyclic AMP-dependent protein kinase (Pierre and Nunez, 1983. Johnson, 1992; Scott et al., 1993), calcium/calmodulin-dependent protein kinase (Yamamoto et al., 1983; Steiner et al., 1990; Johnson, 1992), protein kinase C (Hoshi et al., 1987; Baudier et al., 1987; Correas et al., 1992), casein kinase I (Pierre and Nunez,

TAU

Figure 5. Tau phosphorylation by proline-directed protein kinases. The diagram shows the positions on the tau molecule of motifs containing either S-P (white circles) or T-P (black filled circles) sequences which are consensus targets for proline-directed protein kinases. TBD refers to the tubulin-binding domain of the molecule.

1983), casein kinase II (Correas et al., 1992) and proline-directed protein kinases such as MAP2 kinase (Drewes et al., 1992; Ledesma et al., 1992), cdc2 kinase (Ledesma et al., 1992; Mawal-Dewan et al., 1992; Vulliet et al., 1992; Liu et al., 1993), and glycogen synthase kinase-3 (Hagder et al., 1992; Mandelkow et al., 1992). Some of the residues modified by these protein kinases have been identified. A serine residue located downstream of the repeats in the carboxy terminus of the molecule can be phosphorylated by calcium/calmodulin-dependent protein kinase (Steiner et al., 1990) and cyclic AMP-dependent protein kinase (Scott et al., 1993). The functional consequences of phosphorylation at this site are not yet clear (Johnson, 1992; Drechsel et al., 1992). Serine residues located on the repeats can be phosphorylated by protein kinase C (Correas et al., 1992) and cyclic AMP-dependent protein kinase (Scott et al., 1993); phosphorylation at these residues may decrease the binding of tau to tubulin (Correas et al., 1992; Johnson, 1992; Scott et al., 1993). Several serine and threonine residues located on the proline-rich region can be phosphorylated by proline-directed protein kinases (Ledesma et al., 1992; Drewes et al., 1992; Vulliet et al., 1992; Hanger et al., 1992; Mandelkow et al., 1992; Ishiguro et al., 1992a, b; Arioka et al., 1993; Liu et al., 1993). Phosphorylation at these sites lowers the affinity of tau for tubulin (Gustke et al., 1992; Drechsel et al., 1992), thus reducing the ability of tau to stabilize microtubules (Drechsel et al., 1992). This type of phosphorylation may control microtubule stabilization during axonal growth in developing neurons (Drechsel et al., 1992; Arioka et al., 1993). Abnormal hyperphosphorylation of tau proteins at these sites in mature neurons may contribute to the microtubule destabilization and cytoskeletal abnormalities associated with certain neurodegenerative disorders including Alzheimer's disease (Ledesma et al, 1992; Drewes et al., 1992; Vulliet et al., 1992; Hanger et al., 1992; Mandelkow et al., 1992; Ishiguro et al., 1992a,b; Arioka et al., 1993; Liu et al., 1993; see also: Trojanowski and Lee, Volume IV in this series). As for MAP2, a certain degree of phosphorylation at specific sites on tau protein is required for an efficient binding to tubulin (García de Ancos et al., 1993). However, neither the modified sites nor the protein kinases implicated in promoting tubulin binding have yet been identified.

SUMMARY AND CONCLUSIONS

It is apparent from the above discussion that microtubule nucleation, elongation and stabilization may be controlled by specific MAPs. Subtle structural differences among distinct MAPs may be responsible for the differences observed in microtubule organization and dynamics among different cell types and even among distinct compartments of the same cell. Finally, the ability of MAPs to control microtubule assembly and dynamics may be modulated by posttranslational modifications (mainly phosphorylation and dephosphorylation) at specific sites on MAP molecules. Factors other than MAPs that influence microtubule organization and dynamics probably include several MTBPs as well as certain organelles such as the MTOCs.

REFERENCES

Ahmad, F.J., Pienkowski, T.P., & Baas, P.W. (1993). Regional differences in microtubule dynamics in the axon. J. Neurosci. 13, 856–866.

Aizawa, H., Emori, Y., Mori, A., Murofushi, H., & Sakai, H. Suzuki, K. (1991a). Functional analyses of the domain structure of microtubule-associated protein 4 (MAP-U). J. Biol. Chem. 266, 9841–9846.

Aizawa, H., Kamijo, M., Ohba, Y., Mori, A., Okuhara, K., Kawasaki, H., Murofushi, H., Suzuki, K., & Yasuda, H. (1991b). Microtubule destabilization by cdc2/H1 histone kinase: Phosphorylation of the "pro-rich" region in the microtubule binding domain of MAP4. Biochem. Biophys. Res. Common. 171, 1620–1626.

Aletta, J.M., Lewis, S.A., Cowan, N.J., & Greene, L.A. (1988). Nerve growth factor regulates both the phosphorylation and steady-state levels of microtubule-associated protein 1.2. (MAP1.2). J. Cell Biol. 106, 1573–1581.

Amos, L.A. (1979). Structure of microtubules. In: Microtubules (Roberts, K. & Hyams, J.S. Eds.). pp. 1–64, Academic Press, London.

Aoki, C. & Siekevitz, P.C. (1985). Ontogenetic changes in the cyclic adenosine 3', 5'-monophosphate-stimulatable phosphorylation of cat visual cortex proteins, particularly of microtubule-associated protein 2 (MAP2): Effects of normal dark rearing and of the exposure to light. J. Neurosci. 5, 2465–2483.

Argaraña, C.E., Barra, H.S., & Caputo, R. (1980). Tubulinyl-tyrosine carboxypeptidase from chicken brain: Properties and partial purification. J. Neurochem. 34, 114–118.

Arioka, M., Tsukamoto, M., Ishiguro, K., Kato, R., Sato, K., Imahori, K., & Uchida, T. (1993). Tau protein kinase II is involved in the regulation of the normal phosphorylation state of tau protein. J. Neurochem. 60, 461–468.

Avila, J. (1990). Microtubule dynamics. FASEB J. 4, 3284–3290.

Avila, J. (1991). Microtubule functions. Life Sciences 50, 327–334.

Avila, J. & Díaz-Nido, J. (1991). Phosphorylation of microtubule protein. In: Encyclopedia of Human Biology (Dulbecco, R. Ed.) Vol. 5, pp. 893–902. Academic Press, Boca Raton, Florida.

Baas, P.W. & Ahmad, F.J. (1992). The plus ends of stable microtubules are the exclusive nucleating structures for microtubules in the axon. J.Cell Biol. 116, 1231–1241.

Baas, P.W. & Joshi, H.C. (1992). Gamma-tubulin distribution in the neuron: Implications for the origins of neuritic microtubules. J. Cell Biol. 119, 171–178.

Baas, P.W., Black, M.M., & Banker, G.A. (1989). Changes in microtubule polarity orientation during the development of hippocampal neurons in culture. J. Cell Biol. 109, 3085–3094.

Baas, P.W., Deitch, J.S., Black, M.M. & Banker, G.A. (1988). Polarity orientation of microtubules in hippocampal neurons: Uniformity in the axon and nonuniformity in the dendrite. Proc. Natl. Acad. Sci. USA 85, 8335–8339.

Baas, P.W., Slaughter, T., Brown, A., & Black, M.M. (1991). Microtubule dynamics in axons and dendrites. J. Neurosci. Res. 30, 134–153.

Barnes, G., Lovie, K.A., & Botstein, D. (1992). Yeast proteins associated with microtubules *in vitro* and *in vivo*. Mol. Biol. Cell 3, 29–47.

Barra, H.S., Arce, C.A., & Arañaga, C.E. (1988). Posttranslational tyrosination/detyrosination of tubulin. Mol. Neurobiol. 2, 133–153.

Baudier, J., Lee, S-H., & Cole, R.D. (1987). Separation of the different microtubule-associated tau protein species from bovine brain and their mode II phosphorylation by Ca^{2+}/phospholipid-dependent protein kinase C. J. Biol. Chem. 262, 17584–17590.

Beese, L., Stubbs, G., & Cohen, C. (1987). Microtubule structure at 18Å resolution. J. Mol. Biol. 194, 257–262.

Belmont, L.D., Hyman, A.A., Sawin, K.E., & Mitchison, T. (1990). Real-time visualization of cell cycle-dependent changes in microtubule dynamics in cytoplasmic extracts. Cell 62, 579–589.

Bigot, D., Hunt, S., & Matus, A. (1991). Reorganization of the neuronal cytoskeleton following stimulation with excitatory amino acids. Eur. J. Neurosci. 3, 551–558.

Binder, L.I., Frankfurter, A., & Rebhun, L.I. (1985). The distribution of tau in the mammalian central nervous system. J. Cell Biol. 101, 1371–1378.

Binder, L.I., Frankfurter, A., Kim, H., Cáceres, A., Payne, M.R., & Rebhun, L.I. (1984). Heterogeneity of microtubule-associated protein 2 during rat brain development. Proc. Natl. Acad. Sci. USA 81, 5613–5617.

Black, M.M. (1987). Comparison of the effects of microtubule-associated protein 2 and tau on the packing density of *in vitro* assembled microtubules. Proc. Natl. Acad. Sci. USA 84, 7783–7787.

Bloom, G.S. (1992). Motor proteins for cytoplasmic microtubules. Curr. Opin. Cell. Biol. 4, 66–74.

Bloom, G.S., Luca, F.C., & Vallee, R.B. (1985). Microtubule-associated protein 1B: Identification of a major component of the neuronal cytoskeleton. Proc. Natl. Acad. Sci. USA 82, 5404–5408.

Bloom, G.S., Schoenfeld, T.A., & Vallee, R.B. (1984). Widespread distribution of the major polypeptide component of MAP1A in the nervous system. J. Cell Biol. 98, 320–330.

Bohm, K.J., Vater, W., Fenske, H., & Unger, E.C. (1984). Effect of microtubule-associated proteins on the protofilament number of microtubules assembled *in vitro*. Biochim. Biophys. Acta. 800, 119–126.

Brandt, R. & Lee, G. (1993). Functional organization of microtubule-associated protein tau: Identification of regions which affect microtubule nucleation and bundle formation *in vitro*. J. Biol. Chem. 268, 3414–3419.

Bré, M.-H. & Karsenti, E. (1990). Effects of brain microtubule-associated proteins on microtubule dynamics and nucleation activity of centrosomes. Cell Motil. Cytoskeleton 15, 88–98.

Bré, M.H., Kreis, T.E., & Karsenti, E. (1987). Control of microtubule nucleation and stability in Madin-Darby Canine Kidney cells: The occurrence of noncentrosomal stable detyrosinated microtubules. J. Cell Biol. 105, 1283–1296.

Bré, M.H., Pepperkok, R., Hill, A.M., Levilliers, N., Ansorge, W., Stelzer, E.H.K., & Karsenti, E. (1990). Regulation of microtubule dynamics and nucleation during polarization in MDCK II cells. J. Cell Biol. 111, 3013–3021.

Brion, J.P., Guilleminot, J., Couchie, D., Flament-Durand, J., & Nunez, J. (1988). Both adult and juvenile tau microtubule-associated proteins are axon-specific in the developing and adult rat cerebellum. Neuroscience 25, 139–146.

Brown, A., Slaughter, T.S., & Black, M.M. (1992). Newly assembled microtubules are concentrated in the proximal and distal regions of growing axons. J. Cell Biol. 119, 867–882.

Brugg, B. & Matus, A. (1988). PC12 cells express juvenile microtubule-associated proteins during NGF-induced neurite outgrowth. J. Cell Biol. 107, 643–653.

Brugg, B. & Matus, A. (1991). Phosphorylation determines the binding of microtubule-associated protein (MAP2) to microtubules in living cells. J. Cell Biol. 114, 735–743.

Brugg, B., Reddy, D., & Matus, A. (1993). Attenuation of microtubule-associated protein 1B expressión by antisense oligodeoxynucleotides inhibits initiation of neurite outgrowth. Neurosci. 52, 489–496.

Buendía, B., Draetta, G., & Karsenti, E. (1992). Regulation of the microtubule-nucleating activity of centrosomes in *Xenopus* egg extracts: Role of cyclin A-associated protein kinase. J. Cell Biol. 116, 1431–1442.

Bulinski, J.C. & Borisy, G.G. (1980). Widespread distribution of a 210,000 mol wt microtubule-associated protein in cells and tissues of primates. J. Cell Bio. 87, 802–808.

Bulinski, J.C. & Gundersen, G.G. (1991). Stabilization and post-translational modification of microtubules during cellular morphogenesis. Bioessays 13, 285–293.

Bulinski, J.C., Richards, J.E., & Piperno, G. (1988). Post-translational modifications of α-tubulin: Detyrosination and acetylation differentiate populations of interphase microtubules in cultured cells. J. Cell Biol. 106, 1213–1220.

Burton, P. (1988). Dendrites of mitral neurons contain microtubules of opposite polarity. Brain Res. 473, 107–115.

Butner, K.A. & Kirschner, M.W. (1991). Tau protein binds to microtubules through a flexible array of distributed weak sites. J. Cell Biol. 115, 717–730.

Cáceres, A. & Kosik, K.S. (1990). Inhibition of neurite polarity by tau antisense oligonucleotides in primary cerebellar neurons. Nature 343, 461–463.

Cáceres, A., Banker, G., Steward, O., Binder, L., & Payne, M. (1984). MAP2 is located to the dendrites of hippocampal neurons which develop in culture. Dev. Brain Res. 13, 314–318.

Cáceres, A., Potrebic, S., & Kosik, K.S. (1991). The effects of tau antisense oligonucleotides on neurite formation of cultured cerebellar macroneurons. J. Neurosci. 11, 1515–1523.

Calvert, R. & Anderton, B.H. (1985). A microtubule associated protein (MAP1.X) which is expressed at elevated levels during the development of the rat cerebellum. EMBO J. 4, 1171–1176.

Caplow, M. (1992). Microtubule dynamics. Curr. Opin. Cell. Biol. 4, 58–65.

Carlier, M.F. (1982). Guanosine 5' triphosphate hydrolysis and tubulin polymerization. Mol. Cell. Biochem. 47, 97–113.

Carlier, M.F. & Pantaloni, D. (1981). Kinetics analysis of guanosine 5'-triphosphate hydrolysis associated with tubulin polymerization. Biochemistry 20, 1918–1924.

Cassimeris, L., Pryer, N.K., & Salmon, E. (1988). Real-time observations of microtubule dynamic instability in living cells. J. Cell Biol. 107, 2223–2231.

Centoze, V.E. & Borisy, G.G. (1990). Nucleation of microtubules from mitotic centrosomes is modulated by a phosphorylated epitope. J. Cell Sci. 95, 405–411.

Chapin, S.J. & Bulinski, J.C. (1991). Non-neuronal 210 KD microtubule-associated protein (MAP4) contains a domain homologous to the microtubule-binding domains of neuronal MAP2 and tau. J. Cell Sci. 98, 27–36.

Cheley, S., Kosik, K.S., Paskevich, P., Bakalis, S., & Bayley, H. (1992). Phosphorylated baculovirus p10 is a heat-stable microtubule-associated protein associated with process formation in Sf 9 cells. J. Cell Sci. 102, 739–742.

Chen, J., Kanai, Y., Cowan, N.J., & Hirokawa, N. (1992). Projection domains of MAP2 and tau determine spacings between microtubules in dendrites and axons. Nature 360, 674–677.

Cleveland, D., Hwo, S.Y., & Kirschner, M.W. (1977). Purification of tau, a microtubule associated protein that induces assembly of microtubules from purified tubulin. J. Mol. Biol. 116, 207–225.

Correas, I., Díaz-Nido, J., & Avila, J. (1992). Microtubule-associated protein tau is phosphorylated by protein kinase C on its tubulin-binding domain. J. Biol. Chem. 267, 15721–15728.

Couchie, D., Mavilia, C., Georgieff, I.S., Liem, R.K., Shelanski, M.L., & Nunez, J. (1992). Primary structure of high molecular weight tau present in the peripheral nervous system. Proc. Natl. Acad. Sci. USA 89, 4378–4381.

Cross, D., Dominguez, J., Maccioni, R.B., & Avila, J. (1991). MAP1 and MAP2 binding sites at the C-termini of β tubulin. Studies with synthetic tubulin peptides. Biochemistry 30, 4362–4366.

De Camilli, P., Miller, P.E., Navone, F., Theurkauf, W.E., & Vallee, R.B. (1984). Distribution of microtubule-associated protein 2 in the nervous system of the rat studied by immunofluorescence. Neurosci. 11, 817–846.

Díaz-Nido, J. & Avila, J. (1989). Characterization of proteins immunologically related to brain microtubule-associated protein MAP1B in non-neural cells. J. Cell. Sci. 92, 607–620.

Díaz-Nido, J., Armas-Portela, R., Correas, I., Domínguez, J.E., Montejo, E., & Avila, J. (1991). Microtubule protein phosphorylation in neuroblastoma cells and neurite growth. J. Cell Sci. Suppl. 15, 51–59.

Díaz-Nido, J., Serrano, L., López-Otín, C., Vanderkerckhove, J., & Avila, J. (1990a). Phosphorylation of a neuronal-specific β-tubulin isotype. J. Biol. Chem. 265, 13949–13954.

Díaz-Nido, J., Serrano, L., Hernández, M.A., & Avila, J. (1990b). Phosphorylation of microtubule protein in rat brain at different developmental stages. Comparison with that found in neuronal cultures. J. Neurochem. 54, 211–222.

Díaz-Nido, J., Serrano, L., Méndez, E., & Avila, J. (1988). A casein kinase II-related activity is involved in phosphorylation of microtubule-associated protein MAP1B during neuroblastoma cell differentiation. J. Cell Biol.106, 2057–2065.

Díaz-Nido, J., Montoro, R., López-Barneo, J., & Avila, J. (1993). High external potassium induces an increase in the phosphorylation of the cytoskeletal protein MAP2 and a depression in excitatory synaptic transmission in rat hippocampus. Eur. J. Neurosci. 5, 818–824.

Díez, J.C., de la Torre, J., & Avila, J. (1985). Different association of different microtubule-associated proteins in different in vitro assembly conditions. Biochim. Biophys. Acta. 838, 32–38.

Dinsmore, J.H. & Solomon, F. (1991). Inhibition of MAP2 expression affects both morphological and cell division phenotypes of neuronal differentiation. Cell. 64, 817–826.

Drechsel, D.N., Hyman, A.A., Cobb, M.H., & Kirschner, M.W. (1992). Modulation of the dynamic instability of tubulin assembly by the microtubule-associated protein tau. Mol. Biol. Cell. 3, 1141–1154.

Drewes, G., Lichtenberg, H.B., Doring, F., Mandelkow, E.M., Biernat, J., Goris, J., Doreé, M., & Mandelkow, E., (1992). Mitogen activated protein (MAP) kinase transforms tau protein into an Alzheimer-like state. EMBO J. 11, 2131–2138.

Drubin, D.G. & Kirschner, M.W. (1986). Tau protein function in living cells. J. Cell. Biol. 103, 2739–2746.

Dunphy, W.G., Brizuela, L., Beach, D., & Newport, J. (1988). The Xenopus cdc 2 protein is a component of MPF, a cytoplasmic regulator of mitosis. Cell 54, 423–431.

Dyer, C.A., Hickey, W.F., & Geisert, E.E. (1991). Myelin/oliogodendrocyte-specific protein: A novel surface membrane protein that associates with microtubules. J. Neurosci. Res. 28, 607–613.

Eddé, B., Rossier, J., LeCaer, J.P., Desbruyeres, E., Gros, F., & Denoulet, P. (1990). Post-translational glutamylation of α–tubulin. Science 247, 83–85.

Erikson, H.P. & O'Brien, E.T. (1992). Microtubule dynamic instability and GTP hydrolysis. Ann. Rev. Biophys. Biomol. Struct. 21, 145–166.

Faruki, S., Doreé, M., & Karsenti, E. (1992). Cdc2 kinase-induced destabilization of MAP2-coated microtubules in Xenopus egg extracts. J.Cell Sci. 101, 69–78.

Fischer, E.H. & Krebs, E.G. (1989). Commentary on "The phosphorylase b to a converting enzyme of rabbit skeletal muscle [Biochim. Biophys. Acta (1956) 20, 150–157]". Biochim. Biophys. Acta. 1000, 297–301.

Fischer, I. & Romano-Clarke, G. (1990). Changes in microtubule–associated MAP1B phosphorylation during rat brain development. J. Neurochem. 55, 328–333.

García de Ancos, S., Correas, I., & Avila, J. (1993). Differences in microtubule binding and self association abilities of bovine brain tau isoforms. J. Biol. Chem. 268, 7976–7982.

Gard, D.L. & Kirschner, M.W. (1985). A polymer dependent increase in phosphorylation of β tubulin accompanies differentiation of a mouse neuroblastoma cell line. J. Cell. Biol. 100, 764–774.

Garner, C.C., Tucker, R.P., & Matus, A. (1988). Selective localization of messenger RNA for cytoskeletal protein MAP2 in dendrites, Nature 336, 674–677.

Garner, C.G., Garner, A., Huber, G., Kozak, C., & Matus, A. (1990). Molecular cloning of MAP1A and MAP1B. Identification of distinct genes and their differential expression in developing brain. J. Neurochem. 55, 146–154.

Gautier, J., Norbury, C., Lohka, M., Nurse, P., & Maller, J. (1988). Purified maturation promoting factor contains the product of a *Xenopus* homolog of the fission yeast cell cycle control gene cdc 2+. Cell 54, 433–439.

Georgieff, J.S., Liem, R.K., Mellado, W., Nunez, J., & Shelanski, M.L. (1991). High molecular weight tau: Preferential localization in the peripheral nervous system. J. Cell Sci. 100, 55–60.

Goedert, M. & Jakes, R. (1990). Expression of separate isoforms of tau protein: Correlation with the tau pattern in brain and effects on tubulin polymerization. EMBO J. 9, 4225–4230.

Goedert, M., Crowther, R.A., & Garner, C.C. (1991). Molecular characterization of microtubule-associated proteins tau and MAP2. Trends Neurosci. 14, 193–199.

Goedert, M., Spillantini, M.G., & Crowther, R.A. (1992). Cloning of a big tau microtubule-associated protein characteristic of the peripheral nervous system. Proc. Natl. Acad. Sci. USA 89, 1983–1987.

Gotoh, Y., Nishida, E., Matsuda, S., Shiina, N., Kosako, H., Shiokawa, K., Akiyama, T., Ohta, K., & Skai, H. (1991a). *In vitro* effects on microtubule dynamics of purified *Xenopus* M phase-activated MAP kinase. Nature 349, 251–254.

Gotoh, Y., Moriyama, K., Matsuda, S., Okomura, E., Kishimoto, T., Kawasaki, H., Suzuki, K., Yakara, J., Kasai, H., & Nishida, E. (1991b). *Xenopus* M phase MAP kinase: Isolation of its cDNA and activation by MPF. EMBO J. 10, 2661–2668.

Gozes, I. & Littauer, U.Z. (1978). Tubulin microheterogeneity increase with rat brain maturation. Nature. 276, 411–412.

Greene, L.A., Liem, R.K., & Shelanski, M.L. (1983). Regulation of a high-molecular weight microtubule-associated protein in PC12 cells by nerve growth factor. J. Cell Biol. 96, 76–88.

Griffith, L.M. & Pollard, T.D. (1982). The interaction of actin microfilaments with microtubules and microtubule-associated proteins. J. Biol. Chem. 257, 9143–9151.

Gundersen, G.G., Khawaja, S., & Bulinski, J.C. (1987). Postpolymerization detyrosination of α–tubulin: A mechanism for subcellular differentiation of microtubules. J. Cell Biol. 105, 251–264.

Gustke, N., Steiner, B., Mandelkow, E.M., Biernat, J., Meyer, H.E., Goedert, M., & Mandelkow, E. (1992). The Alzheimer's-like phosphorylation of tau protein reduces microtubule binding and involves Ser-Pro and Thr-Pro motifs. FEBS Lett. 307, 199–205.

Hagestedt, T., Lichtenberg, B., Wille, H., Mandelkow, E-M., & Mandelkow, E. (1989). Tau protein becomes long and stiff upon phosphorylation: Correlation between paracrystalline structure and degree of phosphorylation. J. Cell Biol. 109, 1643–1651.

Halpain, S. & Greengard, P. (1990). Activation of NMDA receptors induces rapid dephosphorylation of the cytoskeletal protein MAP2. Neuron 5, 237–246.

Hamel, E., del Campo, A.A., Lustbader, J., & Lin, C.M. (1983). Modulation of tubulin-nucleotide interactions by microtubule associated proteins. Biochemistry 22, 1271–1279.

Hammarback, J.A., Obar, R.A., Hughers, S.M., & Vallee, R.B. (1991). MAP1B is encoded as a polyprotein that is processed to form a complex N-terminal microtubule-binding domain. Neuron 7, 129–139.

Hanger, D.P., Hughes, K., Woodgett, J.R., Brion, J.P., & Anderton, B.H. (1992). Glycogen synthase kinase 3 induces Alzheimer's disease-like phosphorylation of tau: Generation of paired helical filament epitopes and neuronal localisation of the kinase. Neurosci Lett. 147, 58–62.

Hayden, J.H., Bowser, S.S., & Rieder, C.L. (1990). Kinetochores capture astral microtubules during chromosome attachment to the mitotic spindle: Direct visualization in live newt lung cells. J. Cell Biol. 111, 1039–1045.

Heimann, R., Shelanski, M.L., & Liem, R.K.H. (1985). Microtubule-associated proteins bind specifically to the 70-KDa neurofilament protein. J. Biol. Chem. 260, 12160–12166.

Hemphill, A., Affolter, M., & Seebeck, T. (1992). A novel microtubule-binding motif identified in a high molecular weight microtubule-associated protein from *Trypanosoma brucei*. J. Cell Biol. 117, 95–103.

Hernández, M.A., Avila, J., & Andreu, J.M. (1986). Physicochemical characterization of the heat stable microtubule associated protein MAP2. Eur. J. Biochem. 92, 1–8.

Himmler, A. (1989). Structure of the bovine tau gene: Alternatively spliced transcripts generate a protein family. Mol. Cell Biol. 9, 1389–1396.

Hirokawa, N., Hisanaga, S.L., & Shiomura, Y. (1988a). MAP2 is a component of cross bridges between microtubules and neurofilaments in the neuronal cytoskeleton: Quick-freeze, deep-etch inmunoelectron microscopy and reconstitution studies. J. Neurosci. 8, 2769–2779.

Hirokawa, N., Shiomura, Y., & Okabe, S. (1988b). Tau proteins: The molecular structure and mode of binding on microtubules. J. Cell Biol. 107, 1449–1459.

Horio, T. & Hotani, H. (1986). Visualization of the dynamic instability of individual microtubules by dark-field microscopy. Nature 321, 605–607.

Hoshi, M., Akiyama, T., Shinohara, Y., Miyata, Y., Ogawara, N., Nishida, E., & Sakai, H. (1988). Protein-kinase C catalyzed phosphorylation of the microtubule-binding domain of microtubule-associated protein 2 inhibits its ability to induce tubulin polymerization. Eur. J. Biochem. 174, 225–230.

Hoshi, M., Nishida, E., Miyata, Y., Sakai, H., Miyoshi, T., Ogawara, H., & Akiyama, T. (1987). Protein kinase C phosphorylates tau and induces its functional alterations. FEBS Lett. 217, 237–241.

Hoshi, M., Ohta, K., Gotoh, Y., Mori, A., Murofushi, H., Sakai, H., & Nishida, E. (1992). Mitogen-activated-protein-kinase-catalyzed phosphorylation of microtubule-associated proteins, MAP2 and MAP4, induces an alteration in their function. Eur. J. Biochem. 203, 43–52.

Hotani, T. & Horio, H. (1988). Dynamics of microtubules visualized by dark field microscopy: Treadmilling and dynamic instability. Cell. Motil. Cytoskel. 10, 229–236.

Huber, G. & Matus, A. (1984). Differences in the cellular distribution of two microtubule-associated proteins, MAP1 and MAP2, in rat brain. J. Neurosci. 4, 151–160.

Huber, G. & Matus, A. (1990). Microtubule-associated protein 3 (MAP3) expression in non–neuronal tissues. J. Cell Sci. 95, 237–246.

Ishiguro, K., Omori, A., Takamatsu, M., Sato, K., Arioka, M., Uchida, T., & Imahori, K. (1992a). Phosphorylation sites on tau by tau protein kinase I, a bovine derived kinase generating an epitope of paired helical filaments. Neurosci. Lett. 148, 202–206.

Ishiguro, K., Takamatso, M., Tomizawa, K., Omori, A., Takahasni, M., Arioka, M., Uchida, T., & Imahori, H. (1992b). Tau protein kinase I converts normal tau protein into A68-like component of paired helical filaments. J. Biol. Chem. 267, 10897–10901.

Jameson, L. & Caplow, M. (1981). Modification of microtubule steady state dynamics by phosphorylation of the microtubule-associated proteins. Proc. Natl. Acad. Sci. USA 78, 3413–3417.

Jameson, L., Frey, T., Zeeberg, B., Dalldorf, F., & Caplow, M. (1980). Inhibition of microtubule assembly by phosphorylation of microtubule-associated proteins. Biochemistry 19, 2472–2479.

Job, D., Pabion, M., & Margolis, R.M. (1985). Generation of microtubule stability subclasses by microtubule-associated proteins: Implications for the microtubule "dynamic instability" model. J. Cell Biol. 101, 1680–1689.

Johnson, G.V.W. (1992). Differential phosphorylation of tau by cyclic AMP-dependent protein kinase and Ca^{2+}/Calmolulin-dependent protein kinase II: Metabolic and functional consequences. J. Neurochem. 59, 2056–2062.

Joly, J.C. & Purich, D.L. (1990). Peptides corresponding to the second repeated sequence in MAP2 inhibit binding of microtubule-associated protein to microtubules. Biochemistry 29, 8916–8920.

Joly, J.C., Flynn, G.C., & Purich, D.L. (1989). The microtubule-binding fragments of microtubule-associated protein 2: Localization of the protease-accessible site and identification of an assembly-promoting peptide. J. Cell Biol. 109, 2289–2294.

Joshi, H.C. & Cleveland, D.W. (1990). Diversity among tubulin subunits: Toward what functional end?. Cell. Motil. Cytoskeleton 16, 159–163.

Joshi, H.C., Placios, M.J., McNamara, L., & Cleveland, D.W. (1992). Gamma-tubulin is a centrosomal protein required for cell cycle–dependent microtubule nucleation. Nature 356, 80-83.

Kanai, Y., Chin, J., & Hirokawa, N. (1992). Microtubule bundling by tau proteins *in vivo*: Analysis of functional domains. EMBO J. 11, 3953–3961.

Kanai, Y., Takamura, R., Oshima, T., Mori, H., Ihara, Y., Yanagisawa, M., Masaki, T., & Hirokawa, N. (1989). Expression of multiple tau isoforms and microtubule bundle formation in fibroblasts transfected with a single cDNA clone. J. Cell Biol. 109, 1173–1181.

Karsenti, E. (1991). Mitotic spindle morphogenesis in animal cells. In: Seminars in Cell Biology (Picknett, Ed. T.M.) Vol. 2 W.B. Saunders Co., Pa.

Karsenti, E. & Maro, B. (1986). Centrosomes and the spatial distribution of microtubules in animal cells. Trends. Biochem. Sci. 11, 460–464.

Karsenti, E., Newport, J., Hubble, R., & Kirschner, M. (1984). Interconversion of metaphase and interphase microtubule arrays, as studied by the injection of centrosomes and nuclei into *Xenopus* eggs. J. Cell Biol. 98, 1730–1745.

Kindler, S., Schulz, B., Goedert, M., & Garner, C.C. (1990). Molecular structure of microtubule-associated protein 2b and 2c from rat brain. J. Biol. Chem. 265, 19679–19684.

Kirsch, J., Langusch, D., Prior, P., Littauer, U.Z., Schmitt, B., & Betz, H. (1991). The 93-KDa glycine receptor-associated protein binds to tubulin. J. Biol. Chem. 266, 2242–2246.

Kirschner, M.W. (1980). Implication of treadmilling for the stability and polarity of actin and tubulin polymers *in vivo*. J. Cell. Biol. 86, 330–334.

Kirschner, M.W. & Mitchison, T. (1986). Beyond self assembly: From microtubules to morphogenesis. Cell. 45, 329–342.

Knops, J., Kosik, K.S., Lee, G., Pardee, J.D., Cohen-Could, L., & McConlogue, L. (1991). Overexpression of tau in a non neuronal cell induces long cellular processes. J. Cell Biol. 114, 725–733.

Kosik, K.S., Orecchio, L.D., Bakalis, S., & Neve, R.L. (1989). Developmentally regulated expression of specific tau sequences. Neuron 2, 1389–1397.

Kumagai, H., Nishida, E., Hotani, S., & Sakai, H. (1986). On the mechanisms of calmodulin-induced inhibition of microtubule assembly *in vitro*. J. Biochem. 99, 521–525.

Kuriyama, R. & Borisy, G.G. (1981). Microtubule-nucleating activity of centrosomes in CHO cells is independent on the centriole cycle, but coupled to the mitotic cycle. J. Cell Biol. 91, 822–826.

Labbé, J.C., Lee, M.G., Nurse, P., Picard, A., & Dorée, M. (1988). Activation at M-phase of a protein kinase encoded by a starfish homologue of the cell cycle control gene cdc 2 +. Nature 335, 251–254.

Langkopf, A., Hammarback, J.A., Muller, R., Vallee, R.B., & Garner, C.C. (1992). Microtubule-associated proteins 1A and LC2. J. Biol. Chem. 267, 16561–16566.

Le Dizet, M. & Piperno, G. (1987). Identification of an acetylation site of *Chlamydomonas* α tubulin. Proc. Natl. Acad. Sci. USA 84, 5720–5724.

Ledesma, M.D., Correas, I., Avila, J., & Díaz-Nido, J. (1992). Implication of brain cdc2 and MAP2 kinases in the phosphorylation of tau protein in Alzheimer's disease. FEBS Lett. 308, 218–224.

Lee, G. (1990). Tau protein: An update on structure and function. Cell Motil. Cytoskeleton 15, 199–203.

Lee, G. (1993). Non-motor microtubule-associated proteins. Curr. Opin. Cell Biol. 5, 88–94.

Lee, G. & Brandt, R. (1992). Microtubule-bundling studies revisited: Is there a role for MAPs? Trends Cell Biol. 2, 286–289.

Lee, G. & Rook, S.L. (1992). Expression of tau protein in non-neuronal cells: Microtubule binding and stabilization. J. Cell Sci. 102, 227–237.

Lee, G., Cowan, N., & Kirschner, M.W. (1988). The primary structure and heterogeneity of tau protein from mouse brain. Science 239, 285–288.

Leterrier, J.F., Liem, R., & Shelanki, M. (1982). Interaction between neurofilaments and microtubule-associated proteins: A possible mechanism for intraorganellar binding. J. Cell Biol. 95, 982–986.

Lewis, S. & Cowan, N.J. (1990). Tubulin genes: Structure, expression and regulation. In Microtubule proteins, (Avila, J. Ed.), pp. 33–66 CRC Press, Boca Raton, Fla.

Lewis, S.A., Wang, D., & Cowan, N.J. (1988). Microtubule associated protein MAP2 shares a microtubule binding motif with tau protein. Science 242, 936–939.

Lim, S.S., Sammak, P.J., & Borisy, G.G. (1989). Progressive and spatially differentiated stability of microtubules in developing neural cells. J. Cell Biol. 109, 253–264.

Linden, M., Nelson, B.D., & Leterrier, J.F. (1989). The specific binding of the microtubule-associated protein 2 (MAP2) to the outer membrane of rat brain mitochondria. Biochem. J. 261, 167–173.

Lindwall, G. & Cole, R.D. (1984a). The purification of tau proteins and the occurrence of two phosphorylation states of tau in brain. J. Biol. Chem. 259, 12241–12245.

Lindwall, G. & Cole, R.D. (1984b). Phosphorylation affects the ability of tau proteins to promote microtubule assembly. J. Biol. Chem. 259, 5301–5305.

Linse, K. & Mandelkow, E.M. (1988). The GTP binding peptide of β tubulin. J. Biol. Chem. 263, 15205–15210.

Littauer, U.Z., Giveon, D., Thierauf, M., Ginzburg, I., & Ponstingl, H. (1988). Common and distinct tubulin binding sites for microtubule associated proteins. Proc. Natl. Acad. Sci. USA 83, 7162–7166.

Liu, W.K., Moore, W.T., Williams, R.T., Hall, F.L., & Yen, S.H. (1993). Application of synthetic phospho- and unphospho-peptides to identify phosphorylation sites in a subregion of the tau molecule, which is modified in Alzheimer's disease. J. Neurosci. Res. 34, 371–376.

Mandelkow, E.M., Drewes, G., Biernat, J., Gustke, N., Van Lint, J., Vandenheede, J.R., & Mandelkow, E. (1992). Glycogen synthase kinase-3 and the Alzheimer-like state of microtubule-associated protein tau. FEBS Lett. 3, 315–321.

Mansfield, S.G., Díaz-Nido, J., Gordon-Weeks, P.R., & Avila, J. (1992). The distribution and phosphorylation of the microtubule-associated protein MAP1B in growth cones. J. Neurocytol. 21, 1007–1022.

Marchesi, V.T. & Ngo, N. (1993). *In vitro* assembly of multiprotein complexes containing α, β and γ tubulin, heat shock protein hsp70 and elongation factor 1α. Proc. Natl. Acad. Sci. USA 90, 3028–3032.

Margolis, R.L. & Wilson, L. (1978). Opposite end assembly and disassembly of microtubules at steady state *in vitro*. Cell 13, 1–8.

Matus, A. (1988). Microtubule associated proteins. Ann. Rev. Neurosci. 11, 29–44.

Mawal-Dewan, M., Se, P.C., Abdel-Ghany, M., Shalloway, D., & Racker, E. (1992). Phosphorylation of tau protein by purified p34 CDC28 and a related protein kinase from neurofilaments. J. Biol. Chem. 267, 19705–19709.

Melki, R., Kerjan, P., Waller, P., Carlier, M.F., & Pantaloni, D. (1991). Interaction of microtubule-associated proteins with microtubules: yeast lysyl-and valyl-tRNA synthetases and tau 218–235 synthetic peptide as model systems. Biochemistry 30, 11536–11545.

Mitchison, T. (1989). Mitosis: basic concepts Curr. Opin. Cell Biol. 1, 67–74.

Mitchison, T. & Kirschner, M. (1984). Dynamic instability of microtubule growth. Nature 312, 237–242.

Mitchison, T.J. (1989). Polewards microtubule flux in the mitotic spindle. J. Cell Biol. 109, 637–652.

Montoro, R., Díaz-Nido, J., Avila, J., & López-Barneo, J. (1993). NMDA stimulates the dephosphorylation of the cytoskeletal protein MAP2 and potentiates excitatory synaptic transmission in rat hippocampal slices. Neuroscience (in press).

Multigner, L., Gagnon, J., Van Dorsselaer, A., & Job, D. (1992). Stabilization of sea urchin flagellar microtubules by histone H1. Nature 360, 33–39.

Murphy, D.B. (1991). Function of tubulin isoforms. Curr. Opin. Cell. Biol. 3, 43–51.

Murphy, D.B. & Borisy, G.G. (1975). Association of high-molecular weight proteins with microtubules and their role in microtubule assembly *in vitro*. Proc. Natl. Acad. Sci. USA 72, 2696–2700.

Murphy, D.B., Johnson, K.A., & Borisy, G.G. (1977a). Role of tubulin-associated proteins in microtubule nucleation and elongation. J. Mol. Biol. 117, 33–52.

Murphy, D.B.,Vallee, R.B., & Borisy, G.G. (1977b). Identity and polymerization-stimulatory activity of the non-tubulin proteins associated with microtubules. Biochemistry 16, 2598–2605.

Murthy, A.S. & Flavin, M. (1983). Microtubule assembly using microtubule-associated protein MAP2 prepared in defined states of phosphorylation with protein kinase and phosphatase. Eur. J. Biochem. 137, 37–46.

Noble, M., Lewis, S.A., & Cowan, N.J. (1989). The microtubule binding domain of microtubule-associated protein MAP1B contains a repeated sequence motif unrelated to that of MAP2 and tau. J. Cell Biol. 109, 3367–3376.

Nunez, J. (1988). Immature and mature variants of MAP2 and tau proteins and neuronal plasticity. Trends Neurosci. 11, 477–479.

Nurse, P. (1990). Universal control mechanism regulating onset of M-phase. Nature 344, 503–508.

Oakley, B.R. (1992). Gamma tubulin: The microtubule organizer? Trends Cell. Biol. 2, 1–5.

Obar, R,A., Dingus, J., Bayley, H., & Vallee, R.B. (1989). The R II subunit of cAMP-dependent protein kinase binds to a common amino-terminal domain in microtubule-associated proteins 2a, 2b and 2c. Neuron 3, 639–645.

Ohta, K., Shiina, N., Okumura, E., Hisanaga, S.J., Kishimoto, J., Endo, S., Goton, Y., Nishida, E., & Sakai, H. (1993). Microtubule nucleating activity of centrosomes in cell-free extracts from *Xenopus* eggs: Involvement of protein phosphorylation and accumulation of pericentriolar material. J. Cell Sci. 104, 125–137.

Okabe, S. & Hirokawa, N. (1988). Microtubule dynamics in nerve cells: Analysis by microinjection of biotinylated tubulin into PC12 cells. J. Cell Biol. 107, 651–664.

Olmsted, J.B. (1991). Non motor microtubule associated proteins. Curr. Opin. Cell. Biol. 3, 52–58.

Olmsted, J.B., Asnes, C.F., Parysek, L.M., Lyon, H.D., & Hildder, G.M. (1986). Distribution of MAP4 in cells and in adult and developing mouse tissues. Ann. N.Y. Acad. Sci. 466, 292–305.

Oosawa, R. & Askura, S. (1975). Thermodynamics of the polymerization of protein. Academic Press, New York.

Padilla, R., Maccioni, R.B., & Avila, J. (1990). Calmodulin binds to a tubulin binding site of the microtubule-associated protein tau. Mol. Cell. Biochem. 97, 35–41.

Pai, E.F., Kabsch, W., Krengel, U., Holmes, K.C., John, J., & Wittinghoter, A. (1989). Structure of the guanine nucleotide binding domain of the Ha-ras oncogene product p21 in the triphosphate conformation. Nature 341, 209–214.

Papasozomenos, S.C. & Binder, J.I. (1987). Phosphorylation determines two distinct species of tau in the central nervous system. Cell Motil. Cytoskel. 8, 210–226.

Pepperkok, R., Bré, M.H., Davoust, J., & Kreis, T. (1990). Microtubules are stabilized in confluent epithelial cells but not in fibroblasts. J. Cell Biol. 111, 3003–3012.

Pierre, P. & Nunez, J. (1983). Multisite phosphorylation of tau proteins from rat brain. Biochem. Biophys. Res. Commun. 119, 212–219.

Ponstingl, H., Krauhs, E., Little, M., Kemptf, T., Hofer Warbinek, R., & Ade, W. (1982). Amino acid sequence of α and β tubulin from pig brain: Heterogeneity and regional similarity to muscle proteins. Cold Spring Harbor. Symp. Quant. Biol. 46, 191–196.

Prior, P., Schmitt, B., Grenningloh, G., Pribilla, I., Multhamp. G., Beyreuther, K., Maulet, Y., Werner, P., Langosch, D., Leirsch, J., & Betz, H. (1992). Primary structure and alternative splice variants of gephyrin, a putative glycine receptor-tubulin linker protein. Neuron 8, 1161–1170.

Pryer, N., Walker, R., Skeen, V.P. Bourn, B.D., Soboeiro, M.F., & Salmon, E.D. (1992). Brain microtubule associated proteins modulate microtubule dynamic instability *in vivo*. J. Cell. Sci. 103, 965–976.

Ray, L.B. & Sturgill, T.W. (1988). Characterization of insulin-stimulated microtubule-associated protein kinase. J. Biol. Chem. 263, 12721–12727.

Rendon, A., Jung, D., & Jancsik, V. (1990). Interaction of microtubules and microtubule-associated proteins (MAPs) with rat brain mitochondria. Biochem. J. 269, 555–556.

Rickard , J.E. & Kreis, T.E. (1991). Binding of pp170 to microtubules is regulated by phosphorylation. J.Biol. Chem. 266, 17597–17605.

Riederer, B., Cohen, R., & Matus, A. (1986). MAP5: A novel brain microtubule-associated protein under strong developmental regulation. J. Neurocytol. 15, 763–775.

Riederer, B.M. (1992). Differential phosphorylation of some proteins of the neuronal cytoskeleton during brain development. Histochem. J. 24, 783–793.

Sackett, D.L., Bhattacharyya, B. and Wolff, J. (1985). Tubulin subunit carboxy-termini determine polymerization efficiency. J. Biol. Chem. 260, 43–48.

Sammak, P.J. & Borisy, G.G. (1988). Direct observation of microtubule dynamics in living cells. Nature 332, 724–726.

Sato-Yoshitake, R., Shiomura, Y., Miyasaka, H., & Hirokawa, N. (1989). Microtubule-associated protein 1B: Molecular structure, localization, and phosphorylation-dependent expression in developing neurons. Neuron 3, 229–238.

Sawin, K.E. & Mitchison, T.J. (1991). Mitotic spindle assembly by two different pathways *in vitro*. J. Cell Biol. 112, 925–940.

Scott, C.W., Spreen, R.C., Herman, J.L., Chow, F.P., Davison, M.D., Young, J., & Caputo, C.B. (1993). Phosphorylation of recombinant tau by cAMP-dependent protein kinase. J. Biol. Chem. 268, 1166–1173.

Schoenfeld, J.A., McKerracher, L., Obar, R., & Vallee, R.B. (1989). MAP1A and MAP1B are structurally related microtubule-associated proteins with distinct developmental patterns in the CNS. J. Neurosci. 9, 1712–1730.

Schulze, E. & Kirschner, M.W. (1988). New features of microtubule behavior observed *in vivo*. Nature 334, 356–359.

Seitz-Tutter, D., Langford, G.M., & Weiss, D.G. (1988). Dynamic instability of native microtubules from squid axons is rare and independent of gliding and vesicle transport. Exp. Cell Res. 33, 504–512.

Selden, S.C. & Pollard, T.D. (1983). Phosphorylation of microtubule-associated proteins regulates their interaction with actin filaments. J. Biol. Chem. 258, 7064–7071.

Serrano, L. & Avila, J. (1985). The interaction between subunits in the tubulin dimer. Biochem. J. 230, 551–556.

Serrano, L., de la Torre, J., Maccioni, R.B., & Avila, J. (1984a). Involvement of the carboxy terminal domain of tubulin in the regulation of its assembly. Proc. Natl. Acad. Sci. USA 81, 5989–5993.

Serrano, L., Avila, J., & Maccioni, R.B. (1984b). Controlled proteolysis of tubulin by subtilisin: Localization of the site for MPA2 interaction. Biochemistry 23, 4675–4681.

Serrano, L., Díaz Nido, J., Wandosell, F., & Avila, J. (1987). Tubulin phosphorylation by casein kinase II is similar to that found *in vivo*. J. Cell Biol. 105, 1731–1739.

Serrano, L., Hernández, M.A., Díaz-Nido, J., & Avila, J. (1989). Association of casein kinase II with microtubules. Exp. Cell Res. 181, 263–272.

Serrano, L., Montejo, E., Hernández, M.A., & Avila, J. (1985). Localization of the tubulin binding site for tau protein. Eur. J. Biochem. 153, 595–600.

Serrano, L., Wandosell, F., de la Torre, J., & Avila, J. (1986a). Proteolytic modification of tubulin. Methods in Enzymol. 134, 179–190.

Serrano, L., Valencia, A., Caballero, R., & Avila, J. (1986b). Localization of the high affinity calcium binding site on tubulin molecule. J. Biol. Chem. 261, 7076–7081.

Shelanski, M.L., Gaskin, F., & Cantor, C.R. (1973). Microtubule assembly in the absence of added nucleotides. Proc. Natl. Acad. Sci. USA 70, 765–768.

Shelden, E. & Wadsworth, P. (1993). Observation and quantification of individual microtubule behavior *in vivo*: Microtubule dynamics are cell-type specific. J. Cell Biol. 120, 935–945.

Shiina, N., Gotoh, Y., & Nishida, E. (1992). A novel homo-oligomeric protein responsible for an MPF-dependent microtubule severing activity. EMBO J. 11, 4723–4731.

Shiomura, Y. & Hirokawa, N. (1987). The molecular structure of microtubule-associated protein 1A (MAP1A) *in vivo* and *in vitro*. An immunoelectron microscopy and quick-freeze, deep-etch study. J. Neurosci. 7, 1461–1469.

Shivanna, B.D., Mejillano, M., Williams, & Himes, R.H. (1993). Exchangeable GTP binding site of β-tubulin. J. Biol. Chem. 268, 127–132.

Sloboda, R.D. & Rosebaum, J.L. (1979). Decoration and stabilization of intact, smooth-walled microtubules with microtubule-associated proteins. Biochemistry 18, 48–55.

Sloboda, R.D., Denther, W.L., & Rosenbaum, J.L. (1976). Microtubule-associated proteins and the stimulation of tubulin assembly *in vitro*. Biochemistry 15, 4497–4505.

Sloboda, R.D., Rudolph, S.A., Rosenbaum, J.L., & Greengard, P. (1975). Cyclic AMP-dependent endogenous phosphorylation of a microtubule-associated protein. Proc. Natl. Acad. Sci. USA 72, 177–181.

Solty, B.J. & Borisy, G.G. (1985). Polymerization of tubulin *in vivo*: Direct evidence for assembly onto microtubule ends and from centrosomes. J. Cell. Biol. 100, 1682–1689.

Steiner, B., Mandelkow, E.M., Biernat, J., Gustke, N., Meyer, H.E., Schmidt, B., Mieskes, G., Söling, M.D., Dreschel, D., Kirschner, M., Goedert, M., & Mandelkow, E., (1990). Phosphorylation of microtubule-associated protein tau: Identification of the sites for Ca^{2t}-calmodulin dependent kinase and relationship with tau phosphorylation in Alzheimer's tangles. EMBO J. 9, 3539–3544.

Takemura, R., Okabe, S., Umeyama, T., Kanai, Y., Cowan, N.J., & Hirokawa, N. (1992). Increased microtubule stability and alpha tubulin acetylation in cells transfected with microtubule-associated proteins MAP1B, MAP2 or tau. J. Cell Sci. 103, 953–964.

Theurkauf, W.E. & Vallee, R.B. (1983). Extensive cAMP-dependent and cAMP-independent phosphorylation of microtubule-associated protein 2. J. Biol. Chem. 258, 7883–7886.

Tombes, R.M., Peloquin, J.G., & Borisy, G.G. (1991). Specific association of an M-phase kinase with isolated mitotic spindles and identification of its substrates as MAP4 and MAP1B. Cell Regul. 2, 861–874.

Tsuyama, S., Terayama, Y., & Matsuyama, S. (1987). Numerous phosphates of microtubule-associated protein 2 in living rat brain. J. Biol. Chem. 262, 10886–10892.

Tucker, R.P. (1990). The roles of microtubule-associated proteins in brain morphogenesis: A review. Brain Res. Rev. 15, 101–120.

Tucker, R.P., Binder, L.I., & Matus, A.I. (1988). Neuronal microtubule-associated proteins in the embryonic avian spinal cord. J. Comp. Neurol. 271, 44–55.

Tucker, R.P., Garner, C.C., & Matus, A. (1989). *In situ* localization of microtubule-associated protein mRNA in the developing and adult rat brain. Neuron 2, 1245–1256.

Ulloa, L., Díaz-Nido, J., & Avila, J. (1993a). Depletion of casein kinase II by antisense oligonucleotide prevents neuritogenesis in neuroblastoma cells. EMBO J. 12, 1633–1640.

Ulloa, L., Avila, J., & Díaz-Nido, J. (1993b). Heterogeneity in the phosphorylation of microtubule-associated protein MAP1B during rat brain development. J. Neurochem. 61, 961–972.

Umeyama, T., Okabe, S., Kanai, Y., & Hirokawa, N. (1993). Dynamics of microtubules bundled by microtubule-associated protein 2C (MAP2C). J. Cell Biol. 120, 451–465.

Vale, R.D. (1991). Severing of stable microtubules by a mitotically activated protein in *Xenopus* egg extracts. Cell 64, 827–839.

Vallee, R.B. (1985). On the use of heat stability as a criterion for the identification of microtubule-associated proteins. Biochem. Biophys. Res. Commun. 133, 128–133.

Vallee, R.B. (1990). Molecular characterization of high molecular weight microtubule associated proteins: Some answers, many questions. Cell Motil. Cytoskeleton 15, 204–209.

Vallee, R.B. & Bloom, G.S. (1984). High molecular weight microtubule associated proteins. Mol. Cell. Biol. 13, 21–76.

Vallee, R.B. & Borisy, G.G. (1977). Removal of the projections from cytoplasmic microtubules *in vitro* by digestion with trypsin. J. Biol. Chem. 252, 377–382.

Vallee, R.B. & Davis, S.E. (1983). Low molecular weight microtubule associated proteins are light chains of microtubule-associated protein 1 (MAP1). Proc. Natl. Acad. Sci. USA 80, 1342–1346.

Vallee, R.B., Bloom, G.S., & Luca, F.C. (1986). Differential structure and distribution of the high molecular weight microtubule-associated proteins, MAP1 and MAP2. Ann. N.Y. Acad. Sci. 466, 134–144.

Vandre, D.D., Centonze, V.E., Peloquin, J., Tombes, R.M., & Borisy, G.G. (1991). Proteins of the mammalian mitotic spindle: Phosphorylation/dephosphorylation of MAP4 during mitosis. J. Cell Sci. 98, 577–588.

Vera, J.C., Rivas, C.I., & Maccioni, R.B. (1988). Heat-stable microtubule protein MAP1 binds to microtubules and induces microtubule assembly. FEBS Lett. 232, 159–162.

Verde, F., Labbé, J.C., Doreé, M., & Karsenti, E. (1990). Regulation of microtubule dynamics by cdc2 protein kinase in cell-free extracts of *Xenopus* eggs. Nature 343, 233–238.

Viereck, C., Tucker, R.P., & Matus, A. (1989). The adult rat olfactory system expresses microtubule-associated proteins found in the developing brain. J. Neurosci. 9, 3547–3557.

Voter, W.A. & Erickson, H.P. (1982). Electron microscopy of MAP2. J. Ultrastruct. Res. 80, 374–382.

Vulliet, R., Halloran, S.M., Braun, R.K., Smith, A.J., & Lee, G. (1992). Proline-directed phosphorylation of human tau protein. J. Biol. Chem. 267, 22570–22574.

Wadsworth, P. & McGrail, M. (1990). Interphase microtubule dynamics are cell type-specific. J. Cell Sci. 95, 23–32.

Walaas, S.I. & Greengard, P. (1991). Protein phosphorylation and neuronal function. Pharmacol. Rev. 43, 299–349.

Walker, R.A., O'Brien, E.T., Pryer, N.K., Soboerio, M.F., Voter, W.A., Erickson, H.P., & Salmon, E.D. (1988). Dynamic instability of individual microtubules analyzed by video light microscopy: Rate constants and transition frequencies. J. Cell Biol. 107, 1437–1448.

Wang, D., Villasante, A., Lewis, S.A., & Cowan, N.J. (1986). The mammalian β tubulin repertoire: Hematopoietic expression of a novel β tubulin isotype. J. Cell. Biol. 103, 1903–1911.

Weingarten, M., Lockwood, A., Hwo, S., & Kirschner, M.W. (1975). A protein factor essential for microtubule assembly. Proc. Natl. Acad. Sci. USA 66, 436–439.

Weisenberg, R.C. & Deery, W.J. (1976). Role of nucleotide hydrolysis in microtubule assembly. Nature 263, 792–793.

Weisshaar, B., Doll, T., & Matus, A. (1992). Reorganisation of the microtubular cytoskeleton by embryonic microtubule-associated protein 2 (MAP2C). Development 116, 1151–1161.

West, R.R., Tenbarge, K.M., & Olmsted, J.B. (1991). A model for microtubule-associated protein 4 structure. J. Biol. Chem. 266, 21886–21896.

Westheimer, F.H. (1987). Why nature chose phosphates. Science 235, 1173–1178.

Wiche, G., Oberkanins, C., & Himmler, A. (1991). Molecular structure and function of microtubule-associated proteins. Int. Rev. Cytol. 124, 217–273.

Wille, H., Mandelkow, E-M., Dingus, J., Vallee, R.B., Binder, L.I., & Mandelkow, E. (1992). Domain structure and antiparallel dimers of microtubule-associated protein 2. J. Struct. Biol. 108, 49–61.

Yaffe, M.B., Farr, G.W., Miklos, D., Horwich, A.L., Sternlicht, M.L., & Sternlicht, H. (1992). TCP-1 complex is a molecular chaperone in tubulin biogenesis. Nature 358, 245–248.

Yamamoto, H.K., Fukunaga, E., Tanaka, E., & Miyamoto, E. (1983). Ca^{2+} and calmodulin-dependent phosphorylation of microtubule-associated protein 2 and tau factor, and inhibition of microtubule assembly. J. Neurochem. 41, 1119–1125.

Yamauchi, P.S., Flynn, G.C., Marsh, R.L., & Purich, D.L. (1993). Reduction in microtubule dynamics *in vitro* by brain microtubule-associated proteins and by a microtubule-associated protein 2 second repeated sequence analogue. J. Neurochem. 60, 817–826.

Zabala, J.C. & Cowan, N.J. (1992). Tubulin dimer formation via the release of α and β tubulin monomers from multimolecules complexes. Cell. Mot. Cytosk. 23, 222–230.

Zhou, R., Oskaesson, M., Paules, R.S., Schulz, N., Cleveland, D., & Vande, G.F. (1991). Ability of the c-mos product to associate with and phosphorylate tubulin. Science 251, 671–675.

MOTOR PROTEINS IN MITOSIS AND MEIOSIS

Tim J. Yen

The Cytoskeleton, Volume 1
Structure and Assembly, pages 87–122.
Copyright © 1995 by JAI Press Inc.
All rights of reproduction in any form reserved.
ISBN: 1-55938-687-8

I. INTRODUCTION

Biologists have held a long fascination with mitosis. Notwithstanding the fact that accurate chromosome segregation is an essential process, the mechanics of chromosome separation poses the cell with a tremendously complex engineering task. At the onset of mitosis, the interphase cytoskeleton and karyoskeleton completely disassemble and the interphase arrays of microtubules reorganize to form the mitotic spindle. Microtubules of uniform polarity (plus-ends away) nucleate radially from each of the duplicated centrosomes to establish a bipolar spindle. There are three major classes of microtubules in the spindle that are categorized according to their location and interactions. *Kinetochore* microtubules link the chromosome to the spindle poles and are essential for chromosome segregation. *Polar* microtubules extend from one pole towards the center of the spindle where they can overlap and interact with microtubules nucleated from the opposite pole. *Astral* microtubules are directed away from the spindle proper, towards the periphery of the cell where they may interact with the cell membrane. The polar and astral microtubules play important roles in establishing and maintaining a bipolar spindle. The three microtubule classes, with their associated motor proteins, play a major role in the chromosome movements of mitosis.

Mitosis is subdivided into five morphologically distinct stages. At *prophase*, the chromatin begins to condense within the nucleus while the duplicated centrosomes separate and move to opposite sides of the nucleus. Microtubules nucleated from each centrosome interact to help establish a bipolar spindle. At *prometaphase* (Figure 1a), the nuclear envelope dissolves and the chromosomes are exposed to a highly dynamic array of microtubules that probe the intracellular space by frequently alternating between growing and shrinking phases. Microtubules that are captured by the kinetochore complex become stabilized and this specifies a series

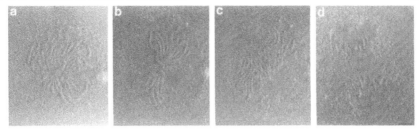

Figure 1. Newt lung epithelial cells at various stages of mitosis, as visualized by differential interference contrast (DIC) microscopy. (a) Early prometaphase; (b) Metaphase; (c) Anaphase; (d) Telophase. Bar = 5µm. Photo courtesy of Dr. Lynn Cassimeris of Lehigh University, Lehigh, PA.

of oscillatory movements that align the chromosomes at the metaphase plate. At *metaphase* (Figure 1b), all the chromosomes align at a position that is equidistant between the two poles. *Anaphase A* commences when the sister chromatids separate and migrate away from each other towards the opposite poles (Figure 1c). After the chromosomes have separated, the spindle poles move apart from each other during anaphase B. At *telophase* (Figure 1d), nuclei begin to reform around the separated chromosomes, the spindle eventually disassembles while cytokinesis bisects the dividing cell into two separate cells.

II. FORCE GENERATION BY THE SPINDLE

The minimum calculated force that is required to move a large chromosome through a cell has been estimated to be approximately 10^{-8} dyn (Nicklas, 1965). However, direct *in vivo* measurements reveal that an anaphase spindle will generate a force that is substantially greater than the calculated force required to move a chromosome (Nicklas, 1983). In these experiments, a calibrated microneedle was hooked to the tip of a separated chromatid undergoing poleward movement. By pulling on the needle, it was experimentally possible to stall movement of this chromosome. By carefully measuring the distance that the needle tip was deflected, the amount of force applied, which must equal the force generated by the spindle, was determined. The results show that the spindle generates a force (7×10^{-5} dyn) that is approximately 10,000 times greater than the calculated force required to move a chromosome. This magnitude of spindle force should cause chromosomes to move significantly faster than the observed velocities. In addition, microtubule-based motors such as flagellar dynein and kinesin can produce velocities of 2 to 10 µm/second, which are nearly 100 times faster than the observed speed of chromosome movement. So the force exerted on the chromosome must be controlled, or *governed*, in some way that prevents chromosomes from moving at the maximal possible speed. The rate of kinetochore microtubule depolymerization could make a good governor because, irrespective of the power of the motor, it should not be any faster than the rate of depolymerization. Alternatively, motors with slower speeds (i.e., smaller step sizes or slower ATPase activity that would increase the time of each mechanochemical cycle) may have evolved to perform this specific task.

A. Chromosomes Encounter Force from Two Directions Within the Spindle

At prometaphase, a single microtubule can make a lateral connection (as opposed to end-on) with the kinetochore, and this is sufficient to allow the mono-oriented chromosome to be pulled polewards at velocities of 0.5 to 1.0 µm/sec (Rieder and Alexander, 1990). This speed is consistent with the use of a minus-end-directed microtubule-based motor such as dynein, which has been localized at kinetochores (Pfarr et al., 1990; Steuer et al., 1990). Although, a mono-oriented chromosome experiences substantial poleward force (Figure 2a), it does not collapse into the pole because of an opposing force that is generated from the vicinity of the spindle pole. This polar-ejection force is thought to be contributed by the frequent and rapid nucleation of microtubules from the poles. The existence of such a force is dramatically demonstrated when a laser beam is used to sever the arms from a mono-oriented chromosome (Rieder et al., 1986). The portion of the chromosome arm that is unattached to the pole is seen to drift away from the pole at 1 to 2 µm/min.

B. Tension Stabilizes Kinetochore-Microtubule Interactions

The initial attachment of a microtubule to one of the two kinetochores at prometaphase orients the sister kinetochore towards the opposite pole so that it can encounter microtubules that originate from that pole. *In vivo*, incorrect connections between kinetochores and microtubules occur (i.e., both kinetochores are attached to microtubules from the same pole) but are highly unstable. In time, the correct connections, whereby each sister kinetochore is connected to the pole that it faces, are made and maintained due to the stability of the interactions. The stability of a bipolar attached chromosome is apparently conferred and maintained by tension. Thus, an unstable interaction between a mono-oriented chromosome and the spindle can be stabilized when a microneedle is used to pull it away from the pole (Ault and Nicklas, 1989). The source of tension is derived from the poleward-directed forces that try to pull the chromosome towards both poles at the same time. Thus, when the microtubule connections are severed between a kinetochore and one of the two poles, the chromosome will move towards the pole that maintained the microtubule attachment (McNeil and Berns, 1981). Likewise, severing the connections between the sister chromatids with a laser beam will result in poleward movement of the separated chromosomes at speeds that are similar to those observed at anaphase (Rieder et al., 1986). The net force exerted on the chromosome is a function of the number of microtubules that is attached to each kinetochore. Reducing the number of microtubules attached to a kinetochore by laser microsur-

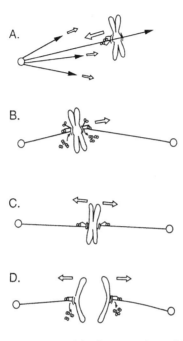

Figure 2. Schematic representation of the forces and possible motors used to translocate chromosomes at different stages of mitosis. (a) At early prometaphase, a mono-
oriented chromosome that makes a lateral connection to a microtubule is pulled
poleward (large arrow) by a minus-end-directed motor such as dynein. The high
frequency of microtubule growth may support a polar ejection force (small arrows)
that opposes the rapid poleward movement of the mono-oriented chromosome. (b) A
bipolar chromosome that has established stable, end-on connections between each
of its sister kinetochores, and microtubules from the opposite poles undergoing
congression. The kinetochore that is furthest from its pole experiences a net poleward
force (large arrow) that is generated by a minus-end motor within the kinetochore.
Plus-end motors at this kinetochore are inactivated (depicted as detached motor). This
poleward force compresses the kinetochore microtubule and stimulates depolymerization from the kinetochore end. Its sister kinetochore responds by inactivating its
minus-end motor (motor that is detached from the microtubule) and activating a
plus-end motor that would translocate towards the end of a growing kinetochore
microtubule. (c) When the forces exerted on the chromosome are balanced, the
chromosomes align equidistant between the poles, at the metaphase plate. (d) At
anaphase, a minus-end motor directs poleward movement of the separated sister
chromatids along a shrinking kinetochore microtubule that is depolymerizing from
the kinetochore end.

gery will displace a chromosome to a position that is closest to the pole with the most kinetochore microtubules (Hays and Salmon, 1990). Likewise, an experimentally derived chromosome with three kinetochores will position itself nearer the pole that faces the two kinetochores (Hays et al., 1982). These results also suggest that the poleward force is proportional to the length of the kinetochore microtubule.

The movement of chromosomes at prometaphase towards the metaphase plate is the result of the complex relationships between poleward forces that are exerted on each kinetochore and microtubule dynamics (Figure 2b). As the kinetochore is pulled towards one pole, the microtubules attached to it must depolymerize, while kinetochore microtubules attached to the opposite pole must simultaneously lengthen. Injection of fluorochrome labeled tubulin subunits into prometaphase cells shows that label is incorporated into the kinetochore microtubules from the kinetochore end (Mitchison, 1988). Examination of cells early after injection shows that label is incorporated into one of the two sister kinetochore microtubule fibers. At longer times, both kinetochore fibers incorporated label but to different extents. These observations suggest that each kinetochore fiber alternates with its sister in its ability to polymerize. Concomitant with the polymerization at one kinetochore, microtubules attached to the sister kinetochore shorten by losing subunits from the kinetochore end. In addition, photobleaching experiments have shown that prometaphase kinetochores will move relative to a marked zone along the kinetochore microtubule (Wise et al., 1991; Cassimeris and Salmon, 1991). These findings, which are similar to the motion of kinetochores during anaphase (Figure 2d; Mitchison et al., 1986; Gorbsky et al., 1987; Nicklas, 1989), suggest that the forces that specify chromosome movement during congression are derived from the kinetochore (Figures 2b). Chromosome alignment is achieved when the spindle is at steady-state and the chromosomes experience no net force from either direction (Figure 2c).

C. Kinetochores Possess Mechanochemical ATPases that Move Microtubules

A closer understanding of kinetochore functions has come from detailed *in vitro* analysis of its interactions with microtubules. By synthesizing microtubules whose minus ends were marked and stabilized against depolymerization, the interaction between the plus-end of the microtubules and kinetochores could be selectively examined. Upon incubation of these microtubules with highly purified chromosomes, the kinetochores captured the plus-ends of the microtubules. When these complexes were diluted to reduce the concentration of unpolymerized tubulin in the reaction, the captured microtubules shrank progressively from the plus-ends, while somehow maintaining attachment to the kinetochore (Koshland et al., 1988). Microtubule depolyermization was rate limiting in these assays, as the extent of microtubule shortening was a function of the final tubulin concentration. Since the depolymerization-driven movement of chromosomes did not require exogenous energy sources (i.e., ATP), sufficient energy must have been released during microtubule depolymerization to translocate the chromosome. In these early experiments, the velocity of depolymerization-driven chromosome movement was estimated to be between 1 to 10 μm/min, a rate consistent with poleward movement of chromosomes at anaphase. More recent advances in technology, which allow real-time observation of this phenomenon, show that the rate of depolymerization-driven chromosome movement (16 μm/min in Coue et al., 1991) can approach the rates (27 μm/min) of rapid depolymerization of the plus-ends of microtubule *in vitro* (Walker et al., 1988). Consistent with the observation that the spindle can generate forces significantly greater than that needed to move a chromosome, the apparent unrestrained use of force in the *in vitro* system will cause chromosomes to move faster than observed *in vivo*.

Kinetochores can move to the plus-ends of microtubules in the presence of ATP (Mitchison and Kirschner, 1985). When microtubules were polymerized from "seeded tubules" captured by kinetochores, the chromosome remained stationary while the microtubules elongated past the kinetochore. Upon addition of ATP, the chromosome translocated towards the plus-end of the newly polymerized microtubule. Since ATP-dependent movement occurred in the absence of microtubule polymerization, the kinetochore must possess a motor-ATPase that specifies plus-end-directed movement along microtubules.

Under experimental conditions whereby microtubule dynamics are inhibited, ATP-dependent motion of kinetochores towards both plus- and minus-ends of microtubules were observed (Hyman and Mitchison, 1991). In this assay, samples were examined in real time by videomicroscopy, so that the rates and direction of movement could be directly observed. The assay was designed so that chromosomes were first adsorbed onto a coverslip, and a highly fluorescent microtubule "seed" added and captured by the kinetochores. Subsequently, a mixture of unlabeled tubulin along with a trace amount of fluorescently tagged tubulin was added to induce polymerization from both ends of the "seed". After stabilizing the polymer with taxol, inspection under low light epifluorescence revealed microtubules that had a very bright region corresponding to the "seed" and then dimly labeled microtubules that extended away in both directions. Plus ends were distinguishable from minus ends because the microtubule length was substantially longer towards the plus end. Since the chromosomes had been immobilized onto the glass surface, motor activity could not be detected by chromosome movement per se, but rather by visualizing a kinetochore-bound microtubule gliding across the surface of the chromosome.

When the kinetochore–microtubule complexes were given ATP, microtubules were seen to be translocated across the kinetochore with their plus ends leading; this would require the activity of a minus-end directed motor. The average speed of 28 μm/min, was similar to the velocity of cytoplasmic dynein, and a comparison of the sensitivity of minus-end movement of the chromosomes to various nucleotide analogs revealed a striking similarity to that of cytoplasmic dynein. These observations suggest that this kinetochore-based movement is due to the motor protein dynein, a view which is consistent with indirect immunocytochemical data showing that anti-dynein antibodies stain the kinetochores. Functionally, the *in vitro* rate of this movement probably reflected the rapid poleward movement of a monovalent chromosome during early prometaphase. On the other hand, this speed is substantially faster than the rates of poleward movement of chromosomes at anaphase A (3–5 μm/min). It remains possible that during anaphase, rate-limiting factors, such as microtubule depolymerization rates, may act as a governor to moderate the velocity of a kinetochore-associated dynein. However, anaphase rates may not have been observed in these experiments since the chromosomes were isolated in a prometaphase-like state after treating cells for prolonged periods with microtubule destabilizing drugs; such chromosomes may lack an anaphase motor or the appropriate modifications that would activate such a cell-cycle-dependent motor.

Plus-end-directed motors that reside at kinetochores have been detected in this *in vitro* system when the minus-end-directed motor is inhibited by phosphorylation (Hyman and Mitchison, 1991). Incubation of chromosomes with ATP-γ-S caused some kinetochore proteins to be thiophosphorylated by a putative kinetochore-associated kinase. Since the thiophosphate group is resistant to phosphatases, the modified proteins, presumably including the motors, are locked into a pseudophosphorylated state. After washing the chromosomes to remove the ATP-γ-S, addition of ATP to these thiophosphorylated kinetochores resulted in microtubules that glided with their minus–ends leading. Characterization of this plus-end motor revealed that it required nearly a 100-fold higher ATP concentration for its activity than that required for the minus-end motor. Furthermore, when compared with a conventional plus-end-directed motor such as kinesin, this kinetochore motor revealed marked differences in nucleotide requirements as well as in significantly slower velocity (3 vs. 30 μm/min). Consistent with this finding, antibodies against sea urchin cytoplasmic kinesin do not stain kinetochores or inhibit mitosis in dividing urchin embryos (Wright et al., 1991; Wright et al., 1993). However, antibodies that are raised against the highly conserved domains within the members of the kinesin superfamily will stain kinetochores (Sawin et al., 1992). This suggests that the novel plus-end motor activity of the kinetochore, although not due to kinesin itself, may be specified by one of these kinesin-related motors.

The presence of a plus-end motor within the kinetochore could be responsible for the "away-from-poles" movements that chromosomes undergo during congression. At the onset of mitosis, the rapid poleward movement of the mono-oriented chromosome is most likely mediated by dynein. However, as the kinetochore becomes attached with increasing numbers of microtubule ends, active plus-end motors, in conjunction with astral exclusion forces, push the chromosome away from the pole. If the astral exclusion force is only limited to a short distance from the pole, the presence of a plus-end motor would be essential for translocating the chromosome far enough away from one pole so that it can interact with microtubules from the opposite pole to establish a bipolar connection.

The distinct pharmacological properties between the plus- and minus-end motor activities of the kinetochore suggest that they are specified by different motor molecules (Hyman and Mitchison, 1991). The role of phosphorylation is unknown but a simple interpretation is that it inhibits the minus-end motor while activating the plus-end motor. The ability of the kinetochore to frequently switch between the two different motors would explain the oscillatory motions of the chromosome during congression. Precisely how these motors are regulated by phosphorylation remains speculative. One attractive hypothesis is that these motor proteins are sensitive to the tension that is exerted on a microtubule (Nicklas, 1983). As a chromosome migrates toward one pole, the kinetochore microtubule attached to that pole is subject to compression, while tension is felt by the kinetochore microtubule that is attached to the opposite pole. Thus, compression may stimulate microtubule shortening through depolymerization, while tension would cause elongation by polymerization (Hill and Kirschner, 1982). Both *in vitro* and *in vivo* data suggest that the changes in microtubule length occur primarily at the more dynamic plus-end that is attached to the kinetochore. Tension and compression may also influence the activities of kinetochore-associated motors in a manner that is consistent with the dynamics of the kinetochore microtubule.

III. THE ROLES OF KINESINS AND DYNEINS IN MITOSIS AND MEIOSIS

The isolation of bona fide microtubule motor proteins such as kinesin and dynein (Vale et al., 1985; Gibbons, 1981), stimulated efforts toward identifying and characterizing, at a molecular level, motor molecules that were predicted to be involved in mitosis. Since these two types of motors moved in opposite directions along microtubules, they were prime candidates for specifying the bidirectional movement of chromosomes during mitosis and meiosis.

A. Kinesin Superfamily

Kinesin is a mechanochemical ATPase that was originally discovered in squid axoplasm as an enzyme complex that moved towards the plus ends of microtubules (Vale et al., 1985). Purifed kinesin exists as a tetrameric complex consisting of two heavy and two light chains (Bloom et al., 1988; Kuznetsov et al., 1988). ATPase and motor activities are specified by the heavy chain. Molecular cloning of the Drosophila kinesin heavy chain cDNA (KHC) revealed a 120-kDa polypeptide consisting of three structural domains (Yang et al., 1989). The domain that specifies motor function resides within the N-terminal 350 amino acids (Yang et al., 1990). This domain is separated from its globular C terminus by an extended α-helical rod region that specifies homodimerization. Drosophila mutants that failed to express KHC died during the early stages of development. The lethal phenotype is not due to defects in cell division but rather to a disruption of their neurological functions (Saxton et al., 1991). This mutant phenotype is consistent with the fact that the Drosophila KHC is a homologue of squid KHC whose main function is axonal transport.

The ability of a Drosophila KHC null mutant to undergo normal cell division suggested that the various motile functions of a cell may be specified by different motor molecules. The identification of motors that are involved in mitosis has come primarily from extensive molecular genetic analysis of large collections of mutant yeasts and Drosophila that exhibited defects in all aspects of cell division. These efforts, along with others, have uncovered a superfamily of kinesin-related motor proteins that have roles in spindle and kinetochore functions. The distinguishing feature among the kinesin superfamily members is a 350-amino-acid motor domain that has invariant subdomains for the nucleotide binding site and regions that are essential for microtubule interactions and mechanical functions. Because the primary structure of the remaining two thirds of the motor domain are not conserved, the overall similarity of this domain among the family members varies between 40% and 65% (Goldstein, 1993). Most members also contain an extended rod

Table 1

Kinesin family member		Possible cellular functions	Motor function[‡] (in vitro motility assays)
	KHC	vesicle transport	+30–40 μm/minute
	bimC	spindle pole	unknown
	cut7	separation and	unknown
BimC	CIN8	maintenance of	unknown
sub	Eg5	bipolar spindle	+2.1 μm/minute
family	KLP6IF		unknown
	KIP1		unknown
		Karyogomy, spindle	
KAR3	KAR3	Kinetochore?	–0.5 μm/minute
sub	ncd	spindle pole separation, MT bundling	–4 μm/minute
family	klpA	spindle pole separation	unknown
	CENP-E	MT bundling (anaphase) kinetochore motor (metaphase)	unknown
	MKLP-1	Antiparallel MT bundling spindle pole separation	+4 μm/minute
	SMY 1	high copy suppressor of myosin null mutant in yeast	unknown

Notes: [‡]+ and – designations denote movement of the motor toward the plus or minus ends of microtubules, respectively.

domain of variable length that separates the motor domain from a highly divergent globular domain. The purpose of this nonconserved domain is believed to provide the ability to recognize a broad variety of "cargo" (i.e., chromosomes, poles, vesicles). This "single motor, many tail hypothesis" explains how kinesins have evolved to carry out multiple cellular functions (Table 1; Vale and Goldstein, 1990).

B. Kinesin-Related Proteins Are Important for Establishing a Bipolar Spindle

In *Aspergillus nidulans*, two groups of cell division mutants, phenotypically distinguished by being *never in mitosis, nim*, and *blocked in mitosis, bim*, allowed identification of a number of genes that have regulatory and structural functions that affect mitosis. The *bimC* gene encodes a 131-kDa kinesin-like protein that was identified by the *bimC4* temperature-sensitive allele (Enos and Morris, 1990). Cells at the restrictive temperature fail to complete mitosis, and the arrested cells are characterized by the presence of a defective spindle that contains two spindle poles at one end. Although microtubules nucleate from this pole, the inability of the duplicated poles to separate blocks the formation of a bipolar spindle. This defect results in the failure of chromosomes to segregate, which in turn, leads to accumulation of multinucleated cells after extended times of incubation at the nonpermissive temperature.

In the fission yeast *Schizosaccharomyces pombe*, temperature-sensitive alleles of the *cut7* gene exhibit a block in spindle pole separation at the restrictive temperature (Hagan and Yanagida, 1990). At mitosis, microtubules do not span between the spindle poles to produce the normal bar-shaped organization but instead are organized into a V shape. If cells are maintained at this temperature for two generations, large cells with two sets of chromosomes and multiple tufts of microtubules accumulate. Molecular characterization of the *cut7* gene revealed that it encodes a member of the kinesin superfamily. Consistent with its *bimC* like phenotype, the motor domain of *cut7* showed more similarity to *bimC* than to KHC (57% vs. 39%).

The notion that there are subfamilies of kinesins that share similar functions has been strengthened by the identification of additional kinesin-related proteins that belong to the *bimC* subfamily. In the budding yeast *Saccharomyces cerevisia*, mutations in the kinesin-related gene *CIN8* lead to increased rate of chromosome loss during normal mitotic growth (Hoyt et al., 1992). When *CIN8* mutants are incubated at elevated temperatures, mononucleate, large-budded cells accumulate, characteristic of a block in mitosis. Examination of these arrested cells by electron microscopy shows the presence of duplicated, but unseparated, spindle poles. This phenotype is consistent with the finding that the motor domain of *CIN8* closely resembles *bimC* and *cut7*. Interestingly, cell viability is unaffected at reduced temperatures when *CIN8* is deleted. This suggests that there are other molecules that have redundant or overlapping functions with *CIN8*. These molecules, however, are unlikely to completely substitute for *CIN8* functions since they are unable to maintain cell viability at higher temperatures. However, suppressor genes, when present in multiple copies, can rescue the temperature sensitive lethal phenotype of *CIN8* mutants (Hoyt et al., 1992). *CIN9*, which encodes a *bimC*like kinesin, is an extragenic suppressor of *CIN8*. The same gene was independently isolated (as *KIP1*) by using degenerate oligoculeotide primers that hybridized to the conserved regions (motor domain) of the kinesin gene to amplify yeast kinesin-related genes by polymerase chain reaction (PCR) (Roof et al., 1992). Consistent with the genetic evidence that *CIN8* and *KIP1/CIN9* genes share overlapping functions is the observation that both gene products are localized immunocytochemically along spindle microtubules. Moreover, because simultaneous deletion of both genes is lethal, it shows that the presence of at least one of these kinesin-related genes is essential for growth.

Genetic analysis indicates that the *bimC* subfamily of kinesins is involved with spindle pole separation at the onset of mitosis. Mechanistically, the separation of duplicated poles that are situated side by side can be brought about either by motors that act along astral microtubules to pull the poles apart, or alternatively, motors may interact (cross-bridge) with the spindle microtubules between the poles to push them apart (Figure 3a; Saunders, 1993). Since *bimC*, *cut7*, *CIN8*, and *CIN9/KIP1* are localized along spindle microtubules, they most likely act by pushing apart the spindle poles. This would require that these motors cross-link the microtubules that nucleate from the adjacent poles. If one end of the motor can bind tightly to one microtubule, when its motor domain tries to move towards the plus-ends of the other microtubule, it will force the cross-linked antiparallel microtubules to slide past each other with the net result of pushing the poles apart. At present the distribution of these putative motors along microtubules remains unknown.

Eg5 is a member of the *bimC* subfamily from the vertebrate *Xenopus* (Le Guellec, 1991) and has been shown *in vitro* to be a plus-end-directed motor that moves at a

A.   B.

Figure 3. Formation and maintenance of a bipolar spindle. (a) Plus-end microtubule motors, such as those that belong to the *bimC* kinesin subfamily, may form cross-bridges between intersecting microtubules that are nucleated from two adjacent poles. Movement of the motor domain towards the plus-end of one microtubule, while its opposite end is rigidly attached to an intersecting microtubule, will push the poles away from each other. (b) A bipolar spindle is maintained by kinesin-like motors that generate opposing forces. *BimC*like kinesins supply the force that push the poles apart. Minus–end motors such as those that belong to the KAR3-subfamily of kinesins provide the counterforce that pulls the poles towards each other. Both families of these kinesin-like motors must somehow cross-bridge the antiparallel microtubules so that force can be exerted in the appropriate direction.

sluggish rate of 3μm/min (Sawin et al., 1992). Eg5 has been localized by im-munofluorescence staining and shown to be in the spindle, with it being concen-trated towards the poles. *In vitro* experiments utilizing *Xenopus* egg extracts to assemble spindles indicate that Eg5 is important for establishing and maintaining spindles. In this system, half-spindles are first formed when chromosomes interact and stabilize centrosome-nucleated microtubules. Subsequent interactions between half-spindles lead to the formation of bipolar spindles. One proposed role of Eg5 is for it to gather the minus-ends of microtubules towards the centrosome by interacting with microtubules and a hypothetical filament system that may be anchored to the poles. If the nonmotor domain of Eg5 can be tethered rigidly to this filament system, its plus-end motor activity would pull the spindle microtubules poleward to stabilize the half-spindle. This model is consistent with data that show that immunodepletion of Eg5 from *Xenopus* egg extracts produced disorganized half-spindles and indirectly blocked formation of bipolar spindles.

C. Bipolar Spindles Are Maintained by a Balance of Opposing Forces

In addition to the role motors play in establishing a bipolar spindle, evidence suggests that motor function is required to maintain pole separation (Figure 3b). Thus, preformed bipolar spindles are disrupted upon incubation with Eg5 antibod-ies. More convincing evidence has derived from the analysis of a temperature-sensitive *CIN8p* and *KIP1Δ* yeast mutant. At the permissive temperature, these cells form a normal bipolar spindle. After the spindle is formed, inactivation of CIN8p by incubation at the restrictive temperature caused the spindle to collapse as the poles retracted back towards each other (Saunders and Hoyt, 1992). The need for the continued presence of *CIN8* and *KIP1* to maintain a bipolar spindle implies that there exists an opposing force that normally is kept in check by the presence of either *CIN8* or *KIP1*. If *CIN8* and *KIP1* are plus-end-directed motors, whose functions are to push the poles apart by cross-linking the overlapping antiparallel microtubules of the spindle, the opposing force must act to pull the poles towards each other. Such a motor, if it resided within the same set of spindle microtubules, would have to specify minus-end-directed movement. The yeast *KAR3* (*kar*yogamy) gene encodes a kinesin-related gene that was predicted to move towards the minus-ends of microtubules, because a defect in this gene inhibited nuclear fusion (Meluh and Rose, 1990). This prediction has been confirmed by the observation that bacterially expressed *KAR3* motor moves towards the minus-ends of microtubules at a rate of 3μm/min (Endow et al., 1994). In cells undergoing normal mitotic growth, the KAR3 protein is localized to the spindle microtubules. Thus *KAR3* might be a motor that pulls the poles together and thus opposes the actions of *CIN8* and *KIP1*. In support of this hypothesis, the lethal temperature-sen-sitive phenotype of the t.s. *CIN8* and *KIP1* double-mutant strain can be partially suppressed by deletion of *KAR3* (Saunders, 1992).

The concept that the bipolar spindles in yeast are stabilized by multiple motors that exert opposing forces (Figure 3b) is further supported by studies in *Aspergillus nidulans. klpA* was identified as a kinesin-related gene by degenerate PCR strategy (O'Connell et al., 1993). Because of its homology with the minus-end kinesins of the *Drosophila* kinesin-related motor called *ncd* and yeast *KAR3*, *klpA* too is predicted to be a minus-end kinesin. All three kinesins share similarities not only within their motor domains but also in the C-terminal location of their motor domains. Although these kinesins do not exhibit obvious similarities in their nonmotor domains, overexpression of *klpA* will complement a *KAR3* null mutant. *klpA* function is not essential for growth as deletion of this gene is not lethal to cells (O'Connell et al., 1993). However, the temperature-sensitive *bimC* mutation, which is blocked in spindle separation, can be rescued when *klpA* is deleted (O'Connell et al., 1993). This result is consistent with the idea that *klpA* and *bimC* have complementary roles in establishing spindles in *Aspergillus*. This hypothesis is further strengthened by the fact that overexpression of *klpA* in otherwise normal cells will cause cells to arrest in mitosis with a monopolar spindle and increased chromosome content. This failure to establish a spindle is due to the elevated levels of *klpA*, which must have overcome a counterforce that is presumably generated by *bimC* and other proteins.

D. Kinesin-Related Proteins That Move Towards the Minus-Ends of Microtubules

The existence of kinesins that can move towards the minus-ends of microtubules could never have been predicted based on sequence comparison. The *in vitro* demonstration that the *Drosophila* kinesin-related protein, *ncd*, moved in the opposite direction (Walker et al., 1990; McDonald et al., 1990) from that of conventional kinesins shattered the dogma that kinesins were all plus-end-directed motors. *Drosophila ncd* belongs to a small subclass of minus-end kinesins that includes *KAR3*, and most likely *klpA*. Like the *bimC* subfamily, the *KAR3* subfamily members exhibit a higher degree of similarity within their motor domains than with other family members. A second shared feature is that the motor domains of this subfamily are not located at the N terminus of the molecule but are deployed at the C terminus. While it made intuitive sense that the "backwards" organization was responsible for the reversal in direction of these motors, testing of this theory proved that directionality was dictated far more subtley by distinct elements in the motor domain. Thus, when the motor domain from conventional KHC was transplanted from the N to the C terminus, this backward motor still moved towards the plus-ends (Stewart et al., 1993).

E. *ncd* and *nod* Are Two *Drosophila* Kinesin-Related Proteins that Are Necessary for Meiosis

The *Drosophila ncd* motor, like its closest relatives, is involved in spindle formation. *ncd* encodes one of two kinesin-related proteins that are important during female meiosis (Endow, 1992; 1993). The *ncd* (*n*onchromosome *d*isjunction, *claret*) gene was originally identified as a mutation that affected chromosome segregation in meiosis of female flies (Lewis and Gencarella, 1952). Because all of the chromosomes undergo high frequency of nondisjunction, the ova from homozygous *ncd* mutant mothers are missing some or all of the chromosomes. Most of the laid eggs are nonviable or are aneuploid. In addition to its role in female meiosis, *ncd* mutation also causes frequent loss of maternal chromosomes during early rounds of mitotic division in the embryo. Abnormal chromosome segregation in *ncd* mutants is due to defects in the establishment of meiotic and mitotic spindles (Hatsumi and Endow, 1992). The spindles of these mutants appear highly disorganized and diffuse, sometimes multipolar, and do not form focused spindle poles. Consistent with these mutant phenotypes, antibodies against *ncd* stain the spindle fibers of normal *Drosophila* oocytes. One potential role for *ncd* is to organize the poles in a meiotic spindle that normally lacks centrioles. The ability of *ncd* to move in the minus direction, coupled with its microtubule-bundling activity (McDonald et al., 1990), suggests that *ncd* might gather and focus the minus-ends of microtubles into spindle poles (Hatsumi and Endow, 1992).

The *nod* gene encodes another kinesin (Zhang et al., 1990) that is also essential for female meiosis (Hawley and Theurkauf, 1993). However, unlike mutations that inactivate *ncd*, *nod* mutants result in nondisjunction of only chromosomes that do not undergo meiotic recombination. In *Drosophila*, the *X* and fourth chromosomes are nonexchange chromosomes and therefore do not form chiasmata. In exchange chromosomes, the chiasmata physically join the paired chromosomes so that they remain together while poleward forces are applied at each sister kinetochore during congression. The distributive segregation system exists to ensure that the nonexchange, achiasmate, chromosomes remain paired during their migration to the metaphase plate. When this mechanism fails, as is the case in *nod* mutants, the nonexchange chromosomes are unable to remain paired, because they are pulled away from each other by poleward forces. These chromosomes fail to align properly and therefore exhibit a high frequency of nondisjunction. Conceptually, the role of *nod* is to counter the poleward forces that are exerted on the nonexchange chromosomes. *Nod* would then be predicted to possess plus-end motor activity that pushes nonexchange chromosomes, toward the metaphase plate (Hawley & Theurkauf, 1993; Endow, 1993). One interesting proposal that has recent experimental support is that *nod* is distributed along chromosome arms (Hawley et al., 1993) and this could explain the correlation between the frequency of chromosome loss and chromosome size in *nod* mutants. *Nod* was found to contain a density that has similarities with human HMG17 that was capable of binding to DNA. On meiotic chromosomes, *nod* was found to be distinguished along the length of the arms (Afshar et al., 1995).

Although both *ncd* and *nod* are required for different aspects of female meiosis, and the loss of these gene functions produce no obvious defects in other cellular events, both mRNAs can be detected in different tissues. This suggests that these two kinesins may have roles other than during female meiosis, and which are redundant with other force generators. This situation would be reminiscent of the redundancy in the spindle motors in budding yeast. Interestingly though, unlike *S. cerevisae*, where loss of one of its *bimC* subfamily members (*CIN8* or *KIP1*) is not lethal, deletion of the *Drosophila bimC* member, *KLP61F* (a.k.a *urchin*), leads to death of the organism (Heck et al., 1993). Typical of the phenotypes of this mutant class of kinesins, *KLP61F* mutants have collapsed, unseparated spindles. Based on the properties of its relatives, *KLP61F* would be predicted to be a plus-end motor. However, it seems illogical that there would be no evolutionary pressure to retain a backup motor as a substitute for *KLP61F*, and indeed, it is possible that *Drosophila* has a second motor that also functions in spindle formation. As the spindle structure became more complex during evolution, multiple *bimC*like motors were required for proper spindle assembly. In yeast, the presence of a simpler spindle may only require one functional *bimC*like motor. An interesting experiment is to test whether the more complex spindles also require opposing motor functions, as was demonstrated by the ability of a loss of *KAR3* function to rescue the lethal phenotype of a *CIN8* and *KIP1* double mutant. By analogy, one could test whether the lethality of a loss-of-function *KLP61F* mutant can be rescued by a compensatory loss of the minus-end *ncd* motor.

F. The Nonmotor Domain of Subfamily Members Share No Obvious Sequence Similarities

The *bimC* and *KAR3* subfamilies share not only similarities within their motor domains, but also have similar functions based on their loss-of-function phenotypes. Similarity in their motor domains may reflect a conservation of the directionality of these two classes of motors, but it may also reflect a conservation of the force-generating capacity. This common feature cannot completely account for their similar functions within the spindle because the portions of the molecules that target these motors to the correct cellular location (i.e., spindle-pole components, microtubules) must play an equally important role. However, within each subclass there is no apparent homology in the regions that lie outside of the motor domain. This could explain the poor efficiency of complementation of a *KAR3* mutant even when *klpA* is overexpressed. The nonmotor domains of these two kinesins may have diverged enough that proteins which normally interact with this region in *KAR3* are unable to recognize the similar functional domain within *klpA*.

G. Dynein

The discovery that *KAR3*, *ncd* and possibly *klpA* are all minus-end kinesins that are important for spindle structure raises the question of the need or importance of dyneins in mitosis. Mammalian cytoplasmic dyneins are multimeric complexes that consist of a pair of 550 kDa heavy chains, together with intermediate and light chains (Vallee, 1991). Although motility has not been directly demonstrated with just the heavy chain, sequence analysis reveals the presence of multiple nucleotide binding sites that would be consistent with its role as a translocator (Gibbons, 1991; Ogawa, 1991). In support of this, antibodies raised against the portion of the heavy chain that contains one such ATP binding site will inhibit *in vitro* motility of dynein complexes. Immunocytochemistry has revealed the presence of dyneins at kineto-chores, as well as throughout the spindle. However, the functional significance of these observations is unknown.

H.   Dynein Is Important for Spindle Pole Separation in Mammalian Cells

The availability of antibodies that block dynein motility has allowed direct testing of the role of this class of microtubule-based motor in spindle functions. PtK1 cells (a marsupial cell line) were microinjected with anti-dynein antibodies during different stages of mitosis (Vaisberg et al., 1993). Injection of antibodies into cells that were in prometaphase, metaphase and anaphase did not block or delay their ability to complete mitosis. These results suggest that the injected antibodies do not interfere with chromosome movement or spindle structure. In contrast, injection of antibodies into prophase cells arrested their progression through mitosis. Exami-nation of these cells revealed that while chromosomes were fully condensed and the nuclear envelope was disassembled, spindle formation was blocked. The chromosomes were arranged around a radial array of microtubules that emanated from unseparated centrosomes much like a monopolar spindle. The formation of radial array of microtubules, with a focused center and chromosomes arranged around the perimeter, indicates that the antibodies do not interfere with the polarity of the microtubules; presumably, the minus-ends of the microtubules are proximal to the centrioles while their plus-ends are attached to kinetochores.

The antibody inhibition experiments suggest that dynein is required during the early stages of spindle pole separation. Once the bipolar spindle is estab-lished, dynein may not be necessary because interactions between the overlap-ping array of spindle microtubules may stabilize this structure. Since PtK1 cells contain centrioles, there may not be a requirement for dynein to gather the minus-ends together as *ncd* has been proposed to do in *Drosophila*. The inhibition of spindle pole separation by blocking dynein motility can occur in two ways (Figure 4a):

1.   Dynein may separate the spindle poles by pulling on the astral array of microtubules. This would require that dynein is anchored to some cytoplasmic matrix, such as the cell cortex, while pulling poleward along the astral microtubule. This application of force would effectively pull the duplicated poles away from each other.

2.   Dynein can also separate the poles by cross-bridging the intersecting micro-tubules that emanate from the two unseparated poles. If one end of dynein remains

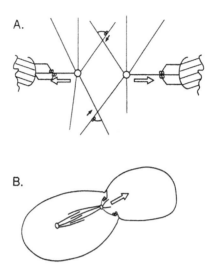

Figure 4. The role of dynein in spindle organization. (a) The minus-end-directed movement of dynein along microtubules provides two possible mechanisms for its role in spindle pole separation in mammalian cells. Dyneins that are fixed to a cellular structure, such as the cortex (hatched), could pull the poles apart by exerting force along astral microtubules (arrows). Dyneins may form cross-bridges between intersecting microtubules (<90°) that have been nucleated from different poles. Poles would be pushed apart by microtubule sliding. (b) In *S. cerevisiae*, dynein functions to position the spindle during mitosis by pulling it through the mother–daughter bud neck (arrows).

fixed to one microtubule, while its motor domain slides towards the minus-end of another microtubule, the net effect is to push apart the spindles. This mode of action for dynein will only work when the angle of the two intersecting microtubules is <90°. Interestingly, this geometric relationship between microtubules is seen at the early stages of spindle pole separation. Once the angle between the microtubules exceeds 90°, as is the case of a bipolar spindle, sliding of the dyneins will pull the poles towards each other.

I.   Dynein Is Important for Positioning the Spindle in Yeast

Isolation of the single dynein heavy chain gene (*DHC1*) in *S. cerevisae* has allowed genetic analysis of dynein function (Li et al., 1993; Eshel et al., 1993). *DHC1* encodes a 2,884-amino-acid protein. Comparison with the dynein heavy chains from *Dictyostelium*, *Drosophila* and rat show that the yeast homologue is 1,700 to 1,900 amino acids smaller. However, yeast dynein is 41% to 42% identical with the other dyneins through a 1,000 amino acid domain that contains the four conserved consensus nucleotide binding sites. Surprisingly, *DHC1* is not essential for mitosis or meiosis because a null mutation is viable. The only discernable defect was a slightly prolonged cell-cycle time. Examination of the ability of these null mutants to carry out nuclear and cell division events showed the accumulation of large-budded cells (size of the bud reflects cell-cycle time; small buds are S phase, large buds are G2 and M phase) with two or more nuclei in the mother bud, while the smaller bud was anucleate. The accumulation of the binucleate phenotype increased from 1% to 15% as cells progressed from the commitment point in G1 to mitosis. However, the *DHC1* null mutants do not accumulate separated anucleate cells. This suggests that the binucleate phenotype can correct itself so that with sufficient time the smaller bud eventually receives its normal complement of chromosomes from the mother. The extra time that the cell takes to reposition the spindle correctly through trial and error may be responsible for the lengthening of the cell division time.

Visualization of the microtubules by immunofluorescence microscopy shows that the binucleate defect is the result of the improper positioning of the spindle in the daughter bud. In normal budded cells, the spindle lies along the mother–bud axis so that chromosomes will segregate into each of the dividing cells. In *DHC1* mutants, the defect is not at the level of spindle formation or chromosome segregation, but due to a failure in positioning the spindle through the bud neck. Since the daughter cells ultimately are nucleated, a redundant mechanism must exist to ensure that the spindle is correctly positioned into the daughter bud. The similarity in the defect between a *DHC1* mutant and a particular tubulin mutant allele that affects astral microtubules suggests that positioning of the spindle through the bud neck requires astral microtubules. Dynein is postulated to be anchored near the bud neck and to bind to the astral array of microtubules that emanate away from the spindle. Exertion of motive force towards the minus-ends of the astral microtubules will pull the spindle towards the neck (Figure 4b). It remains possible that dynein, in conjunction with other proteins (motors), plays important roles in other aspects of spindle function, as well as in other cellular functions. For example, while spindle pole separation is unaffected in a *DHC1* mutant, a second mutation (possibly either *CIN8* or *KIP1*) that eliminates a redundant function may uncover such a role for yeast dynein.

J. Kinetochore Motors in Yeast

Work towards elucidating kinetochore structure and function has come primarily from the analysis of the centromere structure in *S. cerevisae* and mammalian cells. The centromere DNA in *S. cerevisae* was genetically defined as the minimal segment of DNA that would specify mitotic stability to an extrachromosomal plasmid (Carbon and Clarke, 1990). A comparison of a large number of *CEN* sequences (there are 16 chromosomes in a haploid yeast) yielded a consensus centromere element of approximately 125 nucleotides in size that was separated into three domains: CDEI consists of an 8-nucleotide sequence (PuTCACPuTG), CDEII is highly AT-rich and is approximately 80 nucleotides in length, and CDEIII is 26 nucleotides in length and has partial dyad symmetry (TGTTTT/ATGNTTTCCGAAANNNAAAAA). Mutagenesis of these elements revealed that yeast can tolerate changes CDEI and II that reduce centromere function (Cumberledge and Carbon, 1987; Gaudet and Fitzgerald-Hayes, 1987; Hegemann et al., 1988), while point mutations in CDEIII completely abolished centromere function (McGrew et al., 1986; Ng and Carbon, 1987). Genetic analysis pointed to CDEIII as a critical domain that specified interactions with kinetochore proteins.

The factors that interacted with centromere DNA of yeast were identified by using *CEN* DNA as an affinity reagent. Biochemical fractionation of yeast nuclear extracts identified a 240-kDa trimeric complex as the major component which bound specifically to wild-type *CEN* DNA but did not recognize a nonfunctional centromere that contained an inactivating mutation in CDEIII (Lechner and Carbon, 1991). The affinity-purified complex, CBF3, was examined for kinetochore function (i.e., microtubule binding and motility) by first incubating the fraction with beads that were coupled with either wild–type or nonfunctional *CEN* DNA. When polarity-marked microtubules were added to these beads, the wild-type *CEN* DNA–CBF3 complexes were able to move the microtubules in an ATP-dependent fashion. The CBF3 complexes specified minus-end movement with average speeds of 4 µm/min (Hyman et al., 1992). The direction and speed of movement are reminiscent of the poleward movement of the separated chromosomes at anaphase A. Sequence analysis of two of the three genes that encode the subunits of CBF3 do not reveal any significant homologies with any known proteins. Although it is possible that one or all three subunits is the mechanochemical ATPase that moves along microtubules, it is equally plausible that CBF3 itself is not the motor. Analysis of the protein profiles in the affinity-purified CBF3 fractions, which exhibit *CEN* DNA-specific motility, shows the presence of substoichiometric amounts of other proteins. While CBF3 may bind specifically to centromeric DNA, microtubule-based motility could be specified by a motor protein that was simultaneously enriched by the same fractionation procedure. This motor may interact specifically with the CBF3–DNA-bead complex and then translocate the beads along microtubules.

Interestingly, KAR3 was detected in the CBF3 fractions by KAR3-specific antibodies (Middleton and Carbon, 1994). This finding, coupled to the *in vitro* findings that *KAR3* is a minus-end motor that can move at a similar speed (approximately 3 µm/min) as the CBF3-dependent bead motility, implicates this kinesin-related protein as the motor that is responsible for chromosome movement at anaphase A. However, *KAR3* cannot be the sole kinetochore motor in yeast since null mutants are not lethal but exhibit a reduced viability and a delayed mitosis (Meluh and Rose, 1990). In the absence of *KAR3*, *S. cerevesiae* must possess a redundant mechanism to separate its chromosomes.

K. Structure of Mammalian Centromere-Kinetochore Complex

Unlike the simple centromere sequence of *S. cerevesiae*, the centromere of higher eukaryotes is undefined but is estimated to vary between several hundred kilobases to megabases of DNA that consist of highly repetitive elements. Efforts towards characterizing the centromere proteins in mammalian cells were stimulated by the discovery that autoimmune sera from scleroderma patients (CREST variety) recognized three major centromere proteins (Moroi et al., 1980; Brenner et al., 1981; Earnshaw and Rothfield, 1985). In rodents, autoimmune sera stain the trilaminar plates of the kinetochore (Brenner et al., 1981). In humans, the three proteins have been localized to subdomains of the centromere-kinetochore complex. CENP-A is a 17-kDa protein that is a subtype of histone H3 (Sullivan et al., 1994) and is believed to associate with the centromeric heterochromatin (Palmer and Margolis, 1985; Palmer et al., 1987). CENP-B is a highly acidic 80-kDa protein (Earnshaw et al., 1987) that binds specifically to a subset of α-satellite repeats and has been localized by immunoelectron microscopy to the centromeric heterochromatin of metaphase HeLa chromosomes (Cooke et al., 1990). Oligomerization of CENP-B/DNA complexes may also organize the chromatin structure of the region (Matsumoto et al., 1989). CENP-C has been localized to the innermost plate of a trilaminar kinetochore and is also believed to bind DNA (Saitoh et al., 1991; Sugimoto et al., 1994). The sequence, as well as the positions of these three proteins within the centromere, indicates that they are not candidates for generating forces for chromosome motility.

Additional proteins of the centromere-kinetochore complex in human chromosomes have been identified using a biochemical approach. The strategy relied on the assumption that simultaneous enrichment of CENP-B and C which are known to associate with the centromere-kinetochore complex, would also be associated with enrichment of unidentified members of this complex. Monoclonal antibodies specific for these new kinetochore proteins were identified by immunofluorescence microscopy (Yen et al., 1991; Compton et al, 1991). The first such protein to be identified by this approach was CENP-E, a 312-kDa protein that exhibited a number of unexpected properties (Yen et al., 1991). CENP-E is detected in the cytoplasm of a small number of interphase cells that are in the late stages of the cell cycle. CENP-E remains cytoplasmic during prophase and is assembled onto kinetochores at prometaphase, where it remains until late anaphase. In addition to its legalization at kinetochores at anaphase, CENP-E is also found to associate with the overlapping set of microtubules in the spindle interzone. CENP-E is concentrated into the midbody at telophase. The cell-cycle dependent distribution of CENP-E is paralleled by its expression pattern (Yen et al., 1992). CENP-E is found at low levels during the early phases of G1 but increases nearly 10-fold by G2. Pulse-chase experiments show that CENP-E is specifically degraded upon completion of mitosis. This expression pattern suggests that CENP-E functions specifically during mitosis.

L. Mammalian Kinetochore-Associated Motors

Molecular cloning of the CENP-E cDNA revealed it to encode a 312 kDa protein that is the largest member of the kinesin superfamily (Yen et al., 1992). It has a putative motor domain located at the N terminus and is separated from a C-terminal microtubule-binding domain by a huge extended rod that is predicted to self-dimerize. Besides the similarity in size and the presence of the conserved ATP and microtubule-binding sites within the CENP-E motor domain, it does not have any obvious homologies with other family members, suggesting that it could be a prototype of a new subfamily. CENP-E has been localized to the outermost kinetochore plate, a region that should have the most interaction with microtubules (Cooke, Earnshaw, and Yen, unpublished). The pressure of CEMP-E at kinetochores at all stages of mitosis suggest that it may be involved with chromosome alignment during prometaphase (Figure 5a) or chromated separation at anaphase.

Functional analysis of CENP-E has come from microinjection experiments. When prometaphase cells are injected with a monoclonal antibody specific for CENP-E, the chromosomes will align at the metaphase plate but fail to undergo subsequent chromosome separation (Yen et al., 1991). Although an intact spindle is maintained in these metaphase-arrested cells, the chromsomes undergo characteristic oscillations, indicating that the antibody did not interfere with poleward forces. Since the monoclonal antibody recognizes the portion of CENP-E that does not contain any conserved motifs, the mechanism of inhibition remains unclear. Because CENP-E is already kinetochore-bound when cells were injected, it is possible that the antibody interferes with some other essential interactions that may be important for signaling the onset of anaphase.

Another possible role for CENP-E has been suggested by *in vitro* experiments that tested whether microtubule depolymerization releases sufficient energy to translocate chromsomes. Purified chromosomes were attached to a single microtubule at one kinetochore. When microtubules were induced to depolymerize, the chromosome was found to remain attached to the shrinking microtubule and was translocated toward the fixed minus-end of the microtubule independent of an exogenous source of energy. When the interactions between kinetochore components and the depolymerizing end of the microtubule were probed with antibodies against known motor proteins. It was found that CENP-E antibodies directed against the neck region of the motor domain effectively inhibited depolymerization dependent movement. Thus, not only might CENP-E function as a conventional motor that couples ATP hydrolysis to directed chromosome movement, but its motor domain might also have the ability to maintain transient associations with the shrinking end of a kinetochore microtubule. These two properties might be important for a kinetochore motor to remain attached to its microtubule during chromosome alignment, when kinetochores switch rapidly between plus and minus-end directed movements (Skibbens et al., 1994).

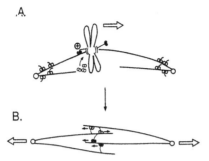

Figure 5. Possible roles for two mammalian spindle-associated kinesin-like motors. (a) At early stages of mitosis, CENP-E is bound to kinetochores and may be important for aligning chromosomes. MKLP-1 is involved with spindle pole separation and perhaps maintenance of the bipolar spindle. (b) At anaphase, CENP-E, along with MKLP-1, are located to the set of overlapping array of microtubules in the spindle midzone. MKLP-1 may function to push the poles apart by applying plus-end motor activity. Arrows show direction of force.

In *S. cerevesiae*, it seems that the anaphase motor for chromosome separation is the minus-end-directed kinesin, KAR3. Dynein (*DHC1*) on the other hand, does not appear to play an essential role in chromosome separation in yeast. The situation is less clear in mammalian cells since dynein is detected, albeit in lower amounts, with kinetochores during anaphase. It remains possible that dynein specifies poleward migration of chromosomes but its speed is somehow regulated. This could occur by modifications to dynein that are introduced at anaphase or by making microtubule depolymerization a rate-limiting process, which would then act as a governor to moderate the velocity of dynein. Given the existence of minus-end kinesins in yeast and *Drosophila*, the use of such a kinesin in poleward chromosome movement in mammalian cells is a strong possibility. The detection of such an activity on chromosomes would be difficult *in vitro,* it would be extremely difficult to obtain chromosomes from anaphase cells.

M. The Smart-Kinetochore Hypothesis

The mechanism of chromosome alignment during congression is significantly more complex than separation because the sister kinetochores must move as a unit despite the presence of opposing forces that want to separate them. The source of this poleward force could be due to a constitutively active anaphase motor within the kinetochore. This possibility is supported by the observation that chromosomes will migrate poleward with anaphase-like speeds when the connections between their kinetochores and one of the pole are severed. In order that the motion of the sister kinetochores do not antagonize one another, the forces that are felt by a kinetochore must somehow be relayed to its sister. How the force that is felt by one kinetochore is sensed by its sister is unknown. One possibility is that as a kinetochore is pulled outwards by a poleward force, its sister kinetochore responds by collapsing inward (Skibbens et al., 1993); depending on how the kinetochore is deformed, different sets of spatially separated motors are postulated to interact with the microtubule. This idea is in line with the "smart kinetochore" hypothesis, which postulates that the kinetochore proteins that specify poleward movement through microtubule depolymerization are spatially separated from those which cause away-from-pole movements (Mitchison, 1989); when a chromosome moves poleward, its kinetochore microtubules are postulated to come under tension and the ends are thought to be pulled out of the kinetochore so that it maintains contacts only with the outermost kinetochore proteins that specify depolymerization-dependent poleward force. It is important to note that only an end-on interaction between the microtubule and the outer kinetochore proteins could specify depolymerization-dependent poleward movement, whereas microtubules that are under compression penetrate far enough into the kinetochore to engage the plus-end-directed motors. The poleward pulling motors would be inoperative because they make lateral instead of end-on interactions along the microtubule wall, and therefore the plus-end-directed motors are active and push the kinetochore away from the pole.

N. Kinesin-Related Proteins That Separate Spindle Poles at Anaphase B

After chromosome separation, the spindle poles migrate away from each other during anaphase B. Spindle pole elongation can be mediated by motors that *push* the poles apart by exerting forces from within the spindle. Spindle elongation has been experimentally reproduced *in vitro* using isolated diatom spindles (Cande and McDonald, 1985, 1986). In this system, spindle elongation is mediated solely by the interdigitating microtubules in the midzone, since the astral microtubules were not preserved during isolation. Antibodies that are specific for the conserved motor domain of kinesin inhibit spindle elongation, demonstrating that members of this superfamily are likely to be the motors that facilitate spindle pole separation (Hogan et al., 1993). An alternative to this mechanism of spindle elongation is to tether the motor to the cell cortex so that it can *pull apart* the poles by acting along astral microtubules. Such a mechanism is thought to exist in the fungus *Fusarium* because laser ablation of the central spindle microtubules does not interfere with, but actually speeds up, pole separation (Aist and Berns, 1981). This would suggest that in some systems, the interactions between the spindle microtubules are to oppose the pulling forces that are used to separate the poles.

In mammalian cells, two kinesins are known to be localized exclusively to the spindle midzone at anaphase (Figure 5a). The localization of CENP-E to the spindle midzone probably reflects its ability to cross-bridge the overlapping and interdigitated microtubules in this region. Cross-bridging activity of CENP-E is supported by the presence of microtubule-binding domains located at the N and C termini of the molecule (Liao et al., 1994), a feature that is shared with the *Drosophila ncd* (Chandra et al., 1993). Biochemical characterization shows that the N-terminal domain of CENP-E, like most kinesins, will bind microtubules in an ATP-sensitive manner. The microtubule-binding property of the C terminus differs from that of both the motor domains and a second class of microtubule-binding proteins called MAPs (Lee, 1993). This domain binds to a different portion of the microtubule lattice than MAPs, and the interaction is resistant to high salt extractions. Although the C terminus can bind to microtubules very tightly, its activity is inhibited by phosphorylation by the mitotic kinase MPF (Liao et al., 1994). Phosphorylation of this domain by MPF could regulate cross-linking activity in a cell-cycle dependent fashion: during the early stages of mitosis, the C-terminal microtubule-binding domain of CENP-E could be inactivated by phosphorylation, thereby preventing microtubule cross-linking; at anaphase, the inactivation of MPF coupled to the dephosphorylation of the C terminus would transform CENP-E a microtubule cross-linker. If the motor domain of CENP-E exerts motive force (plus-ended) along one microtubule while its C terminus remains rigidly attached to another microtubule of opposite polarity, the overall action would force the microtubules past each other. If the minus-ends of these microtubules are attached to the poles, this activity would push apart the spindle poles.

MKLP-1 is another kinesin that is localized exclusively to the spindle midzone at anaphase (Figure 5b). This kinesin-related protein was identified by a monoclonal antibody that was raised against spindle proteins from CHO cells (Selitto and Kuriyama, 1990). Molecular cloning of the human homologue revealed it to be 110-kDa kinesin-related protein that accumulates in the nucleus of interphase cells (Nislow et al., 1992). At mitosis, it is distributed throughout the spindle but is mostly concentrated at the poles. MKLP-1 is relocalized to the spindle midzone at anaphase and to the midbody at telophase. *In vitro*, MKLP-1 forms cross-bridges between microtubules of opposite polarity and exhibits plus-end directed motility with speeds of 4 μm/min (Nislow et al., 1992). These two features are consistent with MKLP-1 as being a motor that is used for spindle pole separation. That both CENP-E and MKLP-1 exist in human cells suggests that they may have overlapping roles in the spindle during anaphase.

O. Motors that May Have Multiple Roles in Mitosis

Like CENP-E, MKLP-1 is thought to have multiple roles during mitosis. Injection of monoclonal anti-MKLP-1 antibodies into PtK1 cells during prophase, prometaphase or metaphase arrests cells at metaphase (Nislow et al., 1990). However, these antibodies do not interfere with mitotic progression once cells have entered anaphase. While it is possible that MKLP-1 plays no role in the spindle midzone at anaphase, it is possibly that the antibody cannot exert its inhibitory effects once MKLP-1 is redistributed into this region. Metaphase-arrested cells appeared to have a normal spindle, although the antibody disrupted the normal distribution of MKLP-1 along spindle fibers to the cytoplasm. Electron microscopy revealed that some chromosomes never reached the metaphase plate, but were located several microns away from the spindle equator. The stray chromosomes may, through a feedback control mechanism, have continued to emit a signal that prevented the onset of anaphase. The spindle microtubules of the arrested cells were not organized into discrete fibers and did not focus at the poles. The loss of microtubule bundles can be explained by the observation that MKLP-1 cross-links microtubules, although, the *in vitro* data show that this molecule cross-links microtubules of opposite polarity instead of parallel microtubules, as would be the case in the metaphase spindle. The absence of MKLP-1 at the poles may deplete the poles of a plus-ended motor whose function is to gather and focus the ends of the spindle microtubules. Such an activity has been proposed for a plus-ended motor from the *bimC* subfamily, Eg5 (Sawin et al., 1992). On the other hand, the use of a minus-end motor, such as *Drosophila ncd*, can theoretically achieve the same result through a different mechanism.

The abrupt changes in the distribution of CENP-E and MKLP-1 during mitosis suggests that the functions of these two proteins may be temporally and spatially regulated. If CENP-E is a kinetochore motor that functions to align chromosomes during prometaphase and metaphase, this activity may antagonize anaphase movements. To allow progression into anaphase, the hypothetical plus-end motor activity of CENP-E would have to be inactivated, perhaps by post-translational modification that occurs at the onset of anaphase releasing it from the kinetochore. Since CENP-E possesses a cross-linking activity, it may have a role in spindle pole separation during anaphase B. Thus, the association of a population of CENP-E at the spindle midzone at anaphase might ensure that spindle pole separation does not occur prematurely. Similarly, MKLP-1 functions during early stages of mitosis may be important for its subsequent functions during anaphase. These molecules, in addition to playing a structural role in spindle mechanics, may have a regulatory role in mitotic progression. The correct spatial positioning of these molecules within a metaphase spindle may be critical for the subsequent formation of a functional anaphase spindle.

P. Reassessment of the "One Motor, Many Tail" Hypothesis

The existence of multiple kinesin-related proteins that specify multiple functions in mitosis forces a reevaluation of the "one motor many tail" hypothesis. The motor domains of all the kinesin family members possess conserved subdomains that are essential for motor function (i.e., ATP and microtubule binding). However, comparison of the growing number of kinesin family members clearly shows that subfamilies can be classified based on similarities in the "variable" regions of the motor domain (Goldstein, 1993). These variable regions are perhaps responsible for the more subtle aspects of motor functions, such as speed, power and direction. The ability of the motor domain of the kinesin gene superfamily to evolve may be a reason why this class of motors can be involved in such a diverse array of cellular functions. The refinements in motor functions of kinesins have coevolved to accommodate the wide array of motile functions within the spindle as well as other cellular processes. In the case of mitosis, the ability of motors to *accurately* segregate chromosomes is clearly more important than speed or power. Consistent with this view, the spindle-associated kinesins, whose motor activities have been characterized, are significantly slower than conventional kinesin, whose role in vesicle transport would require speed.

IV. CYTOKINESIS

The separation and formation of two daughter nuclei during mitosis is coupled to the equal partitioning of the cytoplasm between the dividing cell through a process called cytokinesis. Cell separation is an essential process, as failure will produce polyploid cells. Initiation of cytokinesis, as visualized by the formation of a cleavage furrow at the equatorial cell surface, occurs sometime during anaphase. Electron microscopy of the cleavage furrow of dividing embryos and cells revealed the presence of a circumferential band of microfilaments, arranged in antiparallel fashion, and connected to myosin fibers (Schroeder, 1973; Sanger and Sanger, 1980; Maupin and Pollard 1986). These observations were supported by the concentration of fluorescently tagged actin and myosin, or of myosin II detected by specific antibodies, in the cleavage furrows (Sanger et al., 1989, Fujiwara and Pollard 1976). The force generating mechanism of the actomyosin contractile ring is modeled after the "sliding-filament" mechanism used to describe muscle contraction (Schroeder, 1972; Mabuchi, 1986; Satterwhite and Pollard, 1992). Like muscle, function of the contractile ring is thought to be mediated by the interaction between a bipolar bundle of myosin filaments and two arrays of oppositely oriented actin filaments but, unlike the sarcomere repeats, the barbed ends of the actin filaments of the contractile ring are attached to the plasma membrane. When the myosins within the bundle are translocated towards the barbed ends of each actin filament, this action pulls and constricts the membrane around the equator of the cell (Figures 6a and 6b).

A. Evidence for the Role of Myosin in Cytokinesis

Direct evidence that myosin motors function in cytokinesis has come from genetic investigations and *in vivo* experimentation at the single cell level. In *Dictyostelium*, deletion of its single conventional type II myosin heavy chain gene (*mhcA*) is not lethal to the cell but gives rise to multinucleated cells (Manstein et al., 1989). Since these myosin null mutants are capable of establishing a spindle, but fail to form a cleavage furrow, the multinucleate phenotype most likely arose from a defect in cytokinesis. Similarly, myosin null mutants in *S. cerevesiae* fail to bud, which is indicative of a failure to undergo cytokinesis (Watts et al., 1987; Rodriguez and Paterson, 1990). Independent evidence that further supported the view that a conventional myosin hexameric complex participates in cleavage furrow formation came from the discovery that the mitotic defective *sqh* (spaghetti squash) mutant in *Drosophila* is due to inadequate accumulation of the myosin regulatory light chain (Karess et al., 1991). Depletion of the maternal supply of this

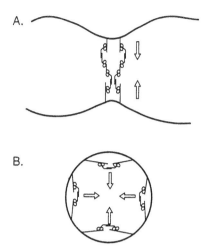

Figure 6. Schematic representation of the role of actomyosin in cytokinesis. (a) Side view of a cleavage furrow and contractile ring. Actin filaments (thin lines) contact the cell cortex through its barbed end. A bipolar array of myosin filaments (two-headed objects) interact with oppositely oriented arrays of actin filaments. Movement of the myosins towards the barbed ends will pull the cell cortex towards the center of the cleavage furrow. (b) Cross-section of a cleavage furrow depicting circumferential arrangement of actin and myosin near the cell surface; force is directed radially towards the center (arrows).

myosin subunit during development produces the late larval lethal phenotype. Inspection of mutant *sqh* larvae shows the presence of a large number of polyploid cells. The defect does not affect mitosis since mutant cells in various stages of normal chromosome segregation can be identified. Since myosin light chain is a regulatory subunit that is essential for motor function, the inability to form a cleavage furrow in *sqh* mutants is presumably a consequence of the loss of myosin function. Antibodies that inhibit myosin motility *in vitro* when injected into echinoderm eggs do not affect mitosis but block cleavage furrow formation (Mabuchi and Okuno, 1977; Kiehart et al., 1982) and this is consistent with the above genetic data.

B. Cleavage Furrow Formation May Be Specified by a 3-Dimensional Network of Contractile Proteins

Although the contractile ring hypothesis is largely accepted as the mechanism for cleavage furrow formation, a growing body of data suggest that the band of actin and myosin filaments may not be the sole source of force. In the early electron microscopy images of cleavage furrows, not only were microfilaments seen parallel to the cell equator, but similar filaments were found randomly distributed throughout the furrow (Zeligs and Wollman, 1979; Maupin and Pollard, 1986). Using high resolution optical sectioning microscopy, the 3-dimensional pattern of actin-filament distribution in dividing rat cells was reconstructed (Fishkind and Wang, 1993); although actin filaments were preferentially oriented parallel to the equator, as expected from the contractile ring hypothesis, the reconstructed images also showed that the equatorial orientation was found on the "noncleaving" ventral side of the cell where it contacts the substratum; in contrast, the distribution of actin filaments on the dorsal and cleavage-active side was significantly less ordered and in addition, actin filaments that were perpendicular to the cell surface were found to traverse through the furrow. The interactions of myosin with the 3-dimensional complex of actin filaments in the furrow may allow greater flexibility in the magnitude and directions of force that can be achieved. In the case of an adherent cell, increased equatorial alignment of actin filaments along the ventral side may signify attempts to overcome the adhesive forces. On the other hand the low level of resistive force on the dorsal side may not require significant filament ordering to achieve cleavage. Such a feedback system could also explain how nonadherent cells can undergo symmetrical cleavage with little ordering of their actin filaments.

C. Regulation of Cleavage Furrow Formation

One of the important issues in cytokinesis is the mechanism of cleavage furrow formation. What are the signals that specify organization of the actin and myosin molecules at the equator of the cell, when are they generated and where do they come from? The timing of cleavage furrow formation may be regulated in part by the mitotic kinase MPF (Maller, 1991). *In vitro*, MPF phosphorylates myosin-regulatory light chain on residues Ser1, Ser2 and Thr9 (Satterwhite et al., 1992), and inhibits the ability of calmodulin-dependent myosin light chain kinase (MLCK) to phosphorylate the activating site at Ser19 (Nishikawa et al., 1984; Bengur et al., 1987). Normally, phosphorylation of Ser19 by MLCK stimulates filament formation and actin-dependent ATPase activity (Nishikawa et al., 1984). Prior to anaphase, phosphorylation of the inhibitory sites at Ser1, Ser2 and Thr9 by MPF prevents phosphorylation of Ser19 by MLCK (Bengur et al., 1987). It is possible that at the onset of anaphase inactivation of MPF, coupled with the dephosphorylation at the inhibitory sites, allows phosphorylation at Ser19 by MLCK and activation of the myosin complex (Satterwhite et al., 1992); localized accumulation of actin filaments near the membrane of the cell equator would then recruit myosin to the region and establish a contractile ring. It is unclear whether the myosin filaments are preactivated before assembling at the contractile ring or the presence of localized phosphatases near the cleavage furrow activates the myosin after their assembly.

The mechanism that specifies furrow formation at the cell equator is unclear. Work performed mostly in echinoderm eggs suggests that the position of the poles and asters specify formation of the cleavage furrow (Rappaport, 1986; Mabuchi, 1986). Micromanipulation experiments showed that when a spindle was moved off-center, the cleavage furrow was formed perpendicular to the new spindle axis. The observation that a cleavage furrow can form in the absence of chromosomes or a visible central spindle strongly suggested that the positioning of the furrow was mediated by a signal that was generated from the poles and asters. The nature of the signal is unknown but is thought to promote tension in the cell membrane; tension may then facilitate alignment of actin filaments into arrays.

Emerging evidence, derived mainly from vertebrate cells, suggests that the central spindle may play an important role in establishing the cleavage furrow. It is thought that components near the spindle midzone may provide a signal that will locally recruit actin filaments at the plasma membrane. The existence of a protein that can interact with both microtubule and actomyosin systems is supported by the discovery that overexpression of the kinesin-related *SMY1* gene in *S. cereviase* will rescue a conditional lethal myosin mutation (Lillie and Brown, 1992). *SMY1* and *MYO2* are believed to interact, or perform similar functions, since deletion of both genes does not produce viable spores. By extending these observations, it is conceivable that a similar protein(s) exists in other cells that can specify interactions between the spindle and the contractile ring. The INCENPs (Earnshaw and Cooke, 1991) and TD-60 (Andreassen et al., 1991) are two types of protein that belong to a growing family of proteins that are localized to the spindle midzone at anaphase and are therefore good candidates for establishing a link between the spindle and the cleavage furrow.

The INCENPs (96- and 110 kDa) are two related chromosome scaffold proteins found in vertebrate cells (McKay et al., 1993) and are the first members of a class of chromosome passenger proteins. This class of proteins is thought to bind transiently to the chromosome so that they can "piggyback" to the spindle midzone when the chromosomes align at the metaphase plate. They are released from the chromosome at the onset of anaphase and become localized to the substructures of the spindle midzone. INCENPs are initially localized to the inner centromere, between the sister chromatids, but are relocalized to the overlapping microtubules in the midzone at anaphase. As anaphase progresses and before cleavage furrow formation is detectable a portion of the INCENPs can be detected near the cell cortex. The concentration of INCENPs at the cell membrane during anaphase may provide part of the signal for subsequent formation of the contractile ring. The possibility that INCENPs might help establish the cleavage furrow is strengthened by the observation that the deduced primary sequence of these proteins contains motifs that are similar to actin and tropomyosin-binding proteins.

TD-60 is a 60-kDa protein of undetermined structure that was identified using an autoimmune serum. Immunofluorescence staining revealed that TD-60 is associated with centromeres from prophase to early anaphase but is redistributed to the spindle midzone at midanaphase, where it is proposed to be a component of a substructure called the *telophase disc* (Andreassen et al., 1991). Confocal microscopy of anaphase cells shows that TD-60 is distributed beyond the boundaries of the spindle and approaches the cell cortex. Furthermore, monoclonal antibodies directed against nonmuscle myosin also stain the telophase disc (Andreassen et al.,

1991). The speculation is that the telophase disc is part of the actomyosin system that may generate a force that is directed radially, as opposed to the tangential force that is generated by a contractile ring.

V. FUTURE DIRECTIONS OF MITOSIS RESEARCH

One of the goals of studying mitosis is to understand, at the molecular level, the mechanisms that specify the variety of processes that occur during cell division. Microscopic examination of dividing cells over the past century has exposed and provided significant insights into the vast and complex arrays of motile events during mitosis. The combined approaches of genetics, biochemistry and molecular biology over the past two decades have identified a series of structural and regulatory proteins that have crucial roles in spindle formation, chromosome alignment and segregation, and cytokinesis. This list, however, is incomplete as there must be additional proteins in the centrioles, kinetochores, centromere, spindle midzone, and cleavage furrow that interact specifically with the existing list of proteins that are localized in these regions. As this list grows, efforts will be turned towards reconstructing this system *in vitro* so that we can understand how these molecules influence the kinetics of the system and attempt to explain the microscopic observations that began 100 years ago.

ACKNOWLEDGMENTS

The author is thankful to K. Truesdale and the secretarial staff for article searches and manuscript preparation. This work was supported in part by the NIH, NCI, Lucille P. Markey Charitable Trust, and an appropriation from the Commonwealth of Pennsylvania. T.J.Y. is a scholar of the L.P. Markey Foundation.

REFERENCES

Afshar, K., Barton, N.R., Hawley, R.S., & Goldstein, L.S.B. (1995). DNA binding and meiotic chromosomal localization of the drosophila and kinesin-like protein. Cell 81, 129–138.

Aist, J.R. & Berns, M.W. (1981). Mechanics of chromosome separation during mitosis in *Fusarium* (Fungi imperfecti). New evidence from ultrastructural and laser microbeam experiments. J. Cell Biol. 91, 446–58.

Andreassen, P.R., Palmer, D.K., Wener, M.H., & Margolis, R.L. (1991). Telophase disc: A new mammalian mitiotic organelle that bisects telophase cells with a possible function in cytokinesis. J. Cell Sci. 99, 523–534.

Ault, J.G. & Nicklas, R.B. (1989). Tension, microtubule rearrangements, and the proper distribution of chromosomes in mitosis. Chromosoma 98, 33–39.

Bengur, A.R., Robinson, E.E., Appella, E., & Sellers, J.R. (1987). Sequence of the sites phosphorylated by protein kinase C in the smooth muscle light chain. J. Biol. Chem. 262, 7613–7617.

Bloom, G.S., Wagner, M.C., Pfister, K.K., & Brady, S.T. (1988). Native structure and physical properties of bovine brain kinesin and identification of the ATP-binding subunit polypeptide. Biochemistry 27, 3409–3416.

Brenner, S., Pepper, D., Berns, M.W., Tan E., & Brinkley, B.R. (1981). Kinetochore structure, duplication, and distribution in mammalian cells: Analysis by human autoantibodies from scleroderma patients. J. Cell Biol. 91, 95–102.

Cande, W.Z. & McDonald, K.L. (1985). In vitro reactivation of anaphase spindle elongation using isolated diatom spindles. Nature 316, 168–170.

Cande, W.Z. & McDonald, K.L. (1986). Physiological and ultrastructural analysis of elongating mitotic spindles reactivated in vitro. J. Cell Biol. 103, 593–604.

Carbon, J. & Clarke, L. (1990). Centromere structure and function in budding and fission yeasts. New Biologist 2, 10–19.

Cassimeris, L. & Salmon, E.D. (1991). Kinetochore microtubules shorten by loss of subunits at the kinetochores of prometaphase chromosomes. J. Cell Sci. 98, 151–58.

Chandra, R., Salmon E.D., Erickson, H.P., Lockhart, A., & Endow S.A. (1993). Structural and functional domains of the Drosophila ncd microtubule motor protein. J. Biol. Chem. 268 (12), 9005–9013.

Compton, D.A., Yen, T.J., & Cleveland, D.W. (1991). Identification of novel centromere/kinetochore associated proteins using monoclonal antibodies generated against human mitotic chromosome scaffolds. J. Cell Biol. (in press).

Cooke, C.A., Heck, M.M.S., & Earnshaw, W.C. (1987). The INCENP antigens: Movement from the inner centromere to the midbody during mitosis. J. Cell Biol. 105, 2053–2067.

Coue, M., Lombillo, V.A., & McIntosh, J.R. (1991). Microtubule depolymerization promotes particle and chromosome movement in vitro. J. Cell Biol. 112, 1165–75.

Cumberledge, S. & Carbon, J. (1987). Mutational analysis of meiotic and mitotic centromere function in Saccharomyces cerevisiae. Genetics 117, 203–12.

Earnshaw, W.C. & Cooke, C.A. (1991). Analysis of the distribution of the INCENPs throughout mitosis reveals the existence of a pathway of structural changes in the chromosomes during metaphase and early events in cleavage furrow formation. J. Cell Sci. 98, 443–461.

Earnshaw, W.C. & Rothfield, N.F. (1985). Identification of a family of human centromere proteins using autoimmune sera from patients with scleroderma. Chromosoma (Berl.) 91, 313–321.

Earnshaw, W.C., Sullivan, K.F., Machlin, P.S., Cooke, C.A., Kaiser, D.A., Pollard, T.D., Rothfield, N.F., & Cleveland, D.W. (1987). Molecular cloning of cDNA for CENP-B, the major human centromere autoantigen. J. Cell Biol. 104, 817–829.

Endow, S.A., Kang, S.J., Satterwhite, L.L., Rose, M.O., Skeen, V.P., Salmon, E.D. (1994). Yeast KAR 3 is a minus-end microtubule motor protein that destabilizes microtubules preferentially at the minus-ends. EMBO J. 13, 2708–2713.

Endow, S.A. (1993). Chromosome distribution, molecular motors and the claret protein. Trends in Genetics 9, 52–55.

Endow, S.A. (1992). Meiotic chromosome distribution in Drosophila oocytes: Roles of two kinesin-related proteins. Chromosoma 102, 1–8.

Enos, A.P. & Morris, N.R. (1990). Mutation of a gene that encodes a kinesin-like protein blocks nuclear division in A. nidulans. Cell 60, 1019–27.

Eshel, D., Urrestarazu, L.A., Vissers, S., Turmioux, J.C., van VlictReedijk, J.C., Planta, R.J., & Gibbons, I.R. (1993). Cytoplasmic Dynein is required for normal nuclear segregation in yeast. Proc. Natl. Acad. Sci. USA 90, 11172–11176.

Fishkind, D.J. & Wang, Y.-L. (1993). Orientation and three-dimensional organization of actin filaments in dividing cells. J. Cell Biol. 123, 837–848.

Fujiwara, K. & Pollard, T.D. (1976). Fluorescent antibody localization of myosin in the cytoplasm, cleavage furrow and mitotic spindle of human cells. J. Cell Biol. 71, 848–875.

Gaudet, A. & Fitzgerald-Hayes, M. (1987). Alterations in the adenine-plus thymine-rich region of CEN3 affect centromere function in Saccharomyces cerevisiae. Mol. Cell. Biol. 7, 68–75.

Gibbons, I.R., Gibbons, B.M., Mocz, G., & Asai, D.J. (1991). Multiple nucleotide-binding sites in the sequence of dynein β heavy chain. Nature 352, 640–643.

Goldstein, L.S.B. (1993). With apologies to Scheherazade: Tails of 1001 kinesin motors. Annu. Rev. Genet. 27, 319–51.

Gorbsky, G.J., Sammak, P.J., & Borisy, G.G. (1987). Chromosomes move poleward in anaphase along stationary microtubules that coordinately disassemble from their kinetochore ends. J. Cell Biol. 104, 9–18.

Hagan, I. & Yanagida, M. (1990). Novel potential mitotic motor protein encoded by fission yeast *cut7+* gene. Nature 347, 563–66.

Hatsumi, M. & Endow, S.A. (1992). Mutants of the microtubule motor protein, nonclaret disjunctional, affect spindle structure and chromosome movement in meiosis and mitosis. J. Cell Sci. 101, 547–59.

Hawley, R.S. & Theurkauf, W.E. (1993). Requiem for distributive segregation: Achiasmate segregation in *Drosophila* females. Trends in Genetics 9 (9), 310–317.

Hawley, R.S., McKim, K.S., & Arbel, T. (1993). Meiotic segregation in *Drosophila Melanogaster* females: Molecules, mechanisms of myths. Annu. Rev. Genet. 27, 281–318.

Hays, T.S. & Salmon, E.D. (1990). Poleward force at the kinetochore in metaphase depends on the number of kinetochore microtubules. J. Cell Biol. 110, 391–404.

Hays, T.S., Wise, D., & Salmon, E.D. (1982). Traction force on a kinetochore at metaphase acts as a linear function of kinetochore fiber length. J. Cell Biol. 93, 374–82.

Heck, M.M.S., Pereira, A., Pesavento, P., Yannoni, Y., Spradling, A.C., & Goldstein, L.S.B. (1993). The kinesin-like protein KLP2 is essential for mitosis in *Drosophila*. J. Cell Biol. (in press).

Hegemann, J.H., Shero, J.H., Cottarel, G., Phillipsen, P., & Hieter, P. (1988). Mutational analysis of the centromere DNA from chromosome VI of *Saccharomyces cerevisiae*. Mol. Cell. Biol. 8, 2523–28.

Hill, T.L. & Kirschner, M.W. (1982). Bioenergetics and kinetics of microtubule and actin filament assembly-disassembly. In: Int. Rev. Cytol. (Bourne, G.H. & Danielli, J.F., Eds.) Vol. 78, pp. 1–123. Academic Press, NY.

Hogan, C.J., Wein, H., Wordeman, L., Scholey, J.M., Sawin, K.E., Cande, W.Z. (1993). Inhibition of anaphase spindle elongation *in vitro* by a peptide antibody that requires kinesin motor domain. PNAS 90, 6611–6615.

Hoyt, M.A., He, L., Loo, K.K., & Sauders, W.S. (1992). Two *Saccharomyces cerevisiae* kinesin-related gene products required for mitotic spindle assembly. J. Cell Biol. 118, 109–20.

Hyman, A.A. & Mitchison, T.J. (1991). Two different microtubule-based motor activities with opposite polarities in kinetochores. Nature 351, 206–11.

Hyman, A.A., Middleton, K., Centola, M., Mitchison, T.J., & Carbon, J. (1992). Microtubule-motor activity of a yeast centromere-binding protein complex. Nature 359, 533–36.

Karess, R.E., Chang, X., Edwards, K.A., Kulkarni, S., & Aguilera, I., (1991). The regulatory light chain of nonmuscle myosin is encoded by *spaghetti-squash*, a gene required for cytokinesis in *Drosophila*. Cell 65, 1177–89.

Kiehart, D.P., Mabuchi, I., & Inoué, S. (1982). Evidence that myosin does not contribute to force production in chromosome movement. J. Cell Biol. 94, 165–78.

Koshland, D., Mitchison, T.J., & Kirschner, M.W. (1988). Polewards chromosome movement driven by microtubule depolymerization *in vitro*. Nature 331, 499–504.

Kuznetsov, S.A., Vaisberg, E.A., Shanina, N.A., Magretova, N.N., & Chernyak, V.Y. (1988). The quaternary structure of bovine brain kinesin. EMBO J. 7, 353–56.

Lechner, J. & Carbon, J. (1991). A 240 kd multi-subunit protein complex, CBF3, is a major component of the budding yeast centromere. Cell 64, 717–25.

Lee, G. (1993). Non-motor microtubule-associated proteins. Curr. Opin. Cell Biol. 5, 88–94.

LeGuellec, R., Paris, J., Couturier, A., Roghi, C., & Philippe, M. (1991). Cloning by differential screening of a *Xenopus* cDNA that encodes a kinesin-related protein. Mol. Cell. Biol. 11, 3395–98.

Lewis, E.B. & Gencarella, W. (1952). Claret and non-disjunction in *Drosophila melanogaster*. Genetics 37, 600–1 (Abstr.).

Li, Y.-Y., Yeh, E., Hays, T., & Bloom, K. (1993). Disruption of mitotic spindle orientation in a yeast dynein mutant. Proc. Natl. Acad. Sci. USA 90, 10096–10100.

Liao, M., Li, G., & Yen, T.J. (1994). Mitotic regulation of microtubule-crosslinking activity of CENP-E kinetochore protein. Science 265, 394–398.

Lillie, S.H. & Brown, S.S. (1992). Suppression of a myosin defect by a kinesin-related gene. Nature 356, 358–61.

Lombillo, V.A., Nislow, C., Yen, T.J., Gelfand, V.I., & McIntosh, J.R. (1995). Antibodies to the kinesin motor domain and CENP-E inhibit microtubule depolymerization-dependent motion of chromosomes in vitro. J. Cell Biol. 128, 107–117.

Mabuchi, I. (1986). Biochemical aspects of cytokinesis. Int. Rev. Cytol. 101, 175–213.

Mabuchi, I. & Okuno, M. (1977). The effect of myosin antibody on the division of starfish blastomeres. J. Cell Biol. 74, 251–263.

Mackay, A.M., Eckley, D.M., Chue, C., & Earnshaw, W.C. (1993). Molecular analysis of the INCENPs (inner centromere proteins): Separate domains are required for association with microtubules during interphase and with the central spindle during anaphase. J. Cell Biol. 123, 373.

Maller, J.L. (1991). Mitotic control. Curr. Opin. Cell Biol. 3, 269–275.

Manstein, D.J., Titus, M.A., De Lozanne, A., & Spudich, J.A. (1989). Gene replacement in Dictyostelium: Generation of myosin null mutants. EMBO J. 8, 923–32.

Masumoto, H., Masukata, H., Muro, Y., Nozaki, N., & Okazaki, T. (1989). A human centromere antigen (CENP-B) interacts with a short specific sequence in alphoid DNA, a human centromeric satellite. J. Cell Biol. 109, 1963–1973.

Maupin, P. & Pollard, T.D. (1986). Arrangement of actin-filaments and myosin-like filaments in the contractile ring and of actin-like filaments in the mitotic spindle of dividing HeLa-cells. J. Ultrastruct. Res. Mol. Struct. 94, 92–103.

McDonald, H.B., Stewart, R.J., & Goldstein, L.S.B. (1990). The kinesin-like ncd protein of Drosophila is a minus end-directed microtubule motor. Cell 63, 1159–65.

McGrew, J., Diehl, B., & Fitzgerald-Hayes, M. (1986). Single base-pair mutations in centromere element III cause chromosome segregation in Saccharomyces cerevisiae. Mol. Cell. Biol. 6, 530–38.

McNeil, P.A. & Berns, M.W. (1981). Chromosome behavior after laser microirradiation of a single kinetochore in mitotic PtK2 cells. J. Cell Biol. 88, 543–53.

Meluh, P.B. & Rose, M.D. (1990). KAR3, a kinesin-related gene required for yeast nuclear fusion. Cell 60, 1029–41.

Middleton, K. & Carbon, J. (1994). KAR3 kinesin is a minus-end-directed motor that functions with centromere binding proteins (CBF3) on an in vitro yeast kinetochore (in press).

Mitchison, T.J. (1988). Microtubule dynamics and kinetochore function in mitosis. Annu. Rev. Cell Biol. 4, 527–50.

Mitchison, T.J. (1989b). Chromosome alignment at mitotic metaphase: Balanced forces or smart kinetochores? In: Cell Motility, Kinesin, Dynein, and Microtubule Dynamics (Warner, F.D. & McIntosh, J.R., Eds.) Vol. 2, pp. 421–30. Liss, NY.

Mitchison, T.J., Evans, L., Schulze, E., & Kirschner, M. (1986). Sites of microtubule assembly and disassembly in the mitotic spindle. Cell 45, 515–27.

Mitchison, T.J. & Kirschner, M.W. (1985). Properties of the kinetochore in vitro. II. Microtubule capture and ATP-dependent translocation. J. Cell Biol. 101, 766–76.

Moroi, Y., Peebles, C., Fritzler, M.J., Steigerwald, J., & Tan, E.M. (1980). Autoantibody to centromere (kinetochore) in scleroderma sera. Proc. Natl. Acad. Sci. USA 77, 1627–1631.

Ng, R. & Carbon, J. (1987). Mutational and in vitro protein-binding studies on centromere DNA from Saccharomyces cerevisae. Mol. Cell. Biol. 7, 4522–4534.

Nicklas. R.B. (1965). Chromosome velocity during mitosis as a function of chromosome size and position. J. Cell Biol. 25, 119–135.

Nicklas, R.B. (1983). Measurements of the force produced by the mitotic spindle in anaphase. J. Cell Biol. 97, 542–48.

Nicklas, R.B. (1989). The motor for poleward chromosome movement in anaphase is at or near the kinetochore. J. Cell Biol. 109, 2245–55.

Nishikawa, M., Sellers, J.R., Adelstein, R.S., & Hidaka, H. (1984). Protein kinase C modulates in vitro phosphorylation of the smooth muscle heavy meromyosin by myosin light chain kinase. J. Biol. Chem. 259, 8808–8814.

Nislow, C., Sellito, C., Kuriyama, R., & McIntosh, J.R. (1990). A monoclonal antibody to a mitotic microtubule-associated protein blocks mitotic progression. J. Cell Biol. 111, 511–22.

Nislow, C., Lombillo, V.A., Kuriyama, R., & McIntosh, J.R. (1992). A plus-end-directed motor enzyme that moves antiparallel microtubules *in vitro* localizes to the interzone of mitotic spindles. Nature 359, 543–47.

O'Connell, M.J., Meluh, P.B., Rose, M.D., & Morris, N.R. (1993). Suppression of the *bimC4* mitotic spindle defect by deletion of *klpA*, a gene encoding a KAR3-related kinesin-like protein in *Aspergillus nidulans*. J. Cell Biol. 120, 153–62.

Ogawa, K. (1991). Four ATP-binding sites in the midregion of the β heavy chain dynein. Nature 352, 643–645.

Palmer, D.K. & Margolis, R.L. (1985). Kinetochore components recognized by human autoantibodies are present on mononucleosomes. Mol. Cell. Biol. 5, 173–186.

Palmer, D.K., O'Day, K., Wener, M.H., Andrews, B.S., & Margolis, R.L. (1987). A 17-kD centromere protein (CENP-A) copurifies with nucleosome core particles and with histones. J. Cell Biol. 104, 805–815.

Pfarr, C.M., Coue, M., Grissom, P.M., Hays, T.S., Porter, M.E., & McIntosh, J.R. (1990). Cytoplasmic dynein is localized to kinetochores during mitosis. Nature 345, 263–65.

Rappaport, R. (1986). Establishment of the mechanisms of cytokinesis in animal cells. Int. Rev. Cytol. 101, 245–281.

Rieder, C.L. & Alexander, S.P. (1990). Kinetochores are transported poleward along a single astral microtubule during chromosome attachment to the spindle in newt lung cells. J. Cell Biol. 110, 81–95.

Rieder, C.L., Davison, E.A., Jensen, L.C.W., Cassimeris, L., & Salmon, E.D. (1986). Oscillatory movements of monooriented chromosomes and their position relative to the spindle pole result from the ejection properties of the aster and half-spindle. J. Cell Biol. 103, 581–91.

Rodriguez, J.R. & Paterson, B.M. (1990). Yeast myosin heavy chain mutant: Maintenance of the cell type specific budding pattern and the normal deposition of chitin and cell wall components requires an intact myosin heavy chain gene. Cell Motil. Cytoskeleton 17, 301–308.

Roof, D., Meluh, P., & Rose, M. (1992). Kinesin-related proteins required for assembly of the mitotic spindle. J. Cell Biol. 118, 95–108.

Saitoh, H., Tomkiel, J., Cooke, C.A., Ratrie III, H., Maurer, M., Rothfield, N.F., & Earnshaw, W.C. (1992). CENP-C, an autoantigen in scleroderma is a component of the human inner kinetochore plate. Cell 70, 115–125.

Sanger, J.M. & Sanger, J.W. (1980). Banding and polarity of actin filaments in interphase and cleaving cells. J. Cell Biol. 86, 568–575.

Sanger, J.M., Mittal, B., Dome, J.S., and Sanger, J.S. (1989). Analysis of cell division using fluorescently labeled actin and myosin in living PtK2 cells. Cell Motil. Cytoskeleton 14, 201–219.

Satterwhite, L.L. & Pollard, T.D. (1992). Cytokinesis. Curr. Opin. in Cell Biol. 4, 43–52.

Satterwhite, L.L., Lohka, M.J., Wilson, K.L., Scherson, T.Y., Cisek, L.J., Corden, J.L., & Pollard, T.D. (1992). Phosphorylation of myosin-II regulatory light chain by cyclin-$\beta34^{cdc2}$: A mechanism for the timing of cytokinesis. JEB 118, 595–604.

Saunders, W.S. (1993). Mitotic spindle pole separation. Ann. Rev. Trends in Cell Biol. 3, 432–437.

Saunders, W. & Hoyt, M.A. (1992). Kinesin-related proteins required for structural integrity of the mitotic spindle. Cell 70, 451–58.

Sawin, K.E., LeGuellec, K., Philippe, M., & Mitchison, T.J. (1992). Mitotic spindle organization by a plus-end-directed microtubule motor. Nature 359, 540–43.

Saxton, W.M., Hicks, J., Goldstein, L.S.B., & Raff, E.C. (1991). Kinesin heavy chain is essential for viability and neuromuscular functions in *Drosophila*, but mutants show no defects in mitosis. Cell 64, 1093–102.

Schroeder, T.E. (1973). Actin in dividing cells. Contractile ring filaments bind heavy meromyosin. Proc. Natl. Acad. Sci. USA 70, 1688–1693.

Schroeder, T.E. (1972). The contractile ring. II. Determining its brief existence, volumetric changes, and vital role in cleaving arbacia eggs. J. Cell Biol. 53, 419–434.

Sellitto, C. & Kuriyama, R. (1988). Distribution of a matrix component of the midbody during the cell cycle in Chinese hamster ovary cells. J. Cell Biol. 106, 431–440.

Skibbens, R.V., Skeen, V.P., & Salmon, E.D. (1993). Directional instability of kinetochore motility during chromosome congression and segregation in mitotic newt lung cells: A push-pull mechanism. J. Cell Biol. 122, 859.

Steuer, E.R., Wordeman, L., Schroer, T.A., & Sheetz, M.P. (1990). Localization of cytoplasmic dynein to mitotic spindles and kinetochores. Nature 345, 266–68.

Stewart, R.J., Thaler, J.P., & Goldstein, L.S.B. (1993). Direction of microtubule movement is on intrinsic property of the motor domains of kinesin heavy chain and Drosophila ncd protein. Proc. Natl. Acad. Sci. USA 90, 5209–5213.

Sugimoto, R., Yata, M., Muso, Y., & Himeno, M. (1994). Human centromere protein (CENP) is a DNA binding protein which possesses a novel DNA binding motif. J. Biochem. 116, 877–881.

Sullivan, K.F., Hechenberger, M., & Masri, K. (1994). Human CENP-A contains a histone H3 related histone fold domain that is required for targeting to the centromere. J. Cell. Biol. 127, 581–592.

Vaisberg, E.A., Koonce, M.P., & McIntosh, J.R. (1993). Cytoplasmic dynein plays a role in mitotic spindle for motion. J. Cell Biol. 123, 849–858.

Vale, R.D. & Goldstein, L.S.B. (1990). One motor, many tails: An expanding repertoire of force-generating enzymes. Cell 60, 883–885.

Vale, R.D., Reese, T.S., & Sheetz, M.P. (1985). Identification of a novel force generating protein, kinesin, involved in microtubule-based motility. Cell 42, 39–50.

Vallee, R. (1991). Cytoplasmic dynein: Advances in microtubule-based motility. Ann. Rev. Trends in Cell Biol. 1, 25–29.

Walker, R.A., Salmon, E.D., & Endow, S.A. (1990). The Drosophila claret segregation protein is a minus-end directed motor molecule. Nature 347, 780–82.

Walker, R.A., O'Brien, E.T., Pryer, N.K., Soboeiro, M.F., Voter, W.A., Erickson, H.P., & Salmon, E.D. (1988). Dynamic instability of individual, MAP free, microtubules analyzed by video light microscopy: Rate constants and transition frequencies. J. Cell Biol. 107, 1437–1443.

Watts, F.Z., Shiels, G., & Orr, E. (1987). The yeast MYO1 gene encoding a myosin-like protein required for cell division. EMBO J. 6, 3499–505.

Wise, D., Cassimeris, L., Rieder, C.L., Wadsworth, P., & Salmon, E.D. (1991). Chromosome fiber dynamics and congression oscillations in metaphase of PtK2 cells at 23°C. Cell Motil. Cytoskeleton 18, 131–42.

Wright, B.D., Terasaki, M., & Scholey, J.M. (1993). Roles of kinesin and kinesin-like proteins in sea urchin embryonic cell division: Evaluation using antibody microinjection. J. Cell Biol. 123 (3), 681–689.

Wright, B.D., Henson, J.H., Wedaman, K.P., Willy, P.J., & Morand, J.N. (1991). Subcellular localization and sequence of sea urchin kinesin heavy chain: Evidence for its association with membranes in the mitotic apparatus and interphase cytoplasm. J. Cell Biol. 113, 817–33.

Yang, J.T., Saxton, W.M., Stewart, R.J., Raff, E.C., & Goldstein, L.S.B. (1990). Evidence that the head of kinesin is sufficient for force generation and motility in vitro. Science 249, 42–47.

Yang, J.T., Laymon, R.A., & Goldstein. L.S.B. (1989). A three-domain structure of kinesin heavy chain revealed by DNA sequence and microtubule binding analyses. Cell 56, 879–89.

Yen, T.J., Compton, D.A., Wise, D., Zinkowski, R.P., & Brinkley, B.R. (1991). CENP-E, a novel human centromere-associated protein required for progression from metaphase to anaphase. EMBO J. 10, 1245–54.

Yen, T.J., Li, G., Schaar, B.T., Szilak, I., & Cleveland, D.W. (1992). CENP-E is a putative kinetochore motor that accumulates just before mitosis. Nature 359, 536–39.

Zeligs, J.D. & Wollman, S.H. (1979). Mitosis in thyroid follicular epithelial cells in vivo. J. Ultra. Res. 66 (3), 288–303.

Zhang, P., Knowles, B.A., Goldstein, L.S.B., & Hawley, R.S. (1990). A kinesin-like protein required for distributive chromosome segregation in Drosophila. Cell 62, 1053–62.

MEMBRANE-CYTOSKELETON

Verena Niggli

The Cytoskeleton, Volume 1
Structure and Assembly, pages 123–168.
Copyright © 1995 by JAI Press Inc.
All rights of reproduction in any form reserved.
ISBN: 1-55938-687-8

I. INTRODUCTION

A stabilizing cortical network of cytoskeletal proteins is indispensable to maintain plasma membrane structure and function, as illustrated by the erythrocyte system (Bennett, 1990). Such a network is also very likely involved in positioning of transmembrane proteins, controlling their mobility in the membrane, and regulating their function. Cytoskeleton-membrane linkage is further important for organelle transport, adhesion of cells to the substrate or to other cells, shape changes, extension of pseudopods and cell migration. These cellular processes are involved in physiological events such as embryonal development, immune defense and wound healing. An understanding of the mechanism and regulation of cytoskeleton-membrane linkage at the molecular level is therefore crucial for our understanding of these events. This review will focus on recent data on the interaction of specific cytoskeletal proteins with membranes, for which evidence has been provided at the molecular level. The role of cytoskeleton-membrane linkage in cell physiology will not be discussed in detail (see article in Vol. II of this Treatise). The literature survey was completed end of July 1993. For previous recent summaries of cytoskeleton-membrane linkage the reader should consult Niggli and Burger (1987); Mangeat (1988); Carraway and Carraway (1989); Bretscher (1991); Luna and Hitt (1992).

The most extensively studied system is represented by the erythrocyte membrane skeleton (Bennett, 1990). Mature erythrocytes lack a transcellular cytoskeletal network, but the cytoplasmic side of the erythrocyte plasma membrane is covered by a closely associated network of a rod-shaped protein, spectrin, crosslinked by short actin filaments. This network serves to stabilize the membrane and also allows and supports shape changes. Molecules closely related to erythrocyte cytoskeletal components are now known to be ubiquitous proteins, so that a supporting membrane skeleton based on spectrin-like proteins appears to be a general feature of eukaryotic cells. Very rarely extensive microtubule-based membrane cytoskeletons have been identified, as for instance in the parasitic protozoon Trypanosoma brucei (Seebeck et al., 1990). Non-erythroid cells also express highly specialized and restricted regions of cytoskeleton-membrane linkage, such as areas of cell-cell contacts (adherens junctions and desmosomes) and cell-substrate adhesion (focal contacts and hemidesmosomes). These regions contain specific proteins usually not present in the spectrin-based network, and which link actin filaments or intermediate filaments to integral membrane proteins. In addition, a number of other putative cytoskeleton-membrane linker proteins, not present in erythrocytes, have been identified, such as MARCKS, ezrin, filamin, filensin, myosin I etc. Such proteins could for instance be responsible for the cytoskeletal anchoring of ion channels, or mediate actin-membrane linkage in motile areas of the cell. A few membrane receptors appear to bind actin filaments directly. The structure and assembly of membrane-associated skeletons in special cell types such as muscle cells, platelets, neutrophils, nerve cells or the motile slime mold Dictyostelium discoideum will not be discussed in detail in this article; however some of these systems will be considered in Vol. III of this Treatise.

II. METHODS USED TO STUDY CYTOSKELETON-MEMBRANE INTERACTION

The study of the molecular mechanism of cytoskeleton-membrane linkage is to a great extent based on *in vitro* interactions of purified cytoskeletal proteins with other peripheral or integral membrane proteins and with membranes and lipid bilayers. Complex formation of cytoskeletal proteins can be studied by (equilibrium) gel filtration chromatography, sucrose density gradient centrifugation, affinity chromatography, binding of proteins to other proteins absorbed to nitrocellulose, etc. (Carraway, 1992). Obviously, putative interactions observed *in vitro* have to be verified in intact cells and it is important to ask the following questions when interpreting *in vitro* data: are the isolated proteins in a native state (not easily assessed in the absence of enzymatic activity); are low-affinity interactions really relevant in intact cells; do the proteins interact directly or could the interaction be mediated by impurities in the preparation? For a discussion of methods used to study the interaction of purified cytoskeletal proteins with artificial lipid bilayers see the article by Isenberg & Goldmann in this volume. For assessment of the interaction of cytoskeletal proteins with purified plasma membranes, these membranes are extracted with non-ionic detergents, and the detergent-insoluble residue, operationally called the membrane skeleton, is isolated by high-speed (100,000 × g) centrifugation. Similarly, intact cells can be extracted with non-ionic detergents, and the insoluble cytoskeleton, containing F-actin, intermediate filaments and microtubules (depending on the extraction conditions, see Schliwa and van Blerkom, 1981) and associated proteins, can be isolated. Enrichment of transmembrane proteins or cytoskeletal components in such residues is usually taken as an indication for a close interaction with the membrane skeleton. However, other factors may also induce retention in detergent-insoluble residues. A membrane protein with a glycosylphosphatidyl inositol anchor for example has been shown to be enriched in a detergent-insoluble fraction very likely due to its interaction with detergent-insoluble glycosphingolipids, rather than with the cytoskeleton (Brown and Rose, 1992). Artefactual association of components during solubilization may also occur.

Immunofluorescence staining of fixed cells, or immunoelectron microscopy for better resolution, are widely used techniques to obtain more information on putative interactions of cytoskeletal proteins *in situ*. Clearly, co-localization observed with these techniques does not prove direct interactions. Fluorescently labelled purified proteins can be microinjected into cells, and the dynamics of association with intracellular structures can be studied. Elegant techniques to assess *in situ* functions of putative cytoskeleton-membrane linkers comprise elimination of specific proteins by microinjection of antibodies into cells or by gene disruption. However, problems can arise due to the existence of a system of functionally overlapping or redundant proteins. Localization of truncated or mutated proteins in transfected cells can lead to the identification of domains important for targeting to cytoskeleton-membrane attachment sites. Studies of cells in which cytoskeletal proteins have been overexpressed may also yield important functional indications. The cloning and sequence analysis of a number of proteins involved in cytoskeleton-membrane linkage during the last few years, combined with the approaches discussed above, is making a major contribution to our understanding of such linkage structures at a molecular level.

III. ACTIN–PLASMA MEMBRANE LINKAGE

A. The Erythrocyte Membrane Cytoskeleton

Recent advances on this extensively studied membrane-stabilizing network have been summarized in a number of recent reviews (Lazarides and Woods, 1989; Bennett, 1990; Liu and Derick, 1992; Gilligan and Bennett, 1993). The main molecules of this network are the rod-shaped proteins α- and β-spectrin, which form anti-parallel α/β heterodimers (100 nm long). The dimers assemble head to head into tetramers of 200 nm. Spectrin tetramers are crosslinked by short β-actin filaments, yielding a polygonal (primarily hexagonal) lattice which covers the entire cytoplasmic side of the erythrocyte plasma membrane. This lattice is attached to the membrane via at least two different links: protein 4.1 binds the end of the spectrin tetramers to the transmembrane glycoprotein glycophorin, and the globular protein ankyrin anchors the β-spectrin subunit to the anion transporter band 3. The latter interaction occurs approximately 80 nm from the ends of the extended spectrin tetramers. Erythrocytes contain exclusively β-actin, an isoform possibly specialized in forming short filaments. Additional proteins associated with this lattice are tropomyosin and tropomodulin that may together determine and maintain the length of the short (12–14 monomers) β-actin filaments. The protein adducin enhances *in vitro* crosslinking of spectrin and actin. It is thought to interact with preformed spectrin-actin complexes and to recruit a second spectrin molecule to this ternary complex. The latter event is inhibited by calmodulin. Increases in cytosolic calcium, possibly occurring when erythrocytes are subjected to shear stress, may thus lead to weakening of the spectrin-actin link. Other proteins included in the lattice are protein 4.2, which may stabilize ankyrin-band 3 interactions, and protein 4.9 (dematin) whose function is as yet unknown. Dematin, a villin-like protein (Rana et al., 1993), bundles actin filaments *in vitro*, but such bundles are lacking in mature erythrocytes. Using immunogold labeling in combination with electron microscopy protein 4.1, dematin and adducin have been visualized as components of the junctional complexes crosslinked by spectrin tetramers and oligomers, whereas ankyrin is localized near the midregion of spectrin (Derick et al., 1992).

The main constituents of the erythrocyte membrane skeleton, that is, α- and β-spectrin, ankyrin, proteins 4.1 and 4.2 and adducin have recently been cloned and sequenced (for refs. see Gallagher and Forget, 1993). Knowledge of the primary structure of these proteins, combined with functional analysis of protein domains, have allowed precise localization of the different protein-protein interaction sites in many cases. Mammalian erythrocyte spectrin (α/β-heterodimer) consists of rodlike, flexible filaments, approximately 100 nm in length. Main features of α- and β-spectrin (240 and 220 kD, respectively) are multiple homologous 106-residue repeats consisting of triple helical segments; twenty segments in α-spectrin and seventeen in β-spectrin. The actin-binding site is located in the NH_2-terminus of β-spectrin. α-Spectrin contains a potential EF-hand (a conserved, calcium-binding structure) at the COOH-terminus (Karinch et al., 1990; Dhermy, 1991; Gallagher and Forget, 1993). α-Spectrin also contains a so-called *src* homology 3 (SH3) domain (segment 10) which is found in a number of diverse cytoplasmic proteins, some of them membrane-associated, such as *src* tyrosine kinases and phospholipase C. SH3 motifs may bring proteins involved in signal transduction in close proximity to their targets at cytoskeletal structures (Musacchio et al., 1992a; Bar-Sagi et al., 1993; Meriläinen et al., 1993; Weng et al., 1993). Structure analysis of the crystallized SH3 domain of brain α-spectrin suggests that such domains could be binding sites for large molecules, probably proteins (Musacchio et al., 1992b). Spectrin heterodimer assembly has been shown to be initiated at a single discrete region in each subunit located near the actin binding end of the dimer. After initial association of the nucleation sites (four contiguous 106 residue repeats in each subunit) the remainder of the subunits have been proposed to associate along their full length in a zipper-like mechanism (Speicher et al., 1992). Using fusion peptides, Kennedy et al., 1991, have located the ankyrin-binding site to the entire 15th repeat segment and to a small part of the 16th repeat of β-spectrin. Proper folding of the repeats appears to be crucial for the interaction. The binding site on spectrin for protein 4.1 has not yet been precisely located, but appears to be in the N-terminal region of β-spectrin. Spectrin dimers are required for formation of spectrin-actin-protein 4.1 complexes (Bennett, 1990; Becker et al., 1990). A small tightly membrane-bound fraction of spectrin interacts possibly directly with the lipid bilayer via a palmitoylated residue on the β-subunit (Mariani et al., 1993).

As reviewed by Conboy (1993) several functional domains have been identified in protein 4.1 (78–80 kD). The N-terminal, 30 kD domain appears to be the site of interaction with glycophorin A, glycophorin C, band 3 and calmodulin. There are some indications that glycophorin C is the main membrane binding site for protein 4.1 under physiological conditions. Residues 407–427, in a highly charged 10 kD domain in the C-terminal portion of protein 4.1, are essential for promoting ternary complexes of protein 4.1, spectrin and actin, via linkage to spectrin (Discher et al., 1993). The 10 kD domain may also be responsible for the loose interaction of myosin with erythrocyte membranes (Pasternak and Racusen, 1989). Using anti-idiotypic antibodies, interaction sequences in the cytosolic domain of band 3 and in the 30 kD domain of protein 4.1 have been identified. Arginine-rich clusters in band 3 (IRRRY/LRRRY at position 343–347) may interact with residues of opposite charge and identical hydrophobicity present in protein 4.1 (LEEDY at position 37–41). Flanking of the charged cluster by hydrophobic residues appears to be a crucial feature for tight binding. Interaction of protein 4.1 with glycophorins possibly involves related motifs (Jöns & Drenckhahn, 1992). Somewhat at variance with these results, Lombardo et al., 1992, found that residues 1–9 of the cytosolic domain of band 3, as well as a possible site near the membrane, are involved in protein 4.1 binding. The extent to which protein 4.1 is linked *in vivo* to the membrane via band 3 as well as glycophorins, and whether it interacts simultaneously with spectrin/actin and membranes, is controversial (Lombardo et al., 1992; Conboy, 1993). Biophysical measurements of Wyatt and Cherry, 1992, introduce additional complexities: according to their results, ankyrin and protein 4.1 together are necessary to restrict the rotational motility of band 3 in stripped ghosts and these data suggest more complex interactions involving possibly slowly rotating clusters of band 3 associated with glycophorin. Various protein kinases such as tyrosine kinases, protein kinase C or cAMP-dependent kinase, phosphorylate protein 4.1 and inhibit binding to band 3 and/or spectrin, thus providing mechanisms for dynamic regulation of protein 4.1-membrane linkages (Conboy, 1993).

The protein ankyrin (202 kD) is a monomeric, globular protein with moderate asymmetry. Three main domains have been identified: a N-terminal 89 kD-domain containing the binding site for band 3; a 62 kD spectrin-binding domain and a 50–55 kD regulatory domain at the C-terminus (reviewed in Bennett, 1992; Lambert and Bennett, 1993; Peters and Lux, 1993). The N-terminus contains a 33-amino acid motif, present in 22 contiguous copies. Related motifs, which may mediate protein–protein interaction (Thompson et al., 1991; Roehl and Kimble, 1993), have been identified in a number of proteins (cell-cycle control proteins, transcription factors, etc.). The repeats are thought not to be folded as a series of independent units in an extended domain. Rather, the N terminus likely has a nearly spherical shape. Repeats 21 and 22 have been identified as being essential for high-affinity interaction between ankyrin and band 3 (Davis et al., 1991a). The ankyrin-binding site on band 3 involves at least 100 amino acids contained in noncontiguous segments of the primary sequence. The spectrin-binding region of ankyrin has been localized to an acidic subdomain enriched in proline residues (Bennett, 1992). The regulatory domain represses binding of ankyrin to spectrin, which may be due to induction of changes in the domain arrangement. When added as a peptide, the regulatory domain reduces the affinity of an ankyrin isoform lacking the domain (protein 2.2) for band 3, supporting the notion that the regulatory domain acts as a repressor (Davis et al., 1992). The regulatory domain is the most variable part of ankyrin, a target for alternative splicing, and may define specificity of interaction of ankyrin isoforms with different targets. As discussed by Davis et al. (1992) this domain, or its reaction partner in ankyrin, may be controlled by phosphorylation. The membrane association of ankyrin may be modulated/facilitated by palmitylation, a modification observed in ankyrin from chicken and rabbit (Staufenbiel, 1987), but not from human erythrocytes (Maretzki et al., 1990).

The complete sequence of the human α- and β-adducin subunits has been obtained by cDNA analysis (Joshi et al., 1991). Alpha- and β-subunits (81 kDa and 80 kDa, respectively) share 49% sequence identity and 76% sequence similarity. Both molecules have a globular 39 kDa N-terminal region and a C-terminal extended rodlike tail. Both subunits form head-to-head and tail-to-tail dimers and head-to-head $\alpha_2\beta_2$-tetramers. The tail domain contains potential protein kinase A and protein kinase C phosphorylation sites and the calmodulin binding site. The tail domain is also required for interaction with spectrin and actin. Adducin does not share any extended sequences with known proteins. A domain in the N terminus of α- and β-adducin is closely related to a region in the highly conserved N-terminal F-actin-binding domains of the spectrin superfamily, especially α-actinin (see below). However, the isolated head domain does not bind to F-actin in cosedimentation assays. The spectrin-binding site has not been identified. Predictively, the flexible tail would be ideally suited for simultaneous association with two large molecules. The basic residues in the C terminus may be involved in F-actin-binding because a related sequence occurs in MARCKS, an actin-bundling protein (see below).

Although protein 4.2, or pallidin (77 kDa), is a major protein of the erythrocyte membrane skeletal network (ca. 5% of the total membrane protein), its function is not well defined (Cohen et al., 1993). Recent description of the amino acid sequence of pallidin (Sung et al., 1990; Korsgren et al., 1990) reveals an unexpected homology with transglutaminases. However, protein 4.2 lacks an obligatory cysteine present in these enzymes and is therefore inactive. Protein 4.2 may protect other proteins from being cross-linked by transglutaminases. This protein interacts tightly with membranes and binds to band 3, ankyrin and protein 4.1. Its covalent modification by myristic acid (Risinger et al., 1992) may be involved in stabilizing these interactions. Another tightly bound membrane-associated protein of unknown function is the palmitoylated protein p55. Interestingly, like spectrin, it contains a SH3-domain (Ruff et al., 1991).

Not only protein–protein interactions, but protein–bilayer interactions are very likely involved in stabilizing the erythrocyte cytoskeleton. Spectrin and protein 4.1 have been shown to interact with phospholipids. These interactions may stabilize the fluid lipid bilayer and reinforce interactions between cytoskeleton and membrane (for an extensive discussion of lipid interactions of cytoskeletal proteins see Isenberg [1991] and the article by Isenberg and Goldmann in this volume). The erythrocyte cytoskeleton is thought to permit considerable deformation of the cells in narrow passages, followed by recovery of the normal form. In the latter form the spectrin subunits are randomly folded in close apposition, whereas the fully extended skeleton has been estimated to cover an area 8 times that of the normal erythrocyte surface area (Liu and Derick, 1992). The functional importance of the components of the erythrocyte membrane skeleton is demonstrated by inherited hemolytic anemias that are based on molecular defects of the membrane skeleton. These defects are due to either deficiencies in spectrin, ankyrin, band 3, protein 4.1 or protein 4.2., decreased spectrin dimer-dimer association or abnormal spectrin-protein 4.1 associations. Glycophorin A and B deficiencies are clinically asymptomatic, and glycophorin C deficiency leads only to mild elliptocytosis and no hemolysis, suggesting that glycophorin C is not the major site for membrane–cytoskeleton linkage. The other defects result in altered cell shape, decreased deformability, loss of surface membrane and generally decreased mechanical stability of the membrane (reviewed in Palek and Sahr, 1992; Peters and Lux, 1993). These findings underline the important functional role of the erythrocyte membrane network and its components.

The biogenesis of the erythrocyte membrane skeleton during erythropoiesis has been studied in detail using transformed chicken or murine erythroid cells that can be induced to differentiate *in vitro* (for reviews see Lazarides and Woods, 1989; Hanspal and Palek, 1992). Based on these studies, formation of a stable membrane skeleton correlates with induction of band 3 synthesis, which appears to be the principal rate limiting step. Band 3 recruits ankyrin onto the skeleton, and ankyrin determines assembly of α- and β-spectrin. Protein 4.1 is assembled last, contributing to the stability of the spectrin network. The peripheral components are first synthesized in excess, turning over rapidly. When the peripheral components are assembled into stable skeletons, their turnover is decreased. Despite our detailed knowledge regarding the erythrocyte membrane skeleton, some open questions remain. We still have to learn more on changes in the network structure during erythrocyte deformation and what is the relevance *in situ* of the interactions of protein 4.1 with glycophorin C and band 3.

B.   Isoforms of Erythroid Proteins in Nonerythroid Tissues

Isoforms of erythrocyte α- and β-spectrin, ankyrin, adducin, protein 4.1 and 4.2 have been found in a variety of other tissues. Nonerythrocyte spectrin (fodrin) is widely distributed and occurs in vertebrates, invertebrates, *Drosophila* and higher plants, suggesting that spectrin-based cortical networks are ubiquitous cellular features. Such networks have been implicated in stabilizing plasma membrane areas, in the formation and maintenance of cell–cell junctions and in the regulation of exocytosis. Spectrin is especially abundant in brain where it may, together with other proteins such as ankyrin, stabilize structures such as nodes of Ranvier and postsynaptic densities. Spectrin is also present in neuromuscular junctions (Bennett, 1990).

Nonerythroid spectrins are closely related to the erythroid form. With few exceptions, nonerythrocyte spectrin is composed of two nonidentical α- and β-subunits consisting of homologous 106-amino-acid repeats. These have retained the capacity to interact with actin, ankyrin and adducin and have acquired the capacity to interact with calmodulin, which is lacking in mammalian erythrocyte and *Drosophila* α-spectrin (Dhermy, 1991; Gallagher and Forget, 1993). Currently, two distinct α-spectrin subunits have been identified: a specialized subunit found only in mammalian erythrocytes and a ubiquitous form present in all other tissues. cDNAs for both types have been cloned and sequenced, revealing an overall identity of 58% (McMahon et al., 1987). The highly conserved nonerythroid α-subunit can combine with different β-isoforms, which may form the basis of functional diversity of spectrins. At least five different types of β-spectrin have been identified: $β_R$, in mammalian erythrocytes; $β_G$, the widely distributed nonerythrocyte isoform; $β_{NM}$, an isoform restricted to neuromuscular junctions; $β_{TW}$, an isoform in the terminal web of avian intestinal epithelium lacking the ankyrin binding site and $β_H$, a *Drosophila* isoform (Coleman et al., 1989; Bennett, 1990). $β_R$ and $β_G$ show 60% sequence identity (Hu et al., 1992). $β_G$ is present in all tissues examined (brain, liver, heart, kidney, lung), whereas $β_R$ is detected only in erythrocyte, brain and heart. $β_R$ is highly enriched in the cerebellum compared with the cortex, whereas $β_G$ is distributed equally between cerebellum and cortex (Hu et al., 1992).

In most tissues, spectrin is closely associated with the plasma membrane, indirectly via ankyrin or directly via integral membrane proteins. Protein 4.1 may be a potential interaction partner. Ankyrin-independent spectrin-plasma membrane interaction in brain may be regulated by calcium/calmodulin, because binding of spectrin to synaptosomal membranes is inhibited by calmodulin in the presence of μM concentrations of calcium, whereas spectrin/ankyrin interaction is not affected (Bennett, 1990). Brain spectrin, but not ankyrin, shows a preferred colocalization with the neuronal cell adhesion molecule N-CAM at cell–cell contact sites and growth cones and coimmunoprecipitates with N-CAM, suggesting a direct interaction (Pollerberg et al., 1987). The epithelial cell adhesion protein E-cadherin likewise coimmunoprecipitates with spectrin and ankyrin (Nelson et al., 1990). Spectrin has also been implicated in regulation of glutamate receptor function, although a direct interaction has not been demonstrated (Siman et al., 1985). Spectrin may be involved in neurotransmission, because it has been shown to bind with high affinity to synapsin I on synaptic vesicles at a site close to the actin-binding site (Sikorski et al., 1991). Spectrin could therefore modulate the docking and fusion of synaptic vesicles. An unusual β-spectrin isoform, distinct from $β_R$ and $β_G$, has been identified in acetylcholine receptor clusters of neuromuscular junctions. Interestingly, α-spectrin could not be identified. Spectrin may be linked indirectly (via the 43-kDa protein) or directly to the acetylcholine receptor and would therefore be involved in receptor positioning (Bloch and Morrow, 1989). One of the few examples of functional regulation of a plasma membrane receptor by a cytoskeletal protein is represented by the interaction of the leucocyte common antigen CD45 with spectrin. This transmembrane protein has been shown to be closely associated with spectrin in lymphocytes, and to interact with high (nM) affinity with purified spectrin, but not with ankyrin. Interestingly, spectrin induces a 7.5-fold increase in V_{max} of the tyrosine phosphatase activity of this receptor. Mitogens increase the amount of CD45/spectrin complex in lymphocytes, which may then trigger an increase in phosphatase activity (Lokeswhar and Bourguignon, 1992a).

In addition to spectrin, other closely related molecules have been identified in non-erythroid tissues. All share spectrin-like repeats and contain highly conserved N-terminal actin binding sites. The proteins belonging to this spectrin superfamily are the actin-crosslinker α-actinin and dystrophin, a muscle-specific protein, both implicated in actin-plasma membrane linkage, see below (Dhermy, 1991).

Ankyrin isoforms are also widely distributed, for example in basolateral domains of epithelial cells, and are especially enriched in the vertebrate nervous system where they are found in the node of Ranvier and in neuromuscular junctions. Three major types can be distinguished: Ankyrin$_R$, the erythrocyte isoform, which is also present in brain, particularly in the cerebellum; ankyrin$_B$, the major brain isoform located in neurons and glial cells; ankyrin$_{NODE}$, an isoform restricted to the cytoplasmic surface of axonal plasma membrane at nodes of Ranvier and axon initial segments. Other isoforms identified on the basis of crossreactivity with anti-ankyrin antibodies have been detected in kidney, epithelial cells, liver and skeletal muscle but have not yet been well characterized (Bennett, 1992; Lambert and Bennett, 1993; Peters and Lux, 1993). Ankyrin$_B$ interacts preferentially with β_G-spectrin, and ankyrin$_R$ with β_R-spectrin (Bennett, 1990). Ankyrin appears to be a particularly versatile linker protein, as it interacts with a variety of transmembrane proteins, as well as with tubulin, intermediate filaments and, of course, spectrin/fodrin (Bennett, 1992). According to recent sequence data the repeated 33-amino-acid motif in the N terminus and the spectrin-binding domain are highly conserved in the brain and erythrocyte ankyrins. Sequence divergence occurs in part of the N-terminal region, in the region linking membrane- and spectrin-binding domains and in the C-terminal region. The latter region contains the regulatory domain (see above), and may be involved in recognition of specific membrane-binding sites by different ankyrin isoforms (Otto et al., 1991). Ankyrin has been shown to bind directly to a variety of integral membrane proteins; to voltage-dependent and to amiloride-sensitive sodium channels (Srinivasan et al., 1992; Smith et al., 1991); to the Na$^+$/K$^+$-ATPase (Nelson and Veshnock, 1987); to gastric parietal cell H$^+$/K$^+$-ATPase (Smith et al., 1993); to the lymphocyte adhesion antigen CD44 (Lokeshwar and Bourguignon, 1992b); to a family of recently described neural adhesion glycoproteins (Davis et al., 1993); to the cardiac Na$^+$-Ca^{2+} exchanger (Li et al., 1993) and to a fraction of the inositol-trisphosphate receptor (Joseph and Samanta, 1993). Interestingly, ankyrin significantly reduces IP$_3$ binding to the receptor and inhibits receptor function in lymphoma light density vesicles (Bourguignon et al., 1993). Some information is already available on putative ankyrin domains involved in these different interactions. Using isolated and reconstituted rat brain sodium channels, Srinivasan et al. (1992) demonstrated by competition assays with purified erythrocyte ankyrin fragments that the sodium channel binding domain is located in the 43-kDa C-terminal portion of the 33-amino-acid repeat domain. This sodium channel binding domain is sufficient to mediate band 3-interaction (Bennett, 1992), and the cytoplasmic domain of band 3 competes for sodium channel binding. The repeat region may also mediate tubulin binding (Davis et al., 1992). In contrast, high-affinity interactions with the Na^+/K^+-ATPase require regions of the spectrin-binding domain of ankyrin, as well as weak interactions with the N-terminal domain (Davis and Bennett, 1990). These data support the contention that ankyrin is involved in maintaining high densities of certain membrane proteins in discrete cellular regions such as nodes of Ranvier or basolateral domains of polarized epithelial cells. It is not clear how ankyrin is targeted to these specific domains and if ankyrin enrichment is the primary event.

Multiple isoforms of protein 4.1 of variable size (30–210 kDa) have been identified in a variety of mammalian nonerythroid cells. However, only a single protein 4.1 gene has been identified. The diversity of protein 4.1 must be due to posttranslational or posttranscriptional modifications (phosphorylation, O-linked glycosylation, deamidation, partial proteolysis) and alternative RNA splicing. Indeed, a number of different mRNAs have been detected in leukocytes, brain, liver, and intestine. These mRNAs show variations, for instance, in N-terminal and spectrin-binding domains. Alternative splicing is regulated in a developmental and tissue-specific manner (Tang et al., 1988; Conboy et al., 1991). Certain isoforms found in lymphocytes, liver and intestine lack a 21-amino-acid sequence essential for promoting spectrin-actin association (Discher et al., 1993). In agreement with these structural findings, immunofluorescence data show colocalization of non-erythroid protein 4.1 along stress fibers, in cell–cell contact areas or even in the nucleus, suggesting functions partly unrelated to participation in a spectrin-based membrane skeleton (for references see Tang et al., 1988; Conboy, 1993). Stevenson et al. (1989) detected most of protein 4.1 and 50% of fodrin in detergent-insoluble plasma membrane fractions of human neutrophils, suggesting that some isoforms of protein 4.1 may be part of a membrane skeleton. Using antisense techniques, Giebelhaus et al. (1988) could demonstrate an important functional role of protein 4.1 in embryonal development of *Xenopus* oocytes.

A family of protein 4.1-related proteins has been identified recently based on sequence data. Members of this family include the proteins talin, ezrin, moesin, radixin and merlin (Rees et al., 1990; Funayama et al., 1991; Lankes and Furthmayr, 1991; Sato et al., 1992; Trofatter et al., 1993; Rouleau et al., 1993). All members of this family have a homologous N-terminal domain followed by an α-helical domain and a highly charged C-terminal region. The N-terminal domain, which mediates PIP_2-sensitive binding of protein 4.1 to glycophorin, is thought to also mediate membrane linkage of the other members of the 4.1 family, possibly to glycophorin-like molecules although the LEEDY-sequence (see above) is missing. The helical rod-like region, which in protein 4.1 mediates spectrin/actin binding, may also be involved in the interaction of talin with other cytoskeletal proteins (see below). The homology of this helical segment is mainly based on predicted secondary structure and not on primary structure. Evidence has recently been provided for the importance of the N-terminal domain in mediating membrane association by expression of truncated ezrin cDNAs in kidney cells: in such transfected cells the N-terminal domain of ezrin is still associated with the plasma membrane but is no longer a component of the detergent-insoluble cytoskeleton, and in contrast the C-terminal domain co-localizes with actin filaments (Algrain et al., 1993).

Two protein tyrosine phosphatases with a N-terminal domain homologous to that of the protein 4.1 family have been identified (Yang and Tonks, 1991; Gu et al., 1991). These enzymes may be in close proximity with members of the protein 4.1-family and possibly regulate their function. The functional role of the closely related molecules ezrin, radixin and moesin (the ERM family; see Sato et al., 1992) is not yet clear. Using antibodies crossreactive with all three proteins, the ERM family has been located in microvilli, but also in surface protrusions, microspikes and filopodia of a wide range of cells, as well as in cleavage furrows and in sites of cell-cell and cell-substrate adhesion (Sato et al., 1992). Experiments with antibodies specific to moesin and ezrin show that these proteins are absent from cell-cell and cell-substrate junctions (Franck et al., 1993). As summarized by Funayama et al. (1991) radixin is an actin barbed-end capping protein. How ezrin interacts with actin filaments is not clear, as it does not interact with F-actin *in vitro* at physiological salt concentrations (for references see Algrain et al., 1993). More information is available on the functional role of talin, a candidate for linkage of actin filaments to the transmembrane receptor family of integrins, as discussed below in detail. The protein 4.1-family thus has the potential to fulfill differential, but related, roles in different plasma membrane areas using related sequences. Interestingly, alterations in a 4.1-related protein (merlin or schwannomin) have been postulated to be the cause of the inherited disease neurofibromatosis-2 (Trofatter et al., 1993; Rouleau et al., 1993).

As reviewed by Gilligan and Bennett, 1993, immunoreactive forms of adducin have been detected in different tissues such as rat brain, kidney, liver, lung and bovine lens, as well as in platelets, lymphocytes and fibroblasts. In the three latter types of cells only α-adducin, but not β-1 adducin (a splice variant) could be detected (Waseem and Palfrey, 1990; Gilligan and Bennett, 1993). Future experiments may show whether α-adducin can function alone, or whether other as yet unidentified subunits are present. Adducin is enriched in calcium-dependent cell-cell junctions of human keratinocytes, and could thus play an important role in establishing these junctions. Bovine brain adducin shows properties comparable to those of red cell adducin: it binds to spectrin-actin complexes, interacts with calmodulin and is a substrate of protein kinase C *in vitro* (Gilligan and Bennett, 1993) and in intact fibroblasts (Waseem and Palfrey, 1990).

Immunoreactive forms of human protein 4.2 have been shown to be tightly associated with plasma membrane fractions of platelets, brain and kidney (Friedrichs et al., 1989). A murine platelet storage pool disease (pallid) is associated with the protein 4.2 gene, suggesting that the non-erythroid protein 4.2 is involved in secretory processes (White et al., 1992).

To conclude, the work described in the above section shows that the widely distributed non-erythroid isoforms of erythrocyte membrane proteins have an important role in modulating plasma membrane functions such as cell–cell contact formation and the functioning and positioning of ion channels and receptors.

C. Special Components of Nonerythroid Cortical Actin Networks

Nonmuscle filamin, or actin-binding protein (ABP), is a widely distributed homodimeric actin cross-linking protein (280 kDa) that also links actin directly to membrane glycoproteins in platelets and myelocytic cells. Based on electron microscopic and sequence analyses, the following structural model for human endothelial ABP has been proposed. The subunits of the ABP-dimers are closely associated at the tail domains, whereas the other ends, containing the actin-binding domains, are free, like a leaf spring (Gorlin et al., 1990). This explains why ABP induces branching and not bundling of actin filaments. ABP interacts directly with the major membrane glycoprotein of platelets, GP Ib-IX (Gorlin et al., 1990; Andrews and Fox, 1992), and the Fc RI-receptor in myelocytic cells (Ohta et al., 1991). The GP Ib-IX-binding domain is located near the self-associating tail of ABP, whereas the ABP-binding domain has been mapped to residues 536–568 of the cytosolic region of GP Ib$_\alpha$ (Andrews and Fox, 1992). The Fc receptor does not contain a comparable region, and its ABP-binding site has not been identified. Interestingly, binding of IgG to the Fc receptor prevents ABP binding, implicating modulation of actin–receptor linkage by receptor occupancy (Ohta et al., 1991). ABP may be a hitherto undetected component of the erythrocyte membrane skeleton, as suggested by recent results of Brown et al. (1993). Indeed, evidence for a cortex-stabilizing role of ABP has been provided by Cunningham et al. (1992). Tumor cell lines lacking this protein show impaired locomotion and blebbing of the plasma membrane. Transfection of ABP into these cells restores locomotion and reduces blebbing. Recent results by Cantiello et al. (1993) using these cell lines, suggest that ABP may be involved in activation of ion channels mediating cell volume regulation.

Certain members of a family of calcium- and phospholipid-binding proteins, the annexins, interact with spectrin and/or F-actin and are thought to be constituents of cortical actin networks. Intact calcium-binding sites of annexin II are a prerequisite for its cortical localization (Thiel et al., 1992). However, the physiological role of annexins is poorly understood. An annexin II-like protein in *Xenopus* has been implicated in DNA replication (Vishwanatha and Kumble, 1993).

Another protein family with a potential for transmitting signals to the actin network is myristoylated alanine-rich protein kinase C substrate (MARCKS). MARCKS consists of rodlike acidic proteins with a myristoylated N-terminal domain and an amphipathic effector site containing phosphorylation sites, as well as actin and calmodulin binding sites. MARCKS binds *in vitro* plasma membranes and calmodulin and cross-links actin. All these activities are inhibited by phosphorylation (for reviews see Aderem, 1992; Blackshear, 1993). Membrane interaction, and calmodulin binding of MARCKS depend on myristoylation. Membrane association may be due to direct bilayer insertion of the fatty acid moiety because it is not disturbed by treatments denaturing putative protein receptors, and association of MARCKS with acidic phospholipid bilayers is inhibited by phosphorylation of the protein (George and Blackshear, 1992; Manenti et al., 1993; Taniguchi and Manenti, 1993). The putative important role of MARCKS in mediating stimulus-dependent changes of actin structure at the membrane will have to be substantiated by further experiments. MARCKS shows some structural and functional similarities to adducin (see above) and to GAP43, a component of the cortical actin network in growth cones (Aderem, 1992).

Dystrophin, a rodlike protein of 427 kDa, closely related to α-actinin and spectrin (see above), is a component of the membrane skeleton of all types of muscle and of some neurons. Dystrophin is thought to form antiparallel homodimers with its N terminus linking to actin filaments and its C terminus interacting with a complex of membrane glycoproteins that are in extracellular contact with laminin (Ervasti and Campbell, 1991; Ibraghimov-Beskrovnaya et al., 1992). The glycoprotein-binding site has recently been mapped to the highly conserved cysteine-rich domain and the first half of the carboxy-terminal domain (Suzuki et al., 1992). Dystrophin may link the cortical actin to the plasma membrane and the extracellular matrix, thereby stabilizing the membrane during muscle contraction (Petrof et al., 1993). Close membrane association of dystrophin is essential for normal muscle function as evidenced by patients suffering from the hereditary disease Duchenne muscular dystrophy, whose muscles exhibit a loss or alteration of dystrophin (Ahn and Kunkel, 1993). Similarly, absence of one of the dystrophin-associated membrane proteins results in muscular dystrophy, supporting the above notion (Matsumara et al., 1992). Actin and glycoproteins may not be the only interaction partners. Dystrophin has recently been shown to interact *in vitro* with high affinity with the protein talin, but not with vinculin (Senter et al., 1993). Talin could thus link dystrophin to the integrins (adhesion molecules; see below). The relevance of this *in vitro* interaction is not clear. In smooth muscle dystrophin is located in distinct plasma membrane domains (calveolae), not overlapping with adherens junctions containing integrins, talin, etc. (North et al., 1993). Several dystrophin-related proteins of unknown function have been detected in nonmuscle tissues (Hugnot et al., 1992; Matsumara et al., 1993).

D. Focal Contacts

Focal contacts or focal adhesions, observed most prominently in cultured cells, represent areas of closest contact of cells with the substrate. In these areas, bundles

Table 1. Proteins Enriched in Focal Contacts

Major structural components	Components of as yet unknown function	Enzymes
Integrins ($\alpha_5\beta_1$, $\alpha_V\beta_3$)	Tensin	Protein kinase C (α)
Actin	Zyxin	Calpain
α-Actinin	Tenuin	Tyrosine kinases (pp125^{FAK}; pp60^{src})
Vinculin	Radixin	
Talin	Paxillin	
	VASP	

of transcellular actin filaments (stress fibers) are physically linked to the extracellular matrix via integrins, transmembrane receptors for specific extracellular components (Wang et al., 1993). This connection allows transfer of information from the extracellular space to the cytoskeleton, and vice versa (for a review on integrins see Hynes, 1992). In this section I will be mainly concerned with molecular aspects of integrin–cytoskeleton linkage (for a more comprehensive review on cell adhesion see the article by K. Burridge in Volume II of this treatise).

A variety of cytoskeletal proteins, together with integrins and certain protein kinases, have been shown by immunofluorescence staining to be enriched in areas of cell adhesion (Table 1; see recent reviews by Turner and Burridge, 1991; Geiger and Ginsberg, 1991; Luna and Hitt, 1992). The peripheral focal contact proteins readily exchange with the cytoplasmic pool of these proteins (Geiger and Ginsberg, 1991). Based on *in vitro* interactions of purified proteins, different modes of linkage of actin to the cytosolic domain of integrins have been proposed. The integrins themselves, widely distributed proteins, are α/β-heterodimeric transmembrane proteins. Of the various isoforms, $\alpha_5\beta_1$ and $\alpha_V\beta_3$ are specifically enriched in focal contacts (Table 1). Three regions in the cytosolic domain of the β_1-subunit appear to be responsible for localization to focal contacts (Reszka et al., 1992). Actin filaments are not thought to be directly linked to this domain. Rather, different cytoskeletal proteins have been suggested to be involved. One possible chain of proteins consists of the spectrinlike actin cross-linking protein α-actinin binding to the protein vinculin, which in turn binds talin that interacts directly with the cytosolic domain of β_1-integrin. Other simpler linkages could be mediated by α-actinin binding directly to the cytosolic domain of β_1-integrin, or by talin, which has recently been shown to interact directly with actin. However, most of these interactions have not been verified *in situ*. As summarized in Table 1, other minor components such as tensin, zyxin, paxillin and tenuin have been localized to focal contacts, though their functional role is not at all clear. Zyxin interacts *in vitro* with α-actinin; paxillin and tensin (an actin-binding protein) bind to vinculin, and the protein VASP interacts with F-actin. The protein radixin (see previous section) may cap the barbed ends of actin filaments near focal contacts (Luna and Hitt, 1992).

Turner and Burridge (1991) suggest a regulatory, non-enzymatic role for these minor components, as paxillin, tensin and zyxin are substrates of tyrosine kinases. Recent evidence suggests a crucial role for tyrosine kinases in focal contact assembly; as a tyrosine kinase, $p125^{FAK}$, is enriched in focal contacts and as clustering of integrins leads to tyrosine phosphorylation and activation of this kinase (Zachary and Rozengurt, 1992). Moreover, tyrosine phosphorylation of paxillin and tensin is increased upon adhesion of cells to the extracellular matrix (Bockholt and Burridge, 1993). Both adhesion-induced tyrosine phosphorylation and focal contact and stress fiber formation are prevented by a tyrosine kinase inhibitor (Burridge et al., 1992). Protein kinase C-dependent phosphorylation has also been implicated, as the α-isoform is enriched in focal contacts of fibroblasts and phorbol esters, activators of protein kinase C, promote focal contact formation in fibroblasts (Woods and Couchman, 1992). It is not clear how increases in tyrosine phosphorylation and other signals are generated at the plasma membrane level and how these signals are transferred to the focal contact constituents. It is noteworthy that the ras-related GTP-binding protein *rho* has recently been implicated in initiation of focal contact and stress fiber formation upon stimulation of Swiss 3T3 cells with growth factors (Ridley and Hall, 1992).

It is not known at present which of the proposed actin-integrin linkages occur *in situ*, whether they occur in parallel or whether temporal and/or spatial changes in the type of linkage occur during cell adhesion, motility, etc. In addition to specific protein–protein interactions, direct protein–bilayer contacts may be important. Talin, α-actinin and vinculin have been shown to interact *in vitro* so closely with bilayers containing acidic phospholipids that they react with an analogue of lecithin carrying a photoactivatable group on its apolar portion (Niggli et al., 1986; Goldmann et al., 1992; Niggli and Gimona, 1993). Vinculin is also labeled by a photoactivatable fatty acid in intact fibroblasts (Niggli et al., 1990). As outlined in Niggli and Burger (1987), this close interaction may serve to concentrate these proteins at the plasma membrane and may facilitate, enhance and/or modify subsequent specific interactions with other focal contact proteins. In the following section I will focus on molecular properties of the main focal contact components.

As reviewed by Blanchard et al. (1989), α-actinin (100 kDa), which cross-links F-actin, is an antiparallel homodimeric rod-shaped protein, 25-nm long, with extensive homologies to spectrin and dystrophin. It consists of an N-terminal actin-binding domain, an extended rod domain with four spectrin-like repeats and two C-terminal calcium-binding motifs. α-Actinin is a widely distributed protein, occurring in skeletal and smooth muscle, as well as in brain, kidney, liver, leuko-cytes, *Dictyostelium discoideum* and *Acanthamoeba* (Blanchard et al., 1989). A variety of skeletal muscle isoforms have been identified, whereas smooth muscle and non-muscle isoforms of α-actinin result from alternative splicing of the transcript from a single gene (Waites et al., 1992). The main difference in nonmuscle and muscle α-actinin has been thought to reside in the different calcium sensitivities of their interaction with F-actin. However, in contrast to previous reports Pacaud and Harricane (1993) did not observe an inhibitory effect of physiological calcium concentrations on F-actin bundling by purified macrophage α-actinin. The functional role of the EF-hands is therefore unclear. α-Actinin is generally concentrated in areas of actin-membrane association in muscle and nonmuscle cells such as focal contacts, Z-lines, dense bodies and cell–cell junctions, and it also shows a periodical distribution along stress fibers. α-Actinin interacts *in vitro* with low affinity with vinculin (K_d 10^{-6}–10^{-7} M; for references see Blanchard et al., 1989), with the cytoplasmic domain of the β_1-subunit of integrin (K_d ca 10^{-8} M; Otey et al., 1990) and with zyxin, a protein that is located in cell-substrate and cell–cell contact areas, as well as along stress fibers near adhesion sites (K_d ca 1 µM; Crawford et al., 1992). α-Actinin can be cleaved by the protease thermolysin into two major fragments of 27 and 53 kDa. Binding sites for actin, vinculin and zyxin have been located to the 27-kDa N-terminal fragment, whereas the integrin-binding site is contained in the 53-kDa rod-like, α-helical fragment (Pavalko and Burridge, 1991; Crawford et al., 1992; Otey et al., 1990). α-Actinin interacts with the isolated cytosolic domain of β-integrin, as well as with intact integrins from chicken smooth muscle or with GP IIb/IIIa from platelets (Otey et al., 1990). Conflicting results have been obtained on the localization of truncated α-actinin molecules or fragments in cultured cells (Pavalko and Burridge, 1991; Hemmings et al., 1992). The *in situ* relevance of the binding sites detected *in vitro* is thus not clear. Interestingly, microinjection of either one of the fragments into fibroblasts led first to disassembly of stress fibers and later to loss of focal contacts (Pavalko and Burridge, 1991). Transfection of a truncated muscle α-actinin lacking the C terminus with the two EF-hands into PtK$_2$ cells had the same effect (Schultheiss et al., 1992). These data point to an important role for α-actinin in the stability of stress fibers and focal contacts. The role for α-actinin–zyxin interaction is not clear. Zyxin, which does not interact with actin, shows no sequence homology to known proteins, except for three so-called LIM domains, which have been identified in proteins involved in transcriptional regulation and cell differentiation. Zyxin exhibits an unusual proline-rich N terminus. Zyxin is thus an interesting candidate to mediate adhesion-stimulated changes in gene expression (Sadler et al., 1992).

Talin, another major component of focal contacts, is a member of the protein 4.1 family (see above). According to cDNA sequencing of murine talin, the 270-kDa protein contains an N-terminal region homologous to that of protein 4.1 (47-kDa fragment), followed by an α-helical rod-domain and a highly charged C-terminal domain (190-kDa fragment). In agreement with these data, talin appears in electron microscopic images as an elongated molecule of 60 nm with a globular head and a flexible tail (Rees et al., 1990). A small fraction of chicken gizzard talin is O-glycosylated in the 190-kDa tail domain, a modification shared by protein 4.1 (Hagmann et al., 1992). Talin is also subject to phosphorylation by tyrosine kinases and protein kinase C (for references see Isenberg and Goldmann, 1992). Talin occurs in focal contacts, ruffles of cultured cells, neuromuscular junctions and intercalated disks of cardiac muscle, but is absent from cell–cell junctions (Beckerle and Yeh, 1990). Talin interacts *in vitro* with vinculin, actin, the cytosolic domain of integrins and acidic phospholipids. Talin contains at least two, possibly three, distinct vinculin binding sites in the 190-kDa fragment, located within residues 498–950 (two adjacent sites) and 1328–2268 (Gilmore et al., 1993). Residues 1653–1848 are recognized by an anti-idiotypic antibody generated by injection of vinculin (Lee et al., 1992). It has been proposed that the 190-kDa fragment contains both the integrin- and actin-binding sites (for references see Isenberg and Goldmann, 1992), but precise locations are unknown. Talin, according to recent reports, is able to nucleate actin polymerization *in vitro*, affects F-actin properties and shows a limited F-actin cross-linking activity (Isenberg and Goldmann, 1992; Ruddies et al., in press; Muguruma et al., 1992). The latter activity is only observed in the presence of α-actinin and does not appear to be due to talin-α-actinin interaction, as the two proteins do not interact, at least not with high affinity (Muguruma et al., 1992). The location of the lipid-binding site is not known; however, talin bound to bilayers is still capable of promoting actin-filament growth and anchoring filaments at the bilayer (Kaufmann et al., 1992). Using an elegant approach, Geiger et al. (1992) have recently provided evidence that the *in vitro* low-affinity interaction of talin with β_1-type integrins occurs *in situ*. Chimeric molecules consisting of the transmembrane and cytoplasmic domain of β_1-integrin, linked to the extracellular domain of the cell–cell adhesion protein N-cadherin, were expressed in CHO and 3T3 cells. The β_1-integrin cytosolic domain was (a) sufficient to locate the chimera to focal contacts where N-cadherin is normally not present and (b) was able to direct talin to cell–cell junctions where it is otherwise lacking. The cytoplasmic domain of integrins is thus able to determine, by direct or indirect interaction, the cellular location of talin. Both the 47-kDa and the 190-kDa domains of talin are able to associate with focal contacts, indicating multiple binding sites with different ligands such as integrins, vinculin and lipids. The 40-kDa fragment is important for specific targeting to sites of cell–substrate adhesion (Nuckolls et al., 1990). Stable association of talin with these sites depends on the presence of the vinculin-binding sites (Gilmore et al., 1993). According to a preliminary report (Nikolai et al., 1993), microinjection of a monoclonal antibody specific to the N-terminal domain of talin into fibroblasts results in disruption of focal contacts and stress fibers.

Vinculin, a 117-kDa protein, consists of a globular head and an extended tail. The N-terminal 90-kDa globular head is linked via a proline-rich region to the basic 27-kDa C-terminal tail (reviewed by Otto, 1990). No obvious sequence homology of vinculin to that of other proteins has been detected, with the exception of a bacterial gene product in *Listeria monocytogenes* required for interaction with F-actin, which shows strong regional homology to the proline-rich region of vinculin (Domann et al., 1992). Vinculin shows some similarity to a 102-kDa cadherin-associated protein (see below; Nagafuchi et al., 1991). Vinculin occurs in areas of cell–cell and cell–substrate contact in neuromuscular junctions and in dense plaques of smooth muscle. Vinculin is widely distributed in cultured cells, brain, leukocytes, platelets, etc. (Otto, 1990). Vinculin is a substrate for protein kinase C and tyrosine kinases; however, changes in the level of tyrosine phosphorylation could not be correlated with changes in focal contact structure (Kellie et al., 1991). Evidence has been provided for the *in vitro* interaction of vinculin with talin, α-actinin, actin, paxillin, tensin and acidic phospholipid bilayers. Residues 167–207, located in the 90-kDa globular head of vinculin, are indispensable but not sufficient, for talin binding. Residues 1–258 are required for a fully functional talin-binding domain. Although residues 167–207 are encoded on a separate exon, no evidence was found for vinculin transcripts lacking the talin binding site (Gilmore et al., 1992). Such special vinculin isoforms would be suitable for cell–cell junctions where talin is lacking. The interaction of vinculin with α-actinin is low affinity as outlined above. We have recently obtained evidence that α-actinin and vinculin may form triple complexes with acidic phospholipid bilayers, resulting in changed interactions of these two proteins with the bilayer (Niggli and Gimona, 1993). Specific α-actinin- and lipid-binding domains have not been identified. The direct interaction of vinculin with F-actin has been disputed, but has received support by recent work. A monoclonal antibody reacting with a region containing residues 587–851 in the globular head of vinculin inhibits binding of F-actin to nitrocellulose-bound vinculin (Ruhnau and Wegner, 1988; Westmeyer et al., 1990). Vinculin also shows a tendency to self-associate, possibly via tail–tail contacts (Otto, 1990). Another interaction partner of vinculin is the 68-kDa protein paxillin, which, like talin, is restricted to sites of cell–substrate adhesion. Paxillin binds specifically to the 27-kDa rod domain of vinculin in blot overlay assays (Turner et al., 1990). Burridge et al. (1992) have shown that paxillin incorporates substantial amounts of phosphate linked to tyrosine upon adhesion of cells to the extracellular matrix. Interestingly, both paxillin phosphorylation and adhesion are prevented by tyrosine kinase inhibitors. Paxillin may somehow regulate vinculin function depending on its state of phosphorylation. Paxillin binds to the SH3 domain of pp60[src] and could thus recruit this kinase to focal contacts (Weng et al., 1993). Another putative link between extracellular signals and the cytoskeleton is the protein tensin (150–200 kDa). Tensin, which binds both actin and vinculin, is present in both cell–cell and cell–substrate junctions and is a substrate for tyrosine kinases (Bockholt et al., 1992; Bockholt and Burridge, 1993). As shown by partial cloning of tensin, this protein contains a *src* homology 2 (SH2) domain. Such sites mediate interaction with phosphotyrosine-containing domains of other proteins (Davis et al., 1991b). Tensin is so far the only cytoskeletal protein known to contain an SH2 domain, which is frequent in proteins directly involved in signal transduction such as phospholipase C or pp60[src].

Bendori et al. (1989) provide evidence that both the 45-kDa N-terminal fragments and the 78-kDa C-terminal portion of vinculin, transfected into COS cells, locate specifically to focal contacts. They propose that vinculin associates with focal contacts through both its interaction with talin and self association. The findings that deletion mutants of vinculin lacking both C-terminal and N-terminal sequences, expressed in COS cells, and antivinculin antibodies, microinjected into fibroblasts, lead to disruption of microfilaments and, in the latter case, change focal contact morphology, support a functional role of vinculin in focal contact assembly and F-actin anchoring (Bendori et al., 1989; Westmeyer et al., 1990).

In addition to vinculin, a higher molecular variant of vinculin, metavinculin, prominent in muscle, has been described. It has been demonstrated for human and chicken metavinculin and vinculin that both proteins are derived from a single gene by alternative splicing of mRNA (Kotelianski et al., 1992; Byrne et al., 1992). Porcine vinculin and metavinculin have been shown to differ by a 68 kDa-residue insert, which is located close to the C-terminal part of the molecule (Gimona et al., 1988; Strasser et al., 1993). No functional difference has been shown for vinculin and metavinculin; both proteins interact with talin, α-actinin and acidic phospholipids (see references in Byrne et al., 1992; Niggli and Gimona, 1993). Further studies with metavinculin-specific antibodies or with the isolated insert itself may yield information on specific metavinculin function.

From the work discussed above, it is clear that vinculin, talin and α-actinin are important elements for focal contact integrity and that talin interacts *in situ* with the cytosolic domain of integrin. According to DePasquale and Izzard (1991) talin accumulates in newly formed focal contacts of fibroblasts before vinculin is detectable. Moreover, studies by Brands et al. (1990) suggest that talin-containing adhesion plaques with attached F-actin can exist at least transiently with little vinculin present, in virally transformed fibroblasts. Talin, as a F-actin nucleating and membrane-binding protein, is especially important in the initial phase of focal contact formation. Vinculin and α-actinin might, in a second phase, stabilize nascent focal contacts by cooperative interactions. Recent findings strengthen this hypothesis because microinjection of talin antibodies into spreading fibroblasts, but not into well-spread cells, leads to disassembly of focal adhesions and to inhibition of spreading and migration (Nuckolls et al., 1992). Overexpression of vinculin or α-actinin in fibroblasts or tumor cells leads to increased adhesion, decreased locomotion and decreased tumorigenicity, confirming an important role of these proteins in stabilizing focal contacts (Fernandez et al., 1992a; Fernandez et al., 1992b; Glück et al., 1993; Samuels et al., 1993). These experiments suggest that focal contact structure is modulated by the cytosolic concentration of these proteins. The question remains open whether low levels of vinculin and α-actinin are necessary for locomotion. α-Actinin appears to be dispensable for locomotion in *Dictyostelium Discoideum* (Schleicher et al., 1988), possibly due to the presence of a set of redundant proteins. Both vinculin and α-actinin are present in the highly motile neutrophils, and vinculin is recruited into focal-contactlike structures in these cells (Figure 1; Yürüker and Niggli, 1992). The role of these structures in neutrophil locomotion is not clear. Certainly, vinculin has been shown to be essential for nematode muscle function (Barstead and Waterston, 1991) and α-actinin for *Drosophila* muscle integrity (Fyrberg et al., 1990). The finding that *Drosophila* embryonic development is not affected by mutations in the α-actinin

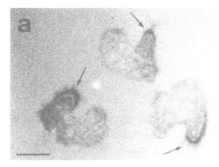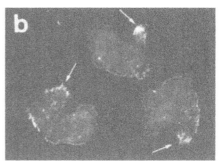

Figure 1. Localization of vinculin in adherent human neutrophils. Isolated human neutrophils were incubated in the presence of 10^{-9} M chemotactic peptide (N-formyl-L-norleucyl-L-leucyl-L-phenylanalyl-L-norleucyl-L-tyrosyl-L-Lysine) for 16 min on glass coverslips at 37°C, in the absence of added protein. Subsequently, the cells were fixed and stained with a monoclonal antivinculin antibody and a second fluorescent antibody (b). The same cells were also visualized with interference reflexion microscopy where areas of very close contact with the substrate appear black (a). For experimental details see Yürüker and Niggli, 1992. Arrows indicate enrichment of vinculin in adhesive areas. Bar, 10 μm (Yürüker and Niggli, unpublished data).

gene could be due either to a redundant role of α-actinin in nonmuscle cells or to encoding of nonmuscle α-actinin by a distantly related gene.

E. Cell–Cell Adherens Junctions

Comparable to areas of cell–substrate contact, areas of cell–cell junctions are physically linked to actin filaments or to intermediate filaments. Actin-based cell–cell junctions correspond to the zonula adherens of polar epithelia, to endothelial junctions or to spot adhesions of other cells; intermediate filament-based junctions are desmosomes of epithelial cells (see below). Actin-based cell–cell and cell substrate adherens junctions share some components; others are specific to one type of junction (Table 2). Both types of adherens junctions are regulated by reversible phosphorylation. Our knowledge of molecular assembly of cell–cell adherens junctions is much less advanced than that of focal contacts (for recent reviews on cell–cell adherens junctions see Geiger and Ginsberg, 1991; Magee and Buxton, 1991; Geiger and Ayalon, 1992). Instead of integrins, cell–cell adherens junctions contain transmembrane proteins of the cadherin family (120–140 kDa). They exhibit a large N-terminal extracellular domain, a single transmembrane region and a C-terminal cytoplasmic domain. Three major subfamilies are represented by N-cadherin (neural cadherin or A-CAM), E-cadherin (epithelial cadherin or uvomorulin) and P-cadherin (placental cadherin). The extracellular domains show calcium-dependent, homophilic interactions. The cytosolic domain is essen-

Table 2. Proteins Enriched in Actin-Linked Cell–Cell
Junctions

Transmembrane adhesion receptors	Peripheral components
Cadherins	Actin
	α-Catenin
	β-Catenin
	γ-Catenin
	Plakoglobin
	Vinculin
	α-Actinin
	Tenuin
	Radixin
	Zyxin

tial for linkage to the actin network and for formation of cell–cell junctions as shown
by experiments with mutated cytoplasmic domains. Cadherins are enriched in
Triton X-100-insoluble fractions of cell extracts, indicative of their interaction with
actin. The actual molecular mechanism of cadherin–actin linkage is not clear.
Proteins called α-, β- and γ-catenin (102, 88 and 80 kDa) have been shown to be
retained in the Triton X-100-insoluble residue and to coimmunoprecipitate with
cadherins. Cross-linking experiments demonstrated a tight, direct, interaction of
β-catenin with E-cadherin (Ozawa and Kemler, 1992). Catenins are thus interesting
candidates for involvement in actin–cadherin linkage (Geiger and Ayalon, 1992).
α-Catenin shows a significant sequence similarity to vinculin. Several domains of
α-catenin show 26% to 34% identity with the corresponding domains of vinculin,
which contain the talin- and paxillin-binding sites. α-Catenin may associate with
itself or with vinculin through a domain containing the self-association site of
vinculin (Nagafuchi et al., 1991). Transfection of a neural isoform of α-catenin into
an α-catenin-deficient and E-cadherin- and β-catenin-containing cell line induces
tight adhesion and epithelial arrangements of these cells, suggesting that α-catenin
is indeed a crucial component of cell–cell junctions (Hirano et al., 1992). β-Caten-
ins belong to a different protein family. They are closely related to the protein
plakoglobin that also interacts with E- and N-cadherin, showing a 68% sequence
identity to plankoglobin. β-Catenin is restricted to adherens junctions, whereas
plakoglobin occurs in intermediate filament-linked cell–cell contacts. Both proteins
are related to the product of the *Drosophila* segment polarity gene armadillo (Peifer
et al., 1992; Butz et al., 1992; Knudsen and Wheelock, 1992). γ-Cadherin has
recently been shown to be distinct from plakoglobin by two-dimensional electrophoresis (Piepenhagen and Nelson, 1993).

   Other cell–cell junction proteins such as α-actinin, vinculin, ezrin, zyxin and
radixin could not be detected in the cadherin-immunoprecipitates (Knudsen and

Wheelock, 1992). However, this may be due to the fact that immunoprecipitation was carried out with the detergent-soluble fraction of cadherin, whereas only the detergent-insoluble cadherin molecules may be linked to the cytoskeleton. The role of these proteins in cadherin–actin linkage is not clear. Cadherins can form complexes with ankyrin and fodrin, possibly an alternative linkage to actin (for references see Geiger and Ayalon, 1992). Another transmembrane adhesion receptor ICAM-1 interacts with the leukocyte integrins LFA$_1$ and Mac-1. Using affinity chromatography, a direct interaction between ICAM-1 with α-actinin has been demonstrated; this involves a region with several positively charged amino acids in the cytoplasmic domain. In contrast, vinculin did not bind to his cytoplasmic domain of ICAM-1 (Carpén et al., 1992).

Cadherin function is important for morphogenesis, and loss of cadherin expression or function in transformed epithelial cells appears to be a key step in the expression of a malignant phenotype (Behrens et al., 1993). Cell–cell junction integrity is regulated by tyrosine phosphorylation because inhibition of tyrosine phosphatases results in disassembly of cell–cell junctions (Geiger and Ayalon, 1992). Interestingly, transformation of chick embryo fibroblasts or MDCK cells with rous sarcoma virus leads to suppression of cadherin-mediated adhesion and concomitantly increases tyrosine phosphorylation of E-cadherin and β-catenin, or of N-cadherin, α- and β-catenin, respectively (Behrens et al., 1993; Hamaguchi et al., 1993). Neither expression nor interaction of N-cadherin with catenins was reduced in the transformed cells, suggesting that other functional aspects may be disturbed by phosphorylation.

F. Direct Interaction of Actin with Integral Membrane Proteins

Very few studies report direct binding of actin filaments to cytosolic domains of transmembrane proteins. Indirect linkages may offer more variety for modulation. Only the epidermal growth factor receptor (EGFR) and ponticulin, an integral membrane protein from the slime mold *Dictyostelium discoideum*, have to date unequivocally been demonstrated to bind directly to actin filaments. Purified EGFR copolymerizes with actin, and an EGFR domain with high homology to the actin-binding domain of profilin is both necessary and sufficient to support this interaction. Only the cytoskeleton-associated receptor fraction exhibits high affinity, suggesting that the cytoskeleton is involved in determining receptor affinity (den Hartigh et al., 1992).

Nucleation of actin polymerization at the plasma membrane could be the driving force for pseudopod extension. Highly purified ponticulin (17 kDa) from *Dictyostelium* has recently been demonstrated to directly accelerate actin filament growth *in vitro*, a process dependent on the presence of *Dictyostelium discoideum* membrane lipids. Ponticulin interacts laterally with actin filaments, leaving both ends free for assembly or disassembly. As a ponticulinlike protein has recently been detected in motile protrusions of human neutrophils, this protein is an interesting candidate for mediating pseudopod protrusion in mammalian cells (Chia et al., 1993).

Actin was shown to be associated in ascites tumor cell microvilli with a complex of transmembrane glycoproteins, which may mediate an indirect linkage to an analog of the EGF receptor (Carraway et al., 1993).

G. Actin-Organelle Linkage

Unconventional myosins (myosin I) represent a novel class of mechanoenzymes, which can interact simultaneously with phospholipid bilayers and support motility of actin filaments. Myosin I exhibits a globular head domain with mechanoenzymatic activity and a positively charged C-terminal tail that interacts *in vitro* directly with acidic phospholipids. The lipid-binding site has been located to the N-terminal part of the tail (residues 701–888) of *Acanthamoeba* myosin I (Doberstein and Pollard, 1992), which also contains an actin-binding site and a SH3 domain (Pollard et al., 1991; Coudrier et al., 1992; Cheney et al., 1993). The electrostatic interaction of myosin I with membranes may concentrate this protein at this location, but cannot direct the protein to specific areas such as filopodia and lamellipodia, or to cytoplasmic vesicles (Wagner et al., 1992). Myosin I is widely distributed in mammalian tissues, occurring in intestinal microvilli where it forms lateral links between the membrane and actin bundles (Bretscher, 1991), as well as in brain, kidney, and fibroblasts, etc. (Wagner et al., 1992). Genes for myosin I isoforms have been cloned and sequenced, all showing considerable homology with brush border myosin I (Ruppert et al., 1993; Sherr et al., 1993). Myosin I may play a crucial role in dynamic processes of intracellular and plasma membranes (Coudrier et al., 1992; Pollard et al., 1991).

Another family of proteins, the synapsins, has been implicated to specifically regulate the efficiency of neurotransmitter release by controlling the fraction of synaptic vesicles available for release. Synapsins interact with synaptic vesicles, at least in part, by direct bilayer insertion. Moreover, they bind to and bundle actin filaments. Distinct domains are thought to be involved in these activities (Greengard et al., 1993). No specific membrane protein receptors have been identified for myosin I or synapsin.

IV. INTERMEDIATE FILAMENT–MEMBRANE LINKAGE

Analogous to the actin-based cell–cell junctions and focal contacts, highly specialized sites of cell–cell and cell–matrix contact intracellularly linked to intermediate filaments (IF) also exist in the desmosomes and hemidesmosomes. Both sites, found mainly in epithelial cells, contain structurally similar dense cytosplasmic plaques that are attached to the IF network. However, on the molecular level, these sites are completely different and more related to their respective, functionally

Table 3. Proteins Involved in Intermediate Filament–Membrane Linkage

a) *Desmosomes*	*b)* *Hemidesmosomes*	*c)* *Other putative intermediate filament- membrane linker proteins*
Integral membrane proteins		Filensin
Desmoglein	Integrin $\alpha_6\beta_4$	Lamin B
Desmocollin I/II	BP 180 antigen	Spectrin
Pemphigus vulgaris antigen		Ankyrin
Peripheral plaque proteins		Filamin
Plakoglobin	BP 230 antigen	A60
Desmocalmin	200 K protein	
Keratocalmin		
Desmoplakin I/II		
Lamin B-like protein		
Plectin		
Band 6 protein		
pp 170		

comparable, actin-associated junctions. Desmosomes contain cadherinlike transmembrane glycoproteins, whereas hemidesmosomes, which mediate adhesion of epithelia to basement membranes, contain a specific integrin isoform (Table 3). In cells not expressing (hemi-)desmosomes, IF-membrane attachment may be mediated via various putative linker proteins, some of which are also part of the actin cortical network, such as ankyrin or filamin (Table 3; for a review see Jones and Green, 1991).

A. Desmosomes

Desmosomes are related to actin-based cell–cell junctions because they contain cadherin-like proteins and because they share at least one component, the protein plakoglobin (for reviews see Magee and Buxton, 1991; Legan et al., 1992). Three types of cadherin-like proteins have been identified in desmosomes: desmoglein (106 kDa), the desmoglein-related protein pemphigus vulgaris antigen (PVA; 130 kDa) and the desmocollins 1 and 2 (85 and 79 kDa) (Legan et al., 1992; Amagai et al., 1991; Kárpáti et al., 1993). These proteins show 30% to 50% identity with N-cadherin. They also mediate calcium-dependent cell adhesion. It is not clear if all three components are directly involved in cell adhesion, and how they interact with each other. The extracellular domains of the desmocollins are very similar to those of cadherins; however, desmoglein has a shortened membrane-proximal domain with an unusual structure not found in the cadherins. As expected from their interaction with two distinct cytoskeletal systems, the cytosolic domains of desmosomal adhesion proteins show some unique features compared with cadherins.

Whereas the cytosolic domain of desmocollin I has a cadherin-like C-terminus, desmoglein has a much longer 400-amino-acid cytoplasmic domain containing a region exhibiting homology to cadherin as well as a unique C-terminus. The desmosomal cadherins represent a complex subfamily with three different types of desmogleins and desmocollins, respectively. Desmocollins I and II are produced by alternative splicing of transcripts from the same gene (Legan et al., 1992). As recently shown, desmoglein subtypes are expressed in a cell-type specific and differentiation-dependent manner (Koch et al., 1992). Troyanovski et al. (1993) have studied the functional role of the cytoplasmic domains of these adhesion proteins using chimeric proteins consisting of the transmembrane domain of the gap junction protein connexin 32 fused to cytosolic tails of desmocollins or desmoglein. These constructs were transfected into carcinoma cells, and formation of gap junction-associated desmosomal plaques was studied. The results show that the last 38 C-terminal amino acids of the tail of desmocollin I are sufficient to form an IF-linked plaque containing plakoglobin and desmoplakin. Despite the marked sequence similarity of this tail to that of cadherins, it must thus contain information for specific linkage of IFs. In contrast, the desmoglein tail did not support plaque formation, and even induced disappearance of desmosomes and detachment of IFs from the plasma membrane. This is surprising, as the long tail of desmoglein would seem to be very suitable to traverse the plaque and link to the cytoskeleton but perhaps this tail domain competes with endogenous desmoglein. The cytosolic domain of desmocollin I thus appears to be a major factor in recruitment of other plaque components and IFs to the membrane.

Although it is not yet known how these filaments are linked to the plaque, the plaque contains several proteins which may possibly be involved (Table 3); plakoglobin co-immunoprecipitates with desmoglein and PVA; the proteins desmocalmin, keratocalmin, the 140 kD lamin-like protein and the band 6 protein (but not the desmoplakins) all interact *in vitro* with IFs (Legan et al., 1992; Magee and Buxton, 1991). Sequence information is only available for a few of these proteins and recent data shows that desmoplakin I (310 kD) and desmoplakin II (238 kD) are closely related proteins generated by alternative splicing. Desmoplakin I is a dumbbell-shaped molecule with a rod-like central domain (132 nm long) flanked by globular C- and N-terminal domains. The rod domain, showing periodicities in charged residues, may mediate self-association or association with related molecules. The C-terminus contains a series of 38 amino acid repeats arranged into three homologous domains, exhibiting a periodicity in acidic and basic residues similar to that of the 1B rod domain of intermediate filaments, and is thus a potential filament-binding site. Plectin, another desmosomal intermediate filament-associated protein, and 230 kD bullous pemphigoid antigen (BP 230 antigen), a component of hemidesmosomes, both contain comparable 38 residue motifs (Virata et al., 1992, and refs. therein). In contrast to these structural implications, no evidence for *in vitro* interaction of desmoplakin with IFs has been obtained (Legan et al., 1992) and this may be due to a lack of accessory proteins or to denaturation. Transfection of the gene coding for the C-terminus of desmoplakin into cultured cells causes the protein to co-localize with, or if overexpressed, to disrupt this network (Stappenbeck and Green, 1992), similar to results obtained with the C-terminus of plectin (see below).

Plakoglobin (82 kD) is the only component known so far which is shared by both cell–cell adherens junctions and desmosomes. Plakoglobin is a globular, dimeric protein, which occurs both in a soluble cytosolic form and also in a closely membrane-associated form in desmosomes and in the zonula adherens (Franke et al., 1989). As discussed in a previous section, plakoglobin is closely related to β-catenin, a component of actin-based cell–cell junctions, and to the segment polarity gene armadillo of *Drosophila* (Peifer et al., 1992; Butz et al., 1992). Plakoglobin has been shown to interact directly with E- and N-cadherin as well as with desmoglein. Interaction with the latter protein occurs possibly via the short region of cadherin-homology present in the cytosolic tail of desmosomal cadherins (Knudsen and Wheelock, 1992; Legan et al., 1992).

Little is known on the actual assembly of desmosomes during cell-cell contact formation. Besides keratins, microtubules have been thought to be involved. Indeed, the microtubule-binding phosphoprotein pp170 has been shown to accumulate at desmosomal plaques in MDCK cells, accompanying cell-cell junction formation (Wacker et al., 1992). However, desmosome formation also occurs in MDCK cells treated with nocodazole or colchicine, where the microtubule network is disrupted (Pasdar et al., 1992). Microtubule-desmosome linkage may serve to stabilize nascent microtubules.

B. Hemidesmosomes

Hemidesmosomes are IF-associated stable structures involved in anchoring epithelial and mesenchymal cells to the basement structure. In agreement with their functional analogy to focal contacts, they contain a specific type of integrin, $\alpha_6\beta_4$ (Stepp et al., 1990; Sonnenberg et al., 1991). This integrin isoform differs from all the other types due to its strikingly large (>1000 amino acids) cytosolic domain. Such long tails could traverse the plaques and interact directly or indirectly with IFs at the border of the plaques. Integrin $\alpha_6\beta_4$ appears to be a crucial molecule for hemidesmosome assembly, because antibodies directed against this protein prevent formation of hemidesmosomes in a wound model (Kurpakus et al., 1991). The extracellular ligand for $\alpha_6\beta_4$ has not been defined; however components of the anchoring filaments that span the basal lamina and connect the plasma membrane to the lamina lucida, or laminin, may be candidates (Legan et al., 1992; Kurpakus et al., 1991). Intracellularly, no information on the molecular linkage of IFs to integrin $\alpha_6\beta_4$ is available. This integrin is enriched in detergent- and salt-insoluble cellular fractions, suggesting a close interaction with IFs (Gomez et al., 1992). HD$_4$, a 180 kDa bullous pemphigoid antigen with two collagen domains, is another hemidesmosomal transmembrane glycoprotein constituent (Nishizawa et al.,

1993). Several peripheral plaque-associated components have been identified (Table 3). As discussed above, the intracellular 230 kDa bullous pemphigoid antigen (Ishiko et al., 1993) shows some sequence homology with the desmosomal component desmoplakin (Sawamura et al., 1991).

C. Intermediate Filament–Membrane Linkage Outside of Desmosomes or Hemidesmosomes

As reviewed by Jones and Green (1991) and by Georgatos (1993), there is ample structural evidence for a close association of IFs with plasma membranes (e.g., in lens or avian erythrocytes) and nuclear membranes. These attachment sites may serve as nucleation sites for filament assembly (Georgatos, 1993). Several candidate linker proteins have been discussed (Table 3). There is evidence of a connection of IFs to the spectrin–actin cortical network. The protein plectin may determine the structure of intermediate filament networks and mediate interactions with microtubules and spectrin. Other proteins that may serve as tissue-specific or organelle-specific membrane linkers are the lens protein filensin, the nuclear lamina protein lamin B and the axonal membrane cytoskeletal protein A-60. Finally, vimentin has been shown to interact *in vitro* directly with acidic phospholipid bilayers (for references see Jones and Green, 1991; Georgatos, 1993).

The neurofilament-binding site of brain spectrin has been located to two fragments in the N-terminal domain of β-spectrin (amino acids 43–288 and 104–288) that contain two highly conserved actin-binding sites (Frappier et al., 1992). The 62-kDa-domain of ankyrin binds both spectrin and vimentin (Bennett, 1992; Peters and Lux, 1993). Gyronemin, a 240-kDa protein associated with IFs, was shown recently to be identical with the actin-cross-linking protein filamin. Filamin, which links F-actin to transmembrane glycoproteins (see above), may thus be involved in IF-membrane linkage and serve to cross-link actin with intermediate filaments (Brown and Binder, 1992). As reviewed by Hayes and Baines (1992), another possible interaction partner is the neuronal protein A60, which interacts tightly with the axonal plasma membrane. A60 binds both ankyrin and neurofilaments, but it is not clear whether it can cross-link these two components. A60 may be involved in creating a link between neuronal plasma membranes and neurofilaments, which is crucial for the role attributed to these filaments in determining axonal caliber.

Another interesting interaction partner of IFs is the protein plectin. Recent cloning and sequencing data show that this protein (466 kDa) has a three domain structure with a long (ca 190 nm) central rod domain flanked by globular N- and C-terminal domains. The plectin sequence shows considerable similarities to that of desmoplakin (as discussed above), especially regarding a repeated motif in the C-terminal domain and almost identical periodicities of charged residues in the rod domains (Wiche et al., 1991). Despite their structural similarities, these two proteins appear to be functionally different. Desmoplakin is restricted to desmosomes and does not interact with IFs *in vitro*, whereas plectin interacts *in vitro* with various proteins such as different types of IF proteins, spectrin, MAP 1 and 2, and appears within cells at different locations, such as desmosomes and focal contacts (Seifert et al., 1992). Using monoclonal antibodies inhibitory to the *in vitro* interaction of plectin with vimentin and lamin B, the binding site for these proteins has been mapped to the center of the rod domain of plectin (Wiche et al., 1991). However, other binding sites may be more important under physiological conditions; later studies showed that colocalization with and disruption of IFs in cultured cells transfected with plectin mutant cDNA depends on the presence of at least two thirds of the globular C-terminus (Wiche et al., 1993). Plectin also forms protein networks via self-assembly (Foisner and Wiche, 1991). Plectin, a substrate for various protein kinases, may be functionally regulated by phosphorylation. It remains to be seen whether plectin is a multifunctional protein *in situ*. The presence of plectin in at least part of the focal contacts of cultured cells reported by Seifert et al. (1992) is intriguing and may explain the observation that IFs appear to be associated with focal contacts in certain cell types.

Another candidate for IF-membrane linkage is the protein filensin, a hydrophilic peripheral membrane protein (100–110 kDa) found in lens fiber cells, structurally similar to IF proteins (Gounari et al., 1993). Like plectin, filensin has a tendency to self-assemble and to interact with vimentin *in vitro*. Recent data suggest that filensin binds to a putative receptor protein of lens plasma membranes, possibly through its C-terminal domain (Brunkener and Georgatos, 1992).

The nuclear envelope protein lamin B may serve as a nuclear receptor for desmin or vimentin; these proteins have been shown to associate cooperatively with lamin B. Recent data using anti-idiotypic antibodies raised against a synthetic peptide that represents the proximal half of the C-terminal tail domain of peripherin (a neuronal type III intermediate filament subunit) strongly suggest that lamin B is a physiological receptor for the tail domain of peripherin. The anti-idiotypic antibody epitopes are located near the conserved C-terminal tryptophan residues of lamin B, very likely the peripherin-binding site (Djabali et al., 1991). Using polyclonal antibodies to lamin B, Cartaud et al. (1990) have identified a lamin B-related protein of 140 kDa in the desmosomal plaque that binds vimentin *in vitro*. A family of lamin B-like proteins may therefore provide different intermediate filament–membrane attachment points.

Finally, direct contacts of IFs with the phospholipid bilayer are possible, because vimentin has been shown to interact *in vitro* directly with negatively charged phospholipids. This interaction of IFs with phospholipid bilayers is specific, as it is restricted to the positively charged N terminus of vimentin (residues 1–96), which has an amphiphilic character. Moreover, vimentin has a high affinity for phosphatidylinositol lipids. The interaction appears to involve an electrostatic interaction followed by hydrophobic insertion, comparable to that of the protein vinculin (see above; Horkovics-Kovats and Traub, 1990). The physiological relevance of this interaction is not clear.

V. MICROTUBULE–MEMBRANE LINKAGE

In mammalian cells microtubules form mainly transcellular filamentous arrays with attachment points at plasma membranes and cell organelles. Microtubule–plasma membrane interaction can occur via components of the spectrin–actin network or via other specialized proteins that link microtubules to transmembrane receptors. Direct bilayer contacts may also be involved. Microtubule–organelle linkage can be mediated by motor proteins, such as dynein or kinesin, or by nonmotor proteins such as CLIP-170. However, in no case is it clear how these microtubule-associated proteins interact with organelle membranes. Extensive microtubule-based membrane skeletons have been shown to occur only in parasitic protozoa. Marginal bands of ringlike peripheral microtubules occur in nucleated nonmammalian erythrocytes and in platelets.

A. Microtubular Membrane Structures

As reviewed by Seebeck et al. (1990), a characteristic feature of protozoic parasites, such as *Trypanosoma brucei* or *Leishmania*, is a highly ordered layer of microtubules localized beneath the plasma membrane, which determines shape and deformability of cells (Figure 2). These protozoa lack transcellular cytoskeletal structures, and the membrane-associated microtubules appear to be the main cytoskeletal system. *Trypanosomal* microtubules, in contrast to the mammalian filaments, appear to be very stable because they are resistant to cold and microtubule-depolymerizing drugs. Also, *trypanosomal* microtubules remain intact during the cell cycle. This unusual stability may be due to a variety of associated proteins that cross-link these filaments and mediate a tight interaction with the membrane. Indeed, a family of trypanosomal microtubule-associated proteins has recently been characterized in detail. The protein MARP-1 (>300 kDa; microtubule-associated repetitive protein), for example, consists mainly of 38-amino-acid acidic repeat units. This sequence has recently been shown to be a novel microtubule-binding motif, interacting with a domain of tubulin distinct from that which binds to basic repeats of mammalian microtubule-associated proteins. MARP-1 may have a dual function in stabilizing microtubules by making multiple contacts along the axis of the filaments and by mediating plasma membrane contacts. Direct evidence for the latter role is lacking, but MARPs are located exclusively at the membrane-oriented face of the microtubules (Hemphill et al., 1992).

Marginal bands of microtubules are a major structural feature of all non-mammalian vertebrate erythrocytes and platelets. However, they do not appear to contact the plasma membrane directly, but seem to be embedded in actin-based membrane skeletons (Cohen et al., 1982; Nakata and Hirokawa, 1987). Electron-microscopical studies suggest that connections exist between actin filaments and the marginal band of microtubules (Nakata and Hirokawa, 1987). Birgbauer and Solomon (1989)

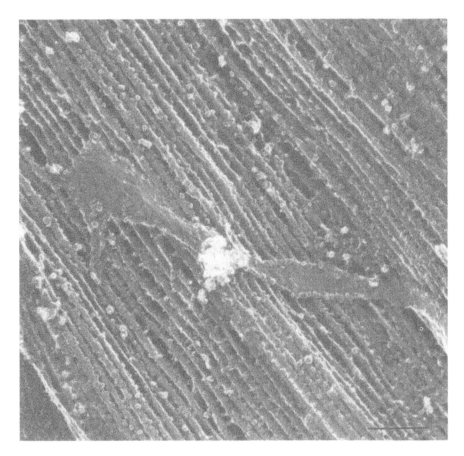

Figure 2. The microtubule-based membrane cytoskeleton of Trypanosomes. Closely packed array of microtubules of the Trypanosome membrane skeleton, with a remaining fragment of the cell membrane. Bar, 200 nm (courtesy of Prof. Thomas Seebeck, University of Bern, Switzerland).

have identified a candidate, an ezrinlike protein, to mediate these linkages. The function of marginal bands is not clear, as erythrocyte morphology is not affected by selective disassembly of microtubules in the cold (Cohen et al., 1982).

B. Microtubule–Plasma Membrane Interaction

One possible mechanism of tubulin–membrane linkage involves components of the spectrin–actin cortical network. Tubulin interacts with repeat regions of ankyrin, including regions that do not interact with band 3 (Davis et al., 1991a). Moreover, protein 4.1 copolymerizes with tubulin *in vitro*. However, this interaction may not mediate membrane linkage, because protein 4.1 appears to be located along microtubules and in the mitotic spindle rather than at the plasma membrane (Correas and Avila, 1988).

Another putative linker protein is gephyrin, a peripheral membrane protein (93 kDa) that associates with postsynaptic glycine receptors and copolymerizes with tubulin (Kirsch et al., 1991). Gephyrin, which exhibits no extensive homology with other cytoskeletal proteins, appears to be a widely distributed protein, not restricted to the central nervous system, and therefore may represent a general microtubule–membrane receptor linker protein (Prior et al., 1992). Gephyrin has been postulated to modulate receptor function by clustering; coexpression of gephyrin with α_2 glycine receptors in kidney cells produced an increase in agonist affinity and a decrease in antagonist affinity (Takagi et al., 1992).

Microtubules may directly modulate signal transduction processes, as high-affinity binding between tubulin dimers and G proteins (α-subunits of G_s and G_{il}) have been demonstrated. This interaction is very likely not involved in microtubule formation at the plasma membrane because these G proteins inhibit tubulin polymerization. Rather, tubulin dimers appear to modulate G-protein function (Wang and Rasenick, 1991). Moreover, tubulin binds to and regulates CD_2, a transmembrane protein involved in T-cell signaling (Offringa and Bierer, 1993).

Similar to intermediate filaments and a variety of actin-associated proteins, tubulin has been shown to interact *in vitro* directly with phospholipid bilayers, a process enhanced by phosphorylation of tubulin by the calmodulin-dependent kinase (Hargreaves et al., 1986). Phosphorylation induces a conformational change revealing hydrophobic domains (Hargreaves et al., 1986). Direct bilayer interaction could explain the presence of tightly bound tubulin in membrane fractions of nerve cells, platelets, cilia, etc., but the exact mechanism of interaction and the physiological role of membrane tubulin is to date unknown (Stephens, 1986).

C. Microtubule-Organelle Interactions

As organelle transport by microtubules is the subject of an article in Volume II of this treatise, the superfamilies of microtubule motor proteins identified in the last years, including kinesin and dynein, will not be discussed here. The mechanism of reversible membrane interaction of kinesin and dynein, which are involved in protein sorting, intracellular membrane trafficking and organization, is as yet unknown (reviewed by Vale, 1992).

A novel, putative endocytic vesicle-microtubule linker protein, CLIP-170 (157 kDa), is a dimeric rodlike protein possessing a basic N-terminal domain and an acidic C-terminal domain with a potential metal-binding domain. By construction of defined mutants, the microtubule-binding site of CLIP-170 has been located to the basic N terminus. CLIP-170 shows no striking sequence homology with other proteins, except for a similarity of N-terminal repeats to motifs in other proteins required for microtubule function. Also, its overall organization is very similar to

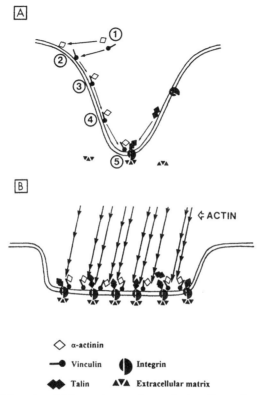

Figure 3. Possible role of bilayer interactions of vinculin and α-actinin in focal contact formation (a speculative model). Cytosolic vinculin and α-actinin (1) interact reversibly with plasma membrane phospholipids (2). Interaction of these two proteins at the bilayer leads to conformational changes in such a way that interaction of vinculin at the bilayer is enhanced and that of α-actinin is decreased (3). The vinculin–α-actinin complexes, as well as talin, are capable of undergoing lateral diffusion along the plasma membrane (4). They may then be targeted to preformed nascent contact sites (containing integrins, talin, paxillin, etc.) to complete formation of a stable actin–membrane linkage cascade (5). The accumulation of vinculin, α-actinin and talin at potential focal adhesion sites may trigger the association of further integrin molecules (6) (model by M. Gimona and V. Niggli). For experimental support see Niggli & Gimona (1993) and Goldmann et al. (1992).

that of the kinesin superfamily, although it does not exhibit a kinesinlike motor domain. How CLIP-170 interacts with the endocytic vesicle membrane is unknown. CLIP-170 may establish initial contacts between microtubules and endocytic vesicles and transiently stabilize peripheral microtubule plus-ends. CLIP-170 also occurs in desmosomes (see above; Pierre et al., 1992).

The structural integrity of the Golgi apparatus depends on functional microtubules. Bloom and Brashear (1989) have identified and purified a putative Golgi-

microtubule linker protein from rat liver. This 58-kDa protein, structurally unrelated to other cytoskeletal proteins, behaves as a peripheral membrane protein that
interacts at low affinity with microtubules and colocalizes with the Golgi. It may
anchor the Golgi in areas where microtubules are densely packed. Mithieux and
Rousset (1989) have identified a 50 kDa integral membrane protein in thyroid
lysosomes, which interacts with microtubules in a manner inhibited by ATP. This
protein may mediate positioning of lysosomes in the cell, a process that depends
on microtubule integrity. The neuronal protein synapsin I (see above) may link
microtubules to synaptic vesicles, because it bundles microtubules *in vitro* (Bennett
and Baines, 1992). Thus, a number of interesting candidate proteins have been
described that could specifically link microtubules to defined organelles.

VI. CONCLUSIONS

Cytoskeleton–membrane linkage can be mediated either by interactions of peripheral proteins with other peripheral or integral proteins or by direct interactions of
peripheral proteins with lipid bilayers via specific protein domains or via covalently
attached fatty acids. Due to the increasing amount of sequence information available, we can now identify key motifs in these proteins that point to specific
interactions and functions. Typical sequences with specific functions have been
identified, such as conserved actin-binding sites in the spectrin superfamily; membrane-binding sites in the talin-protein 4.1 ezrin-moesin-radixin superfamily; phospholipid binding sites in synapsin and myosin I; self-association sites in vinculin
and related proteins; domains pointing to a role in signal transduction such as SH3
domains in spectrin, myosin or SH2 domains in tensin. Despite this increasing
wealth of molecular information, many questions remain to be answered. We do
not yet know why so many proteins with closely related, overlapping functions
exist. It is not clear whether different linkages occur in parallel or at specific sites
at specific time points during processes such as cell adhesion. The relation between
the different types of linker systems, such as focal contact proteins and the
spectrin–ankyrin system, has not been explored. The mechanism of targeting of
cytoskeleton-membrane linkers to specific membrane sites remains unresolved.
Some of the signals and regulatory proteins involved may be specific G proteins,
tyrosine kinases or components of the phosphatidyl-inositol-bisphosphate cycle;
however, the mechanisms whereby these signals are translated into interaction of
membrane receptors with cytoskeletal linker proteins and subsequent receptor
clustering is unknown. We also do not understand the functional relevance of the
interaction of many proteins involved in cytoskeleton–membrane linkage with
membrane phospholipids. In a few cases, lipid binding sites have been identified
suggesting that specific, relevant interactions may occur. Such interactions may
serve to concentrate these proteins at the membrane, converting a three-dimensional
search into a two-dimensional process and facilitating and modifying interactions

with other proteins (Figure 3). Transfection of proteins mutated or truncated in specifically defined protein- or lipid-binding sites into cultured cells, followed by analysis of the *in situ* behavior of these modified proteins, is already providing a promising approach to answer some of the above questions.

NOTE ADDED IN PROOF

After completion of this review article, we have been able to locate the actin binding site of talin in the 200-kDa N-terminal domain, and the lipid binding site in the 47-kDa C-terminal domain (Niggli et al., 1994, Eur. J. Biochem. 224, 951–957). Concerning vinculin, intriguing new evidence shows that intramolecular interactions in vinculin control its binding to α-actinin and actin (Kroemker et al., 1994, FEBS Lett. 355, 259–262; Johnson and Craig, 1995, Nature 373, 261–264).

ACKNOWLEDGMENTS

I would like to thank Dr. E. Sigel for helpful comments on the manuscript, Prof. G. Isenberg for stimulating discussions and Prof. H.U. Keller for continuous support. I also gratefully acknowledge the support by the Swiss National Science Foundation.

REFERENCES

Aderem, A. (1992). The MARCKS brothers: A family of protein kinase C substrates. Cell 71, 713–716.
Ahn, A.H. & Kunkel, L.M. (1993). The structural and functional diversity of dystrophin. Nature Genet. 3, 283–291.
Algrain, M., Turunen, O., Vaheri, A., Louvard, D., & Arpin, M. (1993). Ezrin contains cytoskeleton and membrane binding domains accounting for its proposed role as a membrane-cytoskeletal linker. J. Cell Biol. 120, 129–139.
Amagai, M., Klaus-Kovtun, V., & Stanley, J.R. (1991). Autoantibodies against a novel epithelial cadherin in Pemphigus Vulgaris, a disease of cell adhesion. Cell 67, 869–877.
Andrews, R.K. & Fox, J.E.B. (1992). Identification of a region in the cytoplasmic domain of the platelet membrane glycoprotein Ib-IX complex that binds to purified actin binding protein. J. Biol. Chem. 267, 18605–18611.
Bar-Sagi, D., Rotin, D., Batzer, A., Mandiyan, V., & Schlessinger, J. (1993). SH3 domains direct cellular localization of signalling molecules. Cell 74, 83–91.
Barstead, R.J. & Waterston, R.H. (1991). Vinculin is essential for muscle function in the nematode. J. Cell Biol. 114, 715–724.
Becker, P.S., Schwartz, M.A., Morrow, J.S., & Lux, S.E. (1990). Label-transfer crosslinking demonstrates that protein 4.1 binds to the N-terminal region of beta spectrin and to actin in binary interactions. Eur. J. Biochem. 193, 827–836.
Beckerle, M.C. & Yeh, R. (1990). Talin: Role at sites of cell-substratum adhesion. Cell Motil. Cytoskel. 16, 7–13.
Behrens, J., Vakaet, L., Friis, R., Winterhager, E., Van Roy, F., Mareel, M.M., & Birchmeier, W. (1993). Loss of epithelial differentiation and gain of invasiveness correlates with tyrosine phosphoryla-

tion of the E-cadherin/β-catenin complex in cells transformed with a temperature-sensitive *v-src* gene. J. Cell Biol. 120, 757–766.

Bendori, R., Salomon, D., & Geiger, B. (1989). Identification of two distinct functional domains on vinculin involved in its association with focal contacts. J. Cell Biol. 108, 2383–2393.

Bennett, A.F. & Baines, A.J. (1992). Bundling of microtubules by synapsin 1. Eur. J. Biochem. 206, 783–792.

Bennett, V. (1990). Spectrin-based membrane skeleton: A multipotential adaptor between plasma membrane and cytoplasm. Physiol. Rev. 70, 1029–1065.

Bennett, V. (1992). Ankyrins: Adaptors between diverse plasma membrane proteins and the cytoplasm. J. Biol. Chem. 267, 8703–8706.

Birgbauer, E. & Solomon, F. (1989). A marginal band-associated protein has properties of both microtubule- and microfilament-associated proteins. J. Cell Biol. 109, 1609–1620.

Blackshear, P.J. (1993). The MARCKS family of cellular protein kinase C substrates. J. Biol. Chem. 268, 1501–1504.

Blanchard, A., Vasken, O., & Critchley, D. (1989). The structure and function of α-actinin. J. Muscle Res. Cell Motil. 10, 280–289.

Bloch, R.J. & Morrow, J.S. (1989). An unusual β-spectrin associated with clustered acetylcholine receptors. J. Cell Biol. 108, 481–493.

Bloom, G.S. & Brashear, A. (1989). A novel 58-kDa protein associates with the Golgi apparatus and microtubules. J. Biol. Chem. 264, 16083–16092.

Bockholt, S.M. & Burridge, K. (1993). Cell spreading on extracellular matrix proteins induces tyrosine phosphorylation of tensin. J. Biol. Chem. 268, 14565–14567.

Bockholt, S.M., Otey, C.A., Glenney Jr., J.R., & Burridge, K. (1992). Localization of a 215-kDa tyrosine-phosphorylated protein that cross-reacts with tensin antibodies. Exptl. Cell Res. 203, 39–46.

Bourguignon, L.Y.W., Jin, H., Iida, N., Brandt, N.R., & Zhang, S.H. (1993). The involvement of ankyrin in the regulation of inositol 1,4,5-trisphosphate receptor-mediated internal Ca^{2+} release from Ca^{2+} storage vesicles in mouse T-lymphoma cells. J. Biol. Chem. 268, 7290–7297.

Brands, R., de Boer, A., Feltkamp, C.A., & Roos, E. (1990). Disintegration of adhesion plaques in chicken embryo fibroblasts upon rous sarcoma virus-induced transformation: Different dissociation rates for talin and vinculin. Exptl. Cell Res. 186, 138–148.

Bretscher, A. (1991). Microfilament structure and function in the cortical cytoskeleton. Annu. Rev. Cell. Biol. 7, 337–374.

Brown, D.A. & Rose, J.K. (1992). Sorting of GPI-anchored proteins to glycolipid-enriched membrane subdomains during transport to the apical cell surface. Cell 68, 533–544.

Brown, K.D. & Binder, L.I. (1992). Identification of the intermediate filament-associated protein gyronemin as filamin. J. Cell Sci. 102, 19–30.

Brown, K.D., Zinkowski, R.P., Hays, S.E., & Binder, L.I. (1993). Actin-binding protein is a component of bovine erythrocytes. Cell Motil. Cytoskel. 24, 100–108.

Brunkener, M. & Georgatos, S.D. (1992). Membrane-binding properties of filensin, a cytoskeletal protein of the lens fiber cells. J. Cell Sci. 103, 709–718.

Burridge, K., Turner, C.E., & Romer, L.H. (1992). Tyrosine phosphorylation of paxillin and pp125[FAK] accompanies cell adhesion to extracellular matrix: A role in cytoskeletal assembly. J. Cell Biol. 119, 893–903.

Butz, S., Stappert, J., Weissig, H., & Kemler, R. (1992). Plakoglobin and β-catenin: Distinct but closely related. Science 257, 1142–1144.

Byrne, B.J., Kaczorowski, Y.H., Coutu, M.D., & Craig, S.W. (1992). Chicken vinculin and meta-vinculin are derived from a single gene by alternative splicing of a 207-base pair exon unique to meta-vinculin. J. Biol. Chem. 267, 12845–12850.

Cantiello, H.F., Prat, A.G., Bonventre, J.V., Cunningham, C.C., Hartwig, J.H., & Ausiello, D.A. (1993). Actin-binding protein contributes to cell volume regulatory ion channel activation in melanoma cells. J. Biol. Chem. 268, 4596–4599.

Carpén, O., Pallai, P., Staunton, D.E., & Springer, T.A. (1992). Association of intercellular adhesion molecule-1 (ICAM-1) with actin-containing cytoskeleton and α-actinin. J. Cell Biol. 118, 1223–1234.

Carraway, C.A.C. (1992). Association of cytoskeletal proteins with membranes. In: The Cytoskeleton: A Practical Approach (Carraway, K.L. & Carraway, C.A.C., Eds.), pp. 123–150. IRL Press, Oxford.

Carraway, K.L. & Carraway, C.A.C. (1989). Membrane-cytoskeleton interactions in animal cells. Biochim. et Biophys. Acta 988, 147–171.

Carraway, C.A.C., Carvajal, M.E., Li, Y., & Carraway, K.L. (1993). Association of p185^{neu} with microfilaments via a large glycoprotein complex in mammary carcinoma microvilli. J. Biol. Chem. 268, 5582–5587.

Cartaud, A., Ludosky, M.A., Courvalin, J.C., & Cartaud, J. (1990). A protein antigenically related to nuclear lamin B mediates the association of intermediate filaments with desmosomes. J. Cell. Biol. 111, 581–588.

Cheney, R.E., Riley, M.A., & Mooseker, M.S. (1993). Phylogenetic analysis of the myosin superfamily. Cell Motil. Cytoskel. 24, 215–223.

Chia, C.P., Shariff, A., Savage, S.A., & Luna, E.J. (1993). The integral membrane protein ponticulin acts as a monomer in nucleating actin assembly. J. Cell Biol. 120, 909–922.

Cohen, C.M., Dotimas, E., & Korsgren, C. (1993). Human erythrocyte membrane protein band 4.2 (pallidin). Semin. Hematol. 30, 119–137.

Cohen, W.D., Bartelt, D., Jaeger, R., Langford, G., & Nemhauser, I. (1982). The cytoskeletal system of nucleated erythrocytes. I. Composition and function of major elements. J. Cell. Biol. 93, 828–838.

Coleman, T.R., Fishkind, D.J., Mooseker, M.S., & Morrow, J.S. (1989). Functional diversity among spectrin isoforms. Cell Motil. Cytoskel. 12, 225–247.

Conboy, J.G., Chan, J.Y., Chasis, J.A., Kan, Y.W., & Mohandas, N. (1991). Tissue- and development-specific alternative RNA splicing regulates expression of multiple isoforms of erythroid membrane protein 4.1. Proc. Natl. Acad. Sci. USA 266, 8273–8280.

Conboy, J.G. (1993). Structure, function and molecular genetics of erythroid membrane skeletal protein 4.1 in normal and abnormal red blood cells. Semin. Hematol. 30, 58–73.

Correas, I. & Avila, J. (1988). Erythrocyte protein 4.1 associates with tubulin. Biochem. J. 255, 217–221.

Coudrier, E., Durrbach, A., & Louvard, D. (1992). Do unconventional myosins exert functions in dynamics of membrane compartments? FEBS Lett. 307, 87–92.

Crawford, A.W., Michelsen, J.W., & Beckerle, M.C. (1992). An interaction between zyxin and α-actinin. J. Cell Biol. 116, 1381–1393.

Cunningham, C.C., Gorlin, J.B., Kwiatkowski, D.J., Hartwig, J.H., Janmey, P.A., Byers, R., & Stossel, T.P. (1992). Actin-binding protein requirement for cortical stability and efficient locomotion. Science 255, 325–327.

Davis, J.Q. & Bennett, V. (1990). The anion exchanger and Na$^+$K$^+$-ATPase interact with distinct sites on ankyrin in in vitro assays. J. Biol. Chem. 265, 17252–17256.

Davis, J.Q., McLaughlin, T., & Bennett, V. (1993). Ankyrin-binding proteins related to nervous system cell adhesion molecules: Candidates to provide transmembrane and intercellular connections in adult brain. J. Cell Biol. 121, 121–133.

Davis, L.H., Davis, J.Q., & Bennett, V. (1992). Ankyrin regulation: An alternatively spliced segment of the regulatory domain functions as an intramolecular modulator. J. Biol. Chem. 267, 18966–18972.

Davis, L.H., Otto, E., & Bennett, V. (1991a). Specific 33-residue repeat(s) of erythrocyte ankyrin associate with the anion exchanger. J. Biol. Chem. 266, 11163–11169.

Davis, S., Lu, M.L., Lo, S.H., Lin, S., Butler, J.A., Druker, B.J., Roberts, T.M., An, Q., & Chen, L.B. (1991b). Presence of an SH2 domain in the actin-binding protein tensin. Science 252, 712–715.

Den Hartigh, J.C., van Bergen en Henegouwen, P.M.P., Verkleij, A.J., & Boonstra, J. (1992). The EGF receptor is an actin-binding protein. J. Cell Biol. 119, 349–355.

DePasquale, J.A. & Izzard, C.S. (1991). Accumulation of talin at the edge of the lamellipodium and separate incorporation into adhesion plaques at focal contacts in fibroblasts. J. Cell Biol. 113, 1351–1359.

Derick, L.H., Liu, S.C., Chisthi, A.H., & Palek, J. (1992). Protein immunolocalization in the spread erythrocyte membrane skeleton. Eur. J. Cell Biol. 57, 317–320.

Dhermy, D. (1991). The spectrin superfamily. Biol. Cell 71, 249–254.

Discher, D., Parra, M., Conboy, J.G., & Mohandas, N. (1993). Mechanochemistry of the alternatively spliced spectrin-actin binding domain in membrane skeletal protein 4.1. J. Biol. Chem. 268, 7186–7195.

Djabali, K., Portier, M.M., Gros, F., Blobel, G., & Georgatos, S.D. (1991). Network antibodies identify nuclear lamin B as a physiological attachment site for peripherin intermediate filaments. Cell 64, 109–121.

Doberstein, S. & Pollard, T.D. (1992). Localization and specificity of the phospholipid and actin binding sites on the tail of acanthamoeba myosin IC. J. Cell Biol. 117, 1241–1249.

Domann, E., Wehland, J., Rohde, M., Pistor, S., Hartl, M., Goebel, W., Leimeister-Wächter, M., Wuenscher, M., & Chakraborty, T. (1992). A novel bacterial virulence gene in Listeria monocytogenes required for host cell microfilament interaction with homology to the proline-rich region of vinculin. EMBO J. 11, 1981–1990.

Ervasti, J.M. & Campbell, K.P. (1991). Membrane organization of the dystrophin-glycoprotein complex. Cell 66, 1121–1131.

Fernandez, J.L.R., Geiger, B., Salomon, D., & Ben-Ze'ev, A. (1992a). Overexpression of vinculin suppresses cell motility in BALB/c 3T3 cells. Cell Motil. Cytoskel. 22, 127–134.

Fernandez, J.L.R., Geiger, B., Salomon, D., Sabanay, I., Zöller, M., & Ben-Ze'ev, A. (1992b). Suppression of tumorigenicity in transformed cells after transfection with vinculin cDNA. J. Cell Biol. 119, 427–438.

Foisner, R. & Wiche, G. (1991). Intermediate filament-associated proteins. Curr. Opin. Cell Biol. 3, 75–81.

Franck, Z., Gary, R., & Bretscher, A. (1993). Moesin, like ezrin, colocalizes with actin in the cortical cytoskeleton in cultured cells, but its expression is more variable. J. Cell Sci. 105, 219–231.

Franke, W.W., Goldschmidt, M.D., Zimbelmann, R., Mueller, H.M., Schiller, D.L., & Cowin, P. (1989). Molecular cloning and amino acid sequence of human plakoglobin, the common junctional plaque protein. Proc. Natl. Acad. Sci. USA 86, 4027–4031.

Frappier, T., Derancourt, J., & Pradel, L.A. (1992). Actin and neurofilament binding domain of brain spectrin β subunit. Eur. J. Biochem. 205, 85–91.

Friedrichs, B., Koob, R., Kraemer, D., & Drenckhahn, D. (1989). Demonstration of immunoreactive forms of erythrocyte protein 4.2 in non-erythroid cells and tissues. Eur. J. Cell Biol. 48, 121–127.

Funayama, N., Nagafuchi, A., Sato, N., Tsukita, Sa., & Tsukita, Sh. (1991). Radixin is a novel member of the band 4.1 family. J. Cell Biol. 115, 1039–1048.

Fyrberg, E., Kelly, M., Ball, E., Fyrberg, C., & Reedy, M. (1990). Molecular genetics of Drosophila alpha-actinin: mutant alleles disrupt Z disk integrity and muscle insertions. J. Cell Biol. 110. 1999–2011.

Gallagher, P.G. & Forget, B.G. (1993). Spectrin genes in health and disease. Semin. Hematol. 30, 4–21.

Geiger, B. & Ayalon, O. (1992). Cadherins. Annu. Rev. Cell Biol. 8, 307–332.

Geiger, B. & Ginsberg, D. (1991). The cytoplasmic domain of adherens-type junctions. Cell Motil. Cytoskel. 20, 1–6.

Geiger, B., Salomon, D., Takeichi, M., & Hynes, R. (1992). A chimeric N-cadherin/β1-integrin receptor which localizes to both cell-cell and cell-matrix adhesions. J. Cell Sci. 103, 943–951.

George, D.J. & Blackshear, P.J. (1992). Membrane association of the myristoylated alanine-rich C kinase substrate (MARCKS) protein appears to involve myristate-dependent binding in the absence of a myristoyl protein receptor. J. Biol. Chem. 267, 24879–24885.

Georgatos, S.D. (1993). Dynamics of intermediate filaments. FEBS Lett. 318, 101–107.

Giebelhaus, D.H., Eib, D.W., & Moon, R.T. (1988). Antisense RNA inhibits expression of membrane skeleton protein 4.1 during embryonic development of Xenopus. Cell 53, 601–615.

Gilligan, D.M. & Bennett, V. (1993). The junctional complex of the membrane skeleton. Semin. Hematol. 30, 74–83.

Gilmore, A.P., Jackson, P., Waites, G.T., & Critchley, D.R. (1992). Further characterization of the talin-binding site in the cytoskeletal protein vinculin. J. Cell Sci. 103, 719–731.

Gilmore, A.P., Wood, C., Ohanian, V., Jackson, P., Patel, B., Rees, D.J.G., Hynes, R.O., & Critchley, D.R. (1993). The cytoskeletal talin contains at least two distinct vinculin binding domains. J. Cell Biol. 122, 337–347.

Gimona, M., Small, J.V., Moeremans, M., Van Damme, J., Puype, M., & Vandekerckhove, J. (1988). Porcine vinculin and metavinculin differ by a 68-residue insert located close to the carboxy-terminal part of the molecule. EMBO J. 7, 2329–2334.

Glück, U., Kwiatkowski, D.J., & Ben-Zeev, A. (1993). Suppression of tumorigenicity in simian virus 40-transformed 3T3 cells transfected with α-actinin cDNA. Proc. Natl. Acad. Sci. USA 90, 383–387.

Goldmann, W.H., Niggli, V., Kaufmann, S., & Isenberg, G. (1992). Probing actin and liposome interaction of talin and talin-vinculin complexes: A kinetic, thermodynamic and lipid labeling study. Biochemistry 31, 7665–7671.

Gomez, M., Navarro, P., Quintanilla, M., & Cano, A. (1992). Expression of α6β4 integrin increases during malignant conversion of mouse epidermal keratinocytes: Association of β4 subunit to the cytokeratin fraction. Exp. Cell Res. 201, 250–261.

Gorlin, J.B., Yamin, R., Egan, S., Stewart, M., Stossel, T.P., Kwiatkowski, D.J., & Hartwig, J.H. (1990). Human endothelial actin-binding protein (ABP-280, nonmuscle filamin): A molecular leaf spring. J. Cell Biol. 111, 1089–1105.

Gounari, F., Merdes, A., Quinlan, R., Hess, J., FitzGerald, P.G., Ouzounis, C.A., & Georgatos, S.D. (1993). Bovine filensin possesses primary and secondary structure similarity to intermediate filament proteins. J. Cell Biol. 121, 847–853.

Greengard, P., Valtorta, F., Czernik, A.J., & Benfenati, F. (1993). Synaptic vesicle phosphoproteins and regulation of synaptic function. Science 259, 780–785.

Gu, M., York, J.D., Warshawsky, I., & Majerus, P.W. (1991). Identification, cloning and expression of a cytosolic megakaryocyte protein-tyrosine-phosphatase with sequence homology to cytoskeletal protein 4.1. Proc. Natl. Acad. Sci. USA 88, 5867–5871.

Hagmann, J., Grob, M., & Burger, M.M. (1992). The cytoskeletal protein talin is O-glycosylated. J. Biol. Chem. 267, 14424–14428.

Hamaguchi, M., Matsuyoshi, N., Ohnishi, Y., Gotoh, B., Takeichi, M., & Nagai, Y. (1993). p60[v-src] causes tyrosine phosphorylation and inactivation of the N-cadherin-catenin cell adhesion system. EMBO J. 12, 307–314.

Hanspal, M. & Palek, J. (1992). Biogenesis of normal and abnormal red blood cell membrane skeleton. Semin. Hematol. 29, 305–319.

Hargreaves, A.J., Wandosell, F., & Avila, J. (1986). Phosphorylation of tubulin enhances its interaction with membranes. Nature 323, 827–828.

Hayes, N.V.L. & Baines, A.J. (1992). The axonal membrane cytoskeletal protein A60 and the development of the spectrin/ankyrin-based neuronal membrane skeleton. Biochem. Soc. Trans. 20, 649–652.

Hemmings, L., Kuhlman, P.A., & Critchley, D.R. (1992). Analysis of the actin-binding domain of α-actinin by mutagenesis and demonstration that dystrophin contains a functionally homologous domain. J. Cell Biol. 116, 1369–1380.

Hemphill, A., Affolter, M., & Seebeck, T. (1992). A novel microtubule-binding motif identified in a high molecular weight microtubule-associated protein from *trypanosoma brucei*. J. Cell Biol. 117, 95–103.

Hirano, S., Kimoto, N., Shimoyama, Y., Hirohashi, S., & Takeichi, M. (1992). Identification of a neural α-catenin as a key regulator of cadherin function and multicellular organization. Cell 70, 293–301.

Horkovics-Kovats, S. & Traub, P. (1990). Specific interaction of the intermediate filament protein vimentin and its isolated N-terminus with negatively charged phospholipids as determined by vesicle aggregation, fusion, and leakage measurements. Biochemistry 29, 8652–8657.

Hu, R.-J., Watanabe, M., & Bennett, V. (1992). Characterization of human brain cDNA encoding the general isoform of β-spectrin. J. Biol. Chem. 267, 18715–18722.

Hugnot, J.P., Gilgenkrantz, H., Vincent, N., Chafey, P., Morris, G.E., Monaco, A.P., Berwald-Netter, Y., Koulakoff, A., Kaplan, J.C., Kahn, A., & Chelly, J. (1992). Distal transcript of the dystrophin gene initiated from an alternative first exon and encoding a 75-kDa protein widely distributed in nonmuscle tissues. Proc. Natl. Acad. Sci. USA 89, 7506–7510.

Hynes, R.O. (1992). Integrins: Versatility, modulation, and signalling in cell adhesion. Cell 69, 11–25.

Ibraghimov-Beskrovnaya, O., Ervasti, J.M., Leveille, C.J., Slaughter, C.A., Sernett, S.W., & Campbell, K.P. (1992). Primary structure of dystrophin-associated glycoproteins linking dystrophin to the extracellular matrix. Nature 355, 696–702.

Isenberg, G. (1991). Actin binding proteins-lipid interactions. J. Muscle Res. Cell Motil. 12, 136–144.

Isenberg, G. & Goldmann, W.H. (1992). Actin-membrane coupling: A role for talin. J. Muscle Res. Cell Motil. 13, 587–589.

Ishiko, A., Shimizu, H., Kikuchi, A., Ebihara, T., Hashimoto, T., & Nishikawa, T. (1993). Human autoantibodies against the 230-kD bullous pemphigoid antigen (BPAG1) bind only to the intracellular domain of the hemidesmosome, whereas those against the 180 kD bullous pemphigoid antigen (BPAG2) bind along the plasma membrane of the hemidesmosome in normal human and swine skin. J. Clin. Invest. 91, 1608–1615.

Jones, J.C.R. & Green, K.J. (1991). Intermediate filament-plasma membrane interactions. Curr. Opin. Cell Biol. 3, 127–132.

Jöns, T. & Drenckhahn, D. (1992). Identification of the binding interface involved in linkage of cytoskeletal protein 4.1 to the erythrocyte anion exchanger. EMBO J. 11, 2863–2867.

Joseph, S.K. & Samanta, S. (1993). Detergent solubility of the inositol trisphosphate receptor in rat brain membranes. Evidence for association of the receptor with ankyrin. J. Biol. Chem. 268, 6477–6486.

Joshi, R., Gilligan, D.M., Otto, E., McLaughlin, T., & Bennett, V. (1991). Primary structure and domain organization of human alpha and beta adducin. J. Cell Biol. 115, 665–675.

Karinch, A.M., Zimmer, W.E., & Goodman, S.R. (1990). The identification and sequence of the actin-binding domain of human red blood cell β-spectrin. J. Biol. Chem. 265, 11833–11840.

Kárpáti, S., Amagai, M., Prussick, R., Cehrs, K., & Stanley, J.R. (1993). Pemphigus Vulgaris antigen, a desmoglein type of cadherin, is localized within keratinocyte desmosomes. J. Cell Biol. 122, 409–415.

Kaufmann, S., Käs, J., Goldmann, W.H., Sackmann, E., & Isenberg, G. (1992). Talin anchors and nucleates actin filaments at lipid membranes. FEBS Lett. 314, 203–205.

Kellie, S., Horvath, A.R., & Elmore, M.A. (1991). Cytoskeletal targets for oncogenic tyrosine kinases. J. Cell Sci. 99, 207–211.

Kennedy, S.P., Warren, S.L., Forget, B., & Morrow, J.S. (1991). Ankyrin binds to the 15th repetitive unit of erythroid and nonerythroid β-spectrin. J. Cell Biol. 115, 267–277.

Kirsch, J., Langosch, D., Prior, P., Littauer, U.Z., Schmitt, B., & Betz, H. (1991). The 93-kDa glycine receptor-associated protein binds to tubulin. J. Biol. Chem. 266, 22242–22245.

Knudsen, K.A. & Wheelock, M.J. (1992). Plakoglobin, or an 83-kD homologue distinct from β-catenin, interacts with E-cadherin and N-cadherin. J. Cell Biol. 118, 671–679.

Koch, P.J., Goldschmidt, M.D., Zimbelmann, R., Troyanovsky, R., & Franke, W.W. (1992). Complexity and expression patterns of the desmosomal cadherins. Proc. Natl. Acad. Sci. USA 89, 353–357.

Korsgren, C., Lawler, J., Lambert, S., Speicher, D., & Cohen, C.M. (1990). Complete amino acid sequence and homologies of human erythrocyte membrane protein band 4.2. Proc. Natl. Acad. Sci. USA 87, 613–617.

Koteliansky, V.E., Ogryzko, E.P., Zhidkova, N.I., Weller, P.A., Critchley, D.R., Vancompernolle, K., Vandekerckhove, J., Strasser, P., Way, M., Gimona, M., & Small, J.V. (1992). An additional exon in the human vinculin gene specifically encodes meta-vinculin-specific difference peptide. Eur. J. Biochem. 204, 767–772.

Kurpakus, M.A., Quaranta, V., & Jones, J.C.R. (1991). Surface relocation of Alpha$_6$Beta$_4$ integrins and assembly of hemidesmosomes in an in vitro model of wound healing. J. Cell Biol. 115, 1737–1750.

Lambert, S. & Bennett, V. (1993). From anemia to cerebellar dysfunction. A review of the ankyrin gene family. Eur. J. Biochem. 211, 1–6.

Lankes, W.T. & Furthmayr, H. (1991). Moesin: A member of the protein 4.1-talin-ezrin family of proteins. Proc. Natl. Acad. Sci. USA 88, 8297–8301.

Lazarides, E. & Woods, C. (1989). Biogenesis of the red blood cell membrane-skeleton and the control of erythroid morphogenesis. Ann. Rev. Cell Biol. 5, 427–452.

Lee, S., Wulfkuhle, J.D., & Otto, J.J. (1992). Vinculin binding site mapped on talin with an anti-idiotypic antibody. J. Biol. Chem. 267, 16355–16358.

Legan, P.K., Collins, J.E., & Garrod, D.R. (1992). The molecular biology of desmosomes and hemides-mosomes: "What's in a name?". BioEssays 14, 385–393.

Li, Z., Burke, E.P., Frank, J.S., Bennett, V., & Philipson, K.D. (1993). The cardiac Na$^+$-Ca^{2+} exchanger binds to the cytoskeletal protein ankyrin. J. Biol. Chem. 268, 11489–11491.

Liu, S.C. & Derick, L.H. (1992). Molecular anatomy of the red blood cell membrane skeleton: Structure-function relationship. Semin. Hematol. 29, 231–243.

Lokeswhar, V.B. & Bourguignon, L.Y.W. (1992a). Tyrosine phosphatase activity of lymphoma CD45 (GP 180) is regulated by a direct interaction with the cytoskeleton. J. Biol. Chem. 267, 21551–21557.

Lokeswhar, V.B. & Bourguignon, L.Y.W. (1992b). The Lymphoma transmembrane glycoprotein GP85 (CD44) is a novel guanine nucleotide-binding protein which regulates GP85 (CD44)-ankyrin interaction. J. Biol. Chem. 267, 22073–22078.

Lombardo, C.R., Willardson, B.M., & Low, P.S. (1992). Localization of the protein 4.1-binding site on the cytoplasmic domain of erythrocyte membrane band 3. J. Biol. Chem. 267, 9540–9546.

Luna, E.J. & Hitt, A.L. (1992). Cytoskeleton-plasma membrane interactions. Science 258, 955–964.

Magee, A.I. & Buxton, R.S. (1991). Transmembrane molecular assemblies regulated by the greater cadherin family. Curr. Opin. Cell Biol. 3, 854–861.

Manenti, S., Sorokine, O., Van Dorsselaer, A., & Taniguchi, H. (1993). Isolation of the non-myristoylated form of a major substrate of protein kinase C (MARCKS) from bovine brain. J. Biol. Chem. 268, 6878–6881.

Mangeat, P.H. (1988). Interaction of biological membranes with the cytoskeletal framework of living cells. Biol. Cell 64, 261–281.

Maretzki, D., Mariani, M., & Lutz, H.U. (1990). Fatty acid acylation of membrane skeletal proteins in human erythrocytes. FEBS Lett. 259, 305–310.

Mariani, M., Maretzki, D., & Lutz, H.U. (1993). A tightly membrane-associated subpopulation of spectrin is ^{3}H-palmitoylated. J. Biol. Chem. 268, 12996–13001.

Matsumara, K., Shasby, D.M., & Campbell, K.P. (1993). Purification of dystrophin-related protein (utrophin) from lung and its identification in pulmonary artery endothelial cells. FEBS Lett. 326, 289–293.

Matsumura, K., Tomé, F.M.S., Collin, H., Azibi, K., Chaouch, M., Kaplan, J.-C., Fardeau, M., & Campbell, K.P. (1992). Deficiency of the 50K dystrophin-associated glycoprotein in severe childhood autosomal recessive muscular dystrophy. Nature 359, 320–322.

McMahon, A.P., Giebelhaus, D.H., Champion, J.E., Bailes, J.A., Lacy, S., Carrit, B., Henchman, S.K., & Moon, R.T. (1987). cDNA cloning, sequencing and chromosome mapping of a non-erythroid spectrin, human α-fodrin. Differentiation 34, 68–78.

Meriläinen, J., Palovuori, R., Sormunen, R., Wasenius, V.-M-., & Lehto, V.-P. (1993). Binding of the α-fodrin SH3 domain to the leading lamellae of locomoting chicken fibroblasts. J. Cell Sci. 105, 647–654.

Mithieux, G. & Rousset, B. (1989). Identification of a lysosomal membrane protein which could mediate ATP-dependent stable association of lysosomes to microtubules. J. Biol. Chem. 264, 4664–4668.

Musacchio, A., Gibson, T., Lehto, V.-P., & Saraste, M. (1992a). SH3 - an abundant protein domain in search of a function. FEBS Lett. 307, 55–61.

Musacchio, A., Noble, M., Pauptit, R., Wieringa, R., & Saraste, M. (1992b). Crystal structure of a Src-homology 3 (SH3) domain. Nature 359, 851–855.

Muguruma, M., Matsumura, S., & Fukazawa, T. (1992). Augmentation of α-actinin-induced gelation of actin by talin. J. Biol. Chem. 267, 5621–5624.

Nagafuchi, A., Takeichi, M., & Tsukita, S. (1991). The 102 kd cadherin-associated protein: Similarity to vinculin and posttranslational regulation of expression. Cell 65, 849–857.

Nakata, T. & Hirokawa, N. (1987). Cytoskeletal reorganization of human platelets after stimulation revealed by the quick-freeze deep-etch technique. J. Cell Biol. 105, 1771–1780.

Nelson, W.J., Shore, E.M., Wang, A.Z., & Hammerton, R.W. (1990). Identification of a membrane-cytoskeletal complex containing the cell adhesion molecule uvomorulin (E-cadherin), ankyrin, and fodrin in madin-darby canine kidney epithelial cells. J. Cell Biol. 110, 349–357.

Nelson, W.J. & Veshnock, P.J. (1987). Ankyrin binding to $(Na^{+}+K^{+})$ATPase and implications for the organization of membrane domains in polarized cells. Nature 328, 533–536.

Niggli, V. & Burger, M.M. (1987). Interaction of the cytoskeleton with the plasma membrane. J. Membrane Biol. 100, 97–121.

Niggli, V., Dimitrov, D.P., Brunner, J., & Burger, M.M. (1986). Interaction of the cytoskeletal protein vinculin with bilayer structures analyzed with a photoactivatable phospholipid. J. Biol. Chem. 261, 6912–6918.

Niggli, V. & Gimona, M. (1993). Evidence for a ternary interaction between α-actinin, (meta)vinculin and acidic phospholipid bilayers. Eur. J. Biochem. 213, 1009–1015.

Niggli, V., Sommer, L., Brunner, J., & Burger, M.M. (1990). Interaction in situ of the cytoskeletal protein vinculin with bilayers studied by introducing a photoactivatable fatty acid into living chicken embryo fibroblasts. Eur. J. Biochem. 187, 111–117.

Nikolai, G., Temm-Grove, C.J., Wiegand, C., Wilkinson, J.M., & Jockusch, B.M. (1993). Interaction of junctional proteins at microfilament-membrane attachment sites as probed by antibodies. Eur. J. Cell Biol. 60 (Suppl. 37), 41 (Abstract).

Nishizawa, Y., Uematsu, J., & Owaribe, K. (1993). HD4, a 180 kDa bullous pemphigoid antigen, is a major transmembrane glycoprotein of the hemidesmosome. J. Biochem. 113, 493–501.

North, A.J., Galazkiewicz, B., Byers, T.J., Glenney, Jr., J.R., & Small, J.V. (1993). Complementary distributions of vinculin and dystrophin define two distinct sarcolemma domains in smooth muscle. J. Cell Biol. 120, 1159–1167.

Nuckolls, G.H., Romer, L.H., & Burridge, K. (1992). Microinjection of antibodies against talin inhibits the spreading and migration of fibroblasts. J. Cell Sci. 102, 753–762.

Nuckolls, G.H., Turner, C.E., & Burridge, K. (1990). Functional studies of the domains of talin. J. Cell Biol. 110, 1635–1644.

Offringa, R. & Bierer, B.E. (1993). Association of CD2 with tubulin. J. Biol. Chem. 268, 4979–4988.

Ohta, Y., Stossel, T.P., & Hartwig, J.H. (1991). Ligand-sensitive binding of actin-binding protein to immunoglobulin G Fc receptor I $(Fc_{\gamma}RI)$. Cell 67, 275–282.

Otey, C.A., Pavalko, F.M., & Burridge, K. (1990). An interaction between α-actinin and the β₁ integrin subunit *in vitro*. J. Cell Biol. 111, 721–729.

Otto, E., Kunimoto, M., McLaughlin, T., & Bennett, V. (1991). Isolation and characterization of cDNAs encoding human brain ankyrins reveal a family of alternatively spliced genes. J. Cell Biol. 114, 241–253.

Otto, J.J. (1990). Vinculin. Cell Motil. Cytoskel. 16, 1–6.

Ozawa, M. & Kemler, R. (1992). Molecular organization of the uvomorulin-catenin complex. J. Cell Biol. 116, 989–996.

Pacaud, M. & Harricane, M.C. (1993). Macrophage α-actinin is not a calcium-modulated actin-binding protein. Biochemistry 32, 363–374.

Palek, J. & Sahr, K.E. (1992). Mutations of the red blood cell membrane proteins: From clinical evaluation to detection of the underlying genetic defect. Blood 80, 308–330.

Pasdar M., Li, Z., & Krzeminski, K.A. (1992). Desmosome assembly in MDCK epithelial cells does not require the presence of functional microtubules. Cell Motil. Cytoskel. 23, 201–212.

Pasternak, G.R. & Racusen, R.H. (1989). Erythrocyte protein 4.1 binds and regulates myosin. Proc. Natl. Acad. Sci. USA 86, 9712–9716.

Pavalko, F.M. & Burridge, K. (1991). Disruption of the actin cytoskeleton after microinjection of proteolytic fragments of α-actinin. J. Cell Biol. 114, 481–491.

Peifer, M., McCrea, P.D., Green, K.J., Wieschaus, E., & Gumbiner, B. (1992). The vertebrate adhesive junction proteins β-catenin and plakoglobin and the *Drosophila* segment polarity gene armadillo form a multigene family with similar properties. J. Cell Biol. 118, 681–691.

Peters, L.L. & Lux, S.E. (1993). Ankyrins: Structure and function in normal cells and hereditary spherocytes. Semin. Hematol. 30, 85–118.

Petrof, B.J., Shrager, J.B., Stedman, H.H., Kelly, A.M., & Sweeney, H.L. (1993). Dystrophin protects the sarcolemma from stresses developed during muscle contraction. Proc. Natl. Acad. Sci. USA 90, 3710–3714.

Piepenhagen, P.A. & Nelson, W.J. (1993). Defining E-cadherin-associated protein complexes in epithelial cells: Plakoglobin, β- and γ-catenin are distinct components. J. Cell Sci. 104, 751–762.

Pierre, P., Scheel, J., Rickard, J.E., & Kreis, T.E. (1992). CLIP-170 links endocytic vesicles to microtubules. Cell 70, 887–900.

Pollard, T.D., Doberstein, S.K., & Zot, H.G. (1991). Myosin-I. Annu. Rev. Physiol. 53, 653–81.

Pollerberg, G.E., Burridge, K., Krebs, K.E., Goodman, S.R., & Schachner, M. (1987). The 180-kD component of the neural cell adhesion molecule N-CAM is involved in cell-cell contacts and cytoskeleton-membrane interactions. Cell Tissue Res. 250, 227–236.

Prior, P., Schmitt, B., Grenningloh, G., Pribilla, I., Multhaup, G., Beyreuther, K., Maulet, Y., Werner, P., Langosch, D., Kirsch, J., & Betz, H. (1992). Primary structure and alternative splice variants of gephyrin, a putative glycine receptor-tubulin linker protein. Neuron 8, 1161–1170.

Rana, A.P., Ruff, P., Maalouf, G.J., Spelcher, D.W., & Chishti, A.H. (1993). Cloning of human erythroid dematin reveals another member of the villin family. Proc. Natl. Acad. Sci. USA 90, 6651–6655.

Rees, D.J.G., Ades, S.E., Singer, S.J., & Hynes, R.O. (1990). Sequence and domain structure of talin. Nature 347, 685–689.

Reszka, A.A., Hayashi, Y., & Horwitz, A.F. (1992). Identification of amino acid sequences in the integrin β₁ cytoplasmic domain implicated in cytoskeletal association. J. Cell Biol. 117, 1321–1330.

Ridley, A.J. & Hall, A. (1992). The small GTP-binding protein rho regulates the assembly of focal adhesions and actin stress fibers in response to growth factors. Cell 70, 389–399.

Risinger, M.A., Dotimas, E., & Cohen, C.M. (1992). Human erythrocyte protein 4.2, a high copy number membrane protein, is N-myristylated. J. Biol. Chem. 267, 5680–5685.

Roehl, H. & Kimble, J. (1993). Control of cell fate in C. *elegans* by a GLP-1 peptide consisting primarily of ankyrin repeats. Nature 364, 632–635.

Rouleau, G.A., Merel, P., Lutchman, M., Sanson, M., Zucman, J., Marineau, C., Hoang-Xuan, K., Demczuk, S., Desmaze, C., Plougastel, B., Pulst, S.M., Lenoir, G., Bijlsma, E., Fashold, R.,

Dumanski, J., de Jong, P., Parry, D., Eldridge, R., Aurias, A., Delattre, O., & Thomas, G. (1993). Alteration in a new gene encoding a putative membrane-organizing protein causes neurofibromatosis type 2. Nature 363, 515–521.

Ruff, P., Speicher, D.W., & Husain-Chishti, A. (1991). Molecular identification of a major palmitoylated erythrocyte membrane protein containing the *src* homology 3 motif. Proc. Natl. Acad. Sci. USA 88, 6595–6599.

Ruhnau, K. & Wegner, A. (1988). Evidence for direct binding of vinculin to actin filaments. FEBS Lett. 228, 105–108.

Ruddies, R., Goldmann, W.H., Isenberg, G., & Sackmann, E. (1993). The viscoelasticity of entangled actin networks: The influence of defects and the modulation by talin and vinculin. Eur. Biophysics J. 2, 309–321.

Ruppert, C., Kroschewski, R., & Bähler, M. (1993). Identification, characterization and cloning of myr 1, a mammalian myosin-I. J. Cell Biol. 120, 1393–1403.

Sadler, I., Crawford, A.W., Michelsen, J.W., & Beckerle, M.C. (1992). Zyxin and cCRP: Two interactive LIM domain proteins associated with the cytoskeleton. J. Cell Biol. 119, 1573–1587.

Samuels, M., Ezzell, R.M., Cardozzo, T.J., Critchley, D.R., Coll, J.-L., & Adamson, E.D. (1993). Expression of chicken vinculin complements the adhesion-defective phenotype of a mutant mouse F9 embryonal carcinoma cell. J. Cell Biol. 121, 909–921.

Sato, N., Funayama, N., Nagafuchi, A., Yonemura, S., Tsukita, Sa., & Tsukita, Sh. (1992). A gene family consisting of ezrin, radixin and moesin: Its specific localization at actin filament/plasma membrane association sites. J. Cell Sci. 103, 131–143.

Sawamura, D., Li, K., Chu, M.-L., & Uitto, J. (1991). Human bullous pemphigoid antigen (BPAG1): Amino acid sequences deduced from cloned cDNAs predict biologically important peptide segments and protein domains. J. Biol. Chem. 266, 17784–17790.

Schleicher, M., Noegel, A., Schwarz, T., Wallraff, E., Brink, M., Faix, J., Gerisch, G., & Isenberg, G. (1988). A Dictyostelium mutant with severe defects in α-actinin: Its characterization using cDNA probes and monoclonal antibodies. J. Cell Sci. 90, 59–71.

Schliwa, M. & van Blerkom, J. (1981). Structural interaction of cytoskeletal components. J. Cell Biol. 90, 222–235.

Schultheiss, T., Choi, J., Lin, Z.X., DiLullo, C., Cohen-Gould, L., Fischman, D., & Holtzer, H. (1992). A sarcomeric α-actinin truncated at the carboxyl end induces the breakdown of stress fibers in PtK2 cells and the formation of nemaline-like bodies and breakdown of myofibrils in myotubes. Proc. Natl. Acad. Sci. USA 89, 9282–9286.

Seebeck, T., Hemphill, A., & Lawson, D. (1990). The cytoskeleton of trypanosomes. Parasitology Today 6, 49–52.

Seifert, G.J., Lawson, D., & Wiche, G. (1992). Immunolocalization of the intermediate filament-associated protein plectin at focal contacts and actin stress fibers. Eur. J. Cell Biol. 59, 138–147.

Senter, L., Luise, M., Presotto, C., Betto, R., Teresi, A., Ceoldo, S., & Salviati, G. (1993). Interaction of dystrophin with cytoskeletal proteins: Binding to talin and actin. Biochem. Biophys. Res. Commun. 192, 899–904.

Sherr, E.H., Joyce, M.P., & Greene, L.A. (1993). Mammalian myosin Iα, Iβ, and I_γ: New widely expressed genes of the myosin I family. J. Cell Biol. 120, 1405–1416.

Sikorski, A.F., Terlecki, G., Zagon, I.S., & Goodman, S.R. (1991). Synapsin I-mediated interaction of brain spectrin with synaptic vesicles. J. Cell Biol. 114, 313–318.

Siman, R., Baudry, M., & Lynch, G. (1985). Regulation of glutamate receptor binding by the cytoskeletal protein fodrin. Nature 313, 225–228.

Smith, P.R., Bradford, A.L., Joe, E.-H., Angelides, K.J., Benos, D.J., & Saccomani, G. (1993). Gastric parietal cell H^+-K^+-ATPase microsomes are associated with isoforms of ankyrin and spectrin. Am. J. Physiol. 264, C63–C70.

Smith, P.R., Saccomani, G., Joe, E.-H., Angelides, K., & Benos, D.J. (1991). Amiloride-sensitive sodium channel is linked to the cytoskeleton in renal epithelial cells. Proc. Natl. Acad. Sci. USA 88, 6971–6975.

Sonnenberg, A., Calafat, J., Janssen, H., Daams, H., van der Raaij-Helmer, L.M.H., Falcioni, R., Kennel, S.J., Aplin, J.D., Baker, J., Loizidou, M., & Garrod, D. (1991). Integrin $\alpha6\beta4$ complex is located in hemidesmosomes, suggesting a major role in epidermal cell-basement membrane adhesion. J. Cell Biol. 113, 907–917.

Speicher, D.W., Weglarz, L., & DeSilva, T.M. (1992). Properties of human red cell spectrin heterodimer (side-to-side) assembly and identification of an essential nucleation site. J. Biol. Chem. 267, 14775–14782.

Srinivasan, Y., Lewallen, M., & Angelides, K.J. (1992). Mapping the binding site on ankyrin for the voltage-dependent sodium channel from brain. J. Biol. Chem. 267, 7483–7489.

Stappenbeck, T.S. & Green, K.J. (1992). The desmoplakin carboxyl terminus coaligns with and specifically disrupts intermediate filament networks when expressed in cultured cells. J. Cell Biol. 116, 1197–1209.

Staufenbiel, M. (1987). Ankyrin-bound fatty acid turns over rapidly at the erythrocyte plasma membrane. Mol. Cellul. Biol. 7, 2981–2984.

Stephens, R.E. (1986). Membrane tubulin. Biol. Cell 57, 95–110.

Stepp, M.A., Spurr-Michaud, S., Tisdale, A., Elwell, J., & Gipson, I.K. (1990). $\alpha_6\beta_4$ integrin heterodimer is a component of hemidesmosomes. Proc. Natl. Acad. Sci. USA 87, 8970–8974.

Stevenson, K.B., Clark, R.A., & Nauseef, W.M. (1989). Fodrin and band 4.1 in a plasma membrane-associated fraction of human neutrophils. Blood 74, 2136–2143.

Strasser, P., Gimona, M., Herzog, M., Geiger, B., & Small, J.V. (1993). Variable and constant regions in the C-terminus of vinculin and metavinculin. FEBS Lett. 317, 189–194.

Sung, L.A., Chien, S., Chang, L.-S., Lambert, K., Bliss, S.A., Bouhassira, E.E., Nagel, R.L., Schwartz, R.S., & Rybicki, A.C. (1990). Molecular cloning of human protein 4.2: A major component of the erythrocyte membrane. Proc. Natl. Acad. Sci. USA 87, 955–959.

Suzuki, A., Yoshida, M., Yamamoto, H., & Ozawa, E. (1992). Glycoprotein-binding site of dystrophin is confined to the cysteine-rich domain and the first half of the carboxy-terminal domain. FEBS Lett. 308, 154–160.

Takagi, T., Pribilla, I., Kirsch, J., & Betz, H. (1992). Coexpression of the receptor-associated protein gephyrin changes the ligand binding affinities of α_2 glycine receptors. FEBS Lett. 303, 178–180.

Tang, T.K., Leto, T.L., Correas, I., Alonso, M.A., Marchesi, V.T., & Benz Jr., E.J. (1988). Selective expression of an erythroid-specific isoform of protein 4.1. Proc. Natl. Acad. Sci. USA 85, 3713–3717.

Taniguchi, H. & Manenti, S. (1993). Interaction of myristoylated alanine-rich protein kinase C substrate (MARCKS) with membrane phospholipids. J. Biol. Chem. 268, 9960–9963.

Thiel, C., Osborn, M., & Gerke, V. (1992). The tight association of the tyrosine kinase substrate annexin II with the submembranous cytoskeleton depends on intact p11- and Ca^{2+}-binding sites. J. Cell Sci. 103, 733–742.

Thompson, C.C., Brown, T.A., & McKnight, S.L. (1991). Convergence of Ets- and Notch-related structural motifs in a heteromeric DNA binding complex. Science 253, 762–768.

Trofatter, J.A., MacCollin, M.M., Rutter, J.L., Murrell, J.R., Duyao, M.P., Parry, D.M., Eldrige, R., Kley, N., Menon, A.G., Pulaski, K., Haase, V.H., Ambrose, C.M., Munroe, D., Bove, C., Haines, J.L., Martuza, R.L., MacDonald, M.E., Seizinger, B.R., Short, M.P., Buckler, A.J., & Gusella, J.F. (1993). A novel moesin-, ezrin-, radixin-like gene is a candidate for the neurofibromatosis 2 tumor suppressor. Cell 72, 791–800.

Troyanovsky, S.M., Eshkind, L.G., Troyanovsky, R.B., Leube, R.E., & Franke, W.W. (1993). Contributions of cytoplasmic domains of desmosomal cadherins to desmosome assembly and intermediate filament anchorage. Cell 72, 561–574.

Turner, C.E., Glenney, J.R., & Burridge, K. (1990). Paxillin: A new vinculin-binding protein present in focal adhesions. J. Cell Biol. 111, 1059–1068.

Turner, C.E. & Burridge, K. (1991). Transmembrane molecular assemblies in cell-extracellular matrix interactions. Curr. Opin. Cell Biol. 3, 849–853.

Vale, R.D. (1992). Microtubule motors: Many new models off the assembly line. TIBS 17, 300–304.

Virata, M.L.A., Wagner, R.M., Parry, D.A.D., & Green, K.J. (1992). Molecular structure of the human desmoplakin I and II amino terminus. Proc. Natl. Acad. Sci. USA 89, 544–548.

Vishwanatha, J.K. & Kumble, S. (1993). Involvement of annexin II in DNA replication: Evidence from cell-free extracts of Xenopus eggs. J. Cell Sci. 105, 533–540.

Wacker, I.U., Rickard, J.E., De Mey, J.R., & Kreis, T.E. (1992). Accumulation of a microtubule-binding protein, pp 170, at desmosomal plaques. J. Cell Biol. 117, 813–824.

Wagner, M.C., Barylko, B., & Albanesi, J.P. (1992). Tissue distribution and subcellular localization of mammalian myosin I. J. Cell Biol. 119, 163–170.

Waites, G.T., Graham, I.R., Jackson, P., Millake, D.B., Patel, B., Blanchard, A.D., Weller, P.A., Eperon, I.C., & Critchley, D.R. (1992). Mutually exclusive splicing of calcium-binding domain exons in chick α-actinin. J. Biol. Chem. 267, 6263–6271.

Wang, N., Butler, J.P., & Ingber, D.E. (1993). Mechanotransduction across the cells surface and through the cytoskeleton. Science 260, 1124–1127.

Wang, N. & Rasenick, M.M. (1991). Tubulin-G protein interactions involve microtubule polymerization domains. Biochemistry 30, 10957–10965.

Waseem, A. & Palfrey, H.C. (1990). Identification and protein kinase C-dependent phosphorylation of α-adducin in human fibroblasts. J. Cell Sci. 96, 93–98.

Weng, Z., Taylor, J.A., Turner, C.E., Brugge, J.S., & Seidel-Dugan, C. (1993). Detection of scr homology 3-binding proteins, including paxillin, in normal and v-src-transformed Balb/c 3T3 cells. J. Biol. Chem. 268, 14956–14963.

Westmeyer, A., Ruhnau, K., Wegner, A., & Jockusch, B.M. (1990). Antibody mapping of functional domains in vinculin. EMBO J. 9, 2071–2078.

White, R.A., Peters, L.L., Adkison, L.R., Korsgren, C., Cohen, C.M., & Lux, S.E. (1992). The murine pallid mutation is a platelet storage pool disease associated with the protein 4.2 (pallidin) gene. Nature Genet. 2, 80–83.

Wiche, G., Becker, B., Luber, K., Weitzer, G., Castañón, M.J., Hauptmann, R., Stratowa, C., & Stewart, M. (1991). Cloning and sequencing of rat plectin indicates a 466-kD polypeptide chain with a three-domain structure based on a central alpha-helical coiled coil. J. Cell Biol. 114, 83–99.

Wiche, G., Gromov, D., Donovan, A., Castañón, M.J., & Fuchs, E. (1993). Expression of plectin mutant cDNA in cultured cells indicates a role of COOH-terminal domain in intermediate filament association. J. Cell Biol. 121, 607–619.

Woods, A. & Couchman, J.R. (1992). Protein kinase C involvement in focal adhesion formation. J. Cell Sci. 101, 277–290.

Wyatt, K. & Cherry, R.J. (1992). Both ankyrin and band 4.1 are required to restrict the rotational mobility of band 3 in the human erythrocyte. Biochim. Biophys. Acta 1103, 327–330.

Yang, Q. & Tonks, N.K. (1991). Isolation of a cDNA clone encoding a human protein-tryosine phosphatase with homology to the cytoskeletal-associated proteins band 4.1, ezrin, and talin. Proc. Natl. Acad. Sci. USA 88, 5949–5953.

Yürüker, B. & Niggli, V. (1992). α-Actinin and vinculin in human neutrophils: Reorganization during adhesion and relation to the actin network. J. Cell Sci. 101, 403–414.

Zachary, I. & Rozengurt, E. (1992). Focal adhesion kinase (p125^{FAK}): A point of convergence in the action of neuropeptides, integrins and oncogenes. Cell 71, 891–894.

ACTIN-BINDING PROTEINS–LIPID INTERACTIONS

G. Isenberg and W.H. Goldmann

The Cytoskeleton, Volume 1
Structure and Assembly, pages 169–204.
Copyright © 1995 by JAI Press Inc.
All rights of reproduction in any form reserved.
ISBN: 1-55938-687-8

I. INTRODUCTION

Investigating biological interfaces has gained more interest recently since new biochemical and biophysical techniques have become available to cell biologists. The events at, near or through a plasma membrane are complex, in as much as one does not simply consider a membrane bilayer as a mechanical barrier. From the physicochemical point of view, we have to imagine a multilayered colloidal system consisting of proteins and lipids in various aggregation states. This tri- or tetralayer membrane (Figure 1) extends from inside of a cell to the extracellular environment and involves the cytoskeleton proteins (microfilaments, microtubules, intermediate filaments; cf. Niggli this volume), which are specifically anchored in the fluid mosaic lipid bilayer and interconnected transmembranously with extracellular matrix proteins. In this way signals are transmitted from outside to inside or vice versa. Signals, however, can be generated in the lateral plane by diffusion within the lipid layer or by chemomechanical transduction mechanisms through the interconnected fibrous protein network (plasmalemma undercoat). Furthermore, new microcompartments within or between these multilayers may be generated.

Although the binding of membrane components to the intracellular cytoskeleton and extracellular matrix proteins may be characterized by rapid exchange and low affinity, these may be of considerable importance in determining the order of magnitude of D, the apparent diffusion coefficient of the mobile fraction of membrane proteins (Zhang et al., 1993). The *transient* nature of interactions involving lipid–protein or lipid–protein/protein binding may thus give rise to divergent diffusion constants. A new aspect in this scheme of interactions involves the binding of ectodomains of membrane proteins, the pericellular matrix (PCM), to the large extracellular matrix molecules (Sheetz, 1993). Since some of these PCM proteins are lipid-binding proteins and, in addition, are intracellularly linked to the cytoskeleton, one is confronted with complex cooperative reaction schemes that are restrictive (by inhibiting diffusion) and permissive (by allowing signal transduction).

Before evaluating actin-binding proteins–lipid interactions at the plasma membrane interface in more detail, two aspects should be considered: (1) Generation of intracellular signals or second messengers—besides the classical receptor pathway—may involve mechanical stimuli that are produced by applying physical

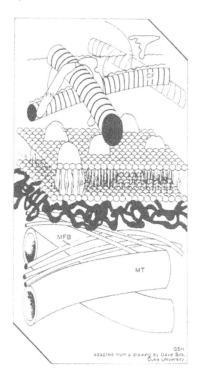

Figure 1. A multi-layer model of the membrane interface. Microfilaments (MF) microtubules (MT) and intermediate filaments (IF) are connected via lipid binding proteins (black) to the membrane bilayer. Pericellular matrix proteins and the extracellular fibronectin (FN)/collagen fiber system (CF) transmit forces through transmembrane receptor proteins (integrins). From Dave Birk, Duke Univ., North Carolina, USA, with permission.

forces to the cell surface. This process of "mechano-transduction" will allow function to follow a given form (P.A. Watson, 1991) and represents a new concept. (2) More than 10 years ago, Lazarides and colleagues proposed that the topogenesis of a membrane skeleton is determined by assembly limiting reaction steps; that is, the stoichiometry of binding would simply be governed by the availability of binding sites independent of the rate of synthesis (Lazarides and Moon, 1984). From these two examples, it is clear that important signals for cell behavior may be generated by specific architecture of the membrane itself.

Hence, actin-binding protein–lipid interactions may play a central role not only in (1) mediating the anchorage of the cytoskeleton in the lipid layer, but also in (2) defining a specific membrane topology. Moreover, the release of actin-binding proteins or certain lipids (e.g., inositolphosphates or diacylglycerol; DAG) from their complex upon stimulation via receptors (see below) may represent a more direct way for a cell to trigger intracellular events through a transmembrane signaling mechanism.

II. THE ROLE OF LIPIDS

The general classification scheme of lipids that interact with actin-binding proteins follows simple criteria:

(1) Fatty acids are distinguished by their number of C atoms and double bonds.
(2) Phospholipids have a common glycerol backbone.
(3) The lipid head groups are linked by a phosphate ester to the glycerol backbone.

Most common phospholipids are: phosphatidylcholine (PC), phosphatidylethanolamine (PE), phosphatidylserine (PS), phosphatidylinositol (PI) and derivatives, phosphatidylinositol-4-monophosphate (PIP), phosphatidylinositol-4,5-biphosphate (PIP-2), inositoltriphosphate (IP3), phosphatidylglycerol (PG), (Figure 2). For the naturally occurring lipid composition of various mammalian cell membranes the reader is referred to the recent article by Zachowsky (1993).

Protein–lipid binding involves hydrogen bonds, van der Waals forces, hydrophobic interactions and electrostatic coupling (Dill, 1990). Different binding phenomena result from distinctly folded protein domains that are properly exposed in the appropriate membrane environment (i.e., after binding to several lipid molecules in protein–lipid complexes and lipid-protein–protein interactions). The actin-binding protein 4.1 serves as an example; its spectrin-binding domain of protein 4.1, which is accessible in aqueous solution, is masked after reconstitution into lipid vesicles, but regains its spectrin-binding capacity in the presence of actin (Cohen and Foley, 1982; Bennett and Stenbuck, 1979). In addition, binding constants for lipid components may change dramatically depending on whether measurements are performed in solution or after reconstitution into a lipid bilayer. For example, the binding constant of protein 4.1 for PIP-2 shifts from low to high affinity after transfer to its "natural" glycophorin-containing membrane environment (Anderson and Marchesi, 1985). Because lipids act as cofactors that modulate the interaction of proteins within the membrane plane and because a selective and high-affinity binding to the lipid bilayer may depend on more than two components, it is essential to study protein–lipid interactions in a system that mimics the natural environment.

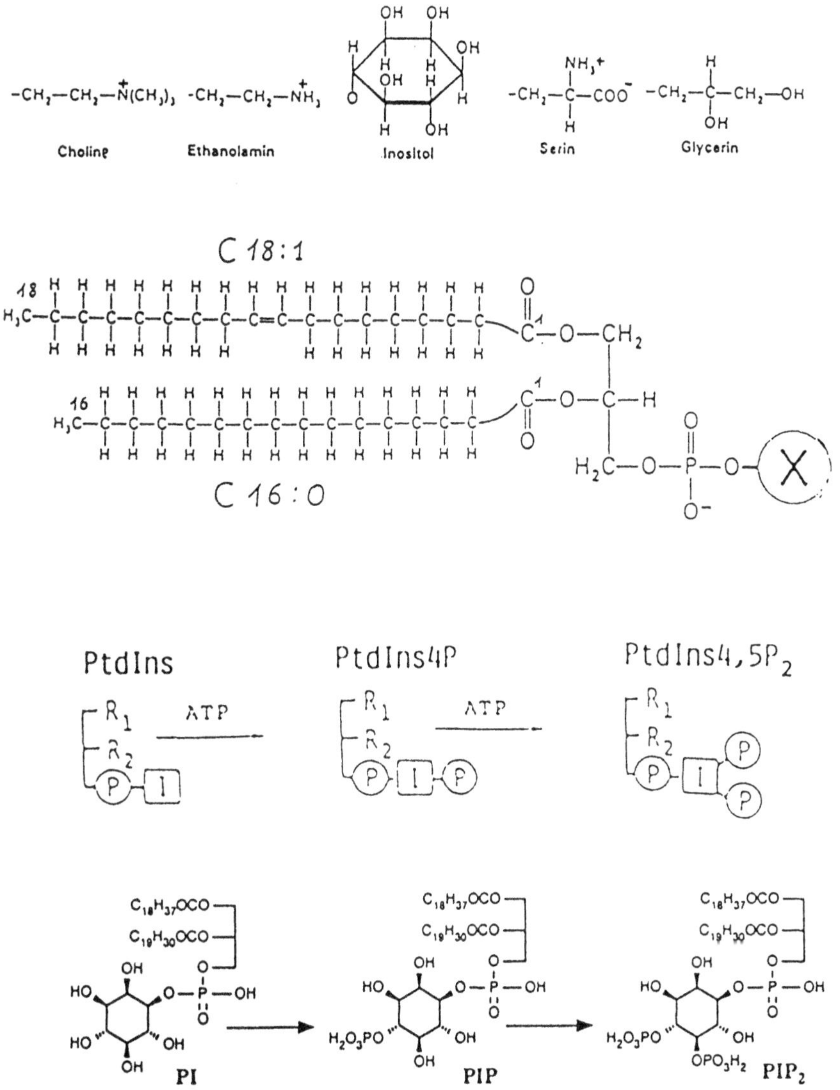

Figure 2. Major lipid components from animal cell membranes. For explanation see chapter II.

III. A SURVEY OF APPLICABLE TECHNIQUES FOR INVESTIGATING PROTEIN–LIPID INTERACTIONS

A number of biophysical techniques, some of these techniques are discussed below in more detail, are now available to investigate lipid–protein interactions within a membrane bilayer. Infrared spectroscopy (Siminovitch et al., 1987) and high-sensitivity differential scanning calorimetry (DSC) (Mason et al., 1981) applied together are powerful tools to provide the thermodynamic and structural information required to characterize lipid conformation and intermolecular interactions. Complementary techniques such as X-ray diffraction (Hui and Huang, 1986), NMR spectroscopy (Ruocco et al., 1985a; 1985b), Raman spectroscopy (O'Leary and Levin, 1984), electron spin resonance spectroscopy (ESR) (Marsh, 1990), and measurements of fluorescence anisotropy using diphenylhexatriene (DPH; Tendian and Lentz, 1990), are becoming routine techniques for studying lipid–protein interactions and to determine the degree of interdigitation in lipid bilayers.

In most laboratories multilayered vesicles, or even better, unilamellar, small (100 nm) and giant liposomes (> 500–1000 nm) serve as membrane models (New, 1990). It should be pointed out that the amount of protein incorporation into liposomes depends on the size of the molecule and the degree of membrane curvature.

Advancement has been made in developing a miniaturized Langmuir technique (Heyn et al., 1991). Two-dimensional lipid monolayers, spread on an air–water interface, can serve as model systems for biological membranes. In mixed lipid–protein films, the lateral packing density of molecules, temperature, ionic strength and pH of the subphase can be controlled in a precise manner. By introducing a small fraction of fluorescent dye molecules, domain formation and mobility of lipids and proteins can be traced individually by epifluorescence light microscopy (Möhwald, 1990; Heyn et al., 1991; Dietrich et al., 1993). An additional advantage is the procedure of transferring lipid–protein monolayers onto planar solid supports, which in the near future, will allow the investigation of pattern formation of lipid–protein complexes in two dimensions by scanning tunneling (Hörber et al., 1988) and atomic force microscopy (Egger et al., 1990; Ohnesorge et al., 1990; Weisenhorn et al., 1991).

IV. ARTEFACTS OF LABELING TECHNIQUES

Interpretation of measurements of actin-binding protein–lipid interactions has to be done with caution: Binding of PI phosphates alone is insufficient to document a stable interaction with membrane lipids. Moreover, as pointed out by Niggli (1993), the labeling of proteins with ^{14}C-PI after SDS-PAGE could be due to oxidative deterioration of polyunsaturated fatty acids and a resulting covalent cross-linking of reactive lipid breakdown products. Application of antioxidants such as 50 mM mercaptoethanol may prevent such artifacts.

It is important to bear in mind that most of the membrane-anchoring proteins are lysed by using detergents and this may lead to structural distortion or an artificial exposure of hydrophobic protein domains. The elution of proteins together with lipid vesicles from gel filtration columns, is indicative of lipid interactions unless the insertion into the hydrophobic layer of lipid molecules has been documented by other techniques. In addition to DSC, hydrophobic labels, which partition into the hydrophobic part of the membrane, are useful tools to document protein–lipid interactions *in situ*. One class of these labels is brominated fatty acids that compete with fatty acids at the hydrophobic-binding sites of spectrin (Isenberg et al., 1981). INA (5-iodonaphtyl-1-azide) is an hydrophobic photoaffinity label used to label membrane-bound α-actinin (Rotman et al., 1982). More recently, the photoactivatable PC analogue [^{3}H]-PTPC-11 was used to document actin-binding protein–lipid interactions (Niggli et al., 1986; Goldmann et al., 1992; Niggli et al., 1994).

The usefulness of this new photolabeling and cross-linking protocol is discussed extensively in a review by Brunner (1993). However, these elegant techniques are not trivial and should be applied with caution. Control experiments are necessary to exclude any unintended labeling of peripheral or soluble proteins by chemical or radiolytic degradation products. Since these probes will also label hydrophobic domains of otherwise electrostatically coupled, hydrophilic proteins, labeling experiments should be carried out at various salt concentrations. Alternatively, scanning calorimetric measurements also help to give a clear answer (see below).

V. FOCAL CONTACT PROTEINS INVOLVED IN CYTOSKELETON–LIPID INTERACTIONS

Specialized adhesion zones of the plasma membrane, which a motile cell uses to establish a connection to a substrate, are called *focal contacts* (for review cf. Burridge et al., 1988; Jockusch and Füchtbauer, 1983; Dunlevy and Couchman, 1993). Contractile stress fibers (Isenberg et al., 1976) terminate in such transmembrane junctions (Samuelsson et al., 1993). Transmission of force requires linkage to the lipid bilayer. Many actin binding proteins anticipated to play a role in anchoring have been localized in focal contacts (see Niggli, this volume), but only a few of them have been shown to interact with lipids directly. Three major actin-binding proteins, talin, vinculin and α-actinin bind to actin and interact with lipids by inserting into the hydrophobic part of membrane leaflets.

A. Talin

A special role for talin, a major protein of focal contacts (Burridge and Connell, 1983a,b; Hock et al., 1989; Beckerle and Yeh, 1990; Burridge et al., 1990), may be in coupling microfilaments to plasma membranes (Isenberg, 1991; Isenberg and Goldmann, 1992). Our laboratory became interested in talin when we applied highly sensitive DSC to actin-binding protein–lipid mixtures. DSC allows the recording of phase transitions of lipids in the absence and presence of proteins. A lipid bilayer will, with rising temperature, undergo these phase transitions by changing its molecular structure from *crystalline* (L_c) in the frozen state to a gel phase ($L_{\beta'}$), the *ripple* phase ($P_{\beta'}$) and to the *fluid* phase (L_α; Figure 3). Phase transitions are endothermic processes, that is, heat is required from the environment. The calorimeter in its reaction chamber compensates for the additional energy by comparing the sample with the reference probe. This energy difference is recorded as a function of temperature. Phase transitions appear as peaks in the profile (thermogram). Integration over the area below the DSC signal yields the difference in enthalpy, ΔH. Assuming a reversible first order phase transition one can calculate the difference in entropy: $\Delta S = \Delta H/T_m$ (T_m = temperature of phase transition). The large change in entropy resulting from the disorder of the C chains in the fluid phase ($T > T_H$) is thus essential for driving the phase transition (Figure 3). T_s^* represents the solidus line, the onset of lipid chain melting, and T_1^* represents the endpoint of chain melting in the fluid phase. Hydrophobic and electrostatic interactions are reflected by a shift of the transition states T_s^* to lower temperatures or T_1^* to higher temperatures, respectively.

DSC measurements have shown that there is a weak but stable hydrophobic interaction of talin with neutral dimyristoylphosphatidylcholine (DMPC) vesicles (Heise et al., 1991). The interaction of talin with DMPC, however, is greatly enhanced in mixed phospholipid bilayers containing negatively charged phospholipids, e.g., dimyristoylphosphatidylglycerol (DMPG) or, (DMPS) (Heise et al., 1991) (see Figure 3). While the weak hydrophobic interaction remains unsaturated up to high protein–lipid molar ratios, the electrostatic interaction reaches steady state conditions rapidly (Figure 3). These are some reasons to assume that talin, like vinculin, (Niggli and Burger, 1987) interacts with lipid membranes in a two-step mechanism: The protein may first be attracted to and fixed at the bilayer surface by electrostatic interactions before it is able to insert its hydrophobic portions into the hydrocarbon bilayer. This mechanism was also proposed for protein 4.1 by Kimelberg and Papahadjopoulos (1971). Covalently bound fatty acids may help to facilitate this insertion (Keenan et al., 1982). If we assume that the major portion of the talin molecule covers the bilayer surface, it would be expected that insertion is maximal at a protein concentration, that leads to steric hindrance at the bilayer surface. This would explain the difference in saturation behavior of hydrophobic and electrostatic interactions (Figure 3).

The selectivity of phospholipid binding of talin and other actin-binding proteins can be demonstrated by Fourier transform infrared spectroscopy (FTIR) (Heise et al., 1991) with one of the lipids being deuterated. Spectra can be taken from DMPC-d_{54} and non-deuterated DMPG in a 1:1 mixture and plotted against temperature (Figure 4). Incorporation of talin almost exclusively affects the DMPG spectra. Moreover, the position and width of the main transition of the noncharged component (i.e., DMPC) are only slightly changed by the protein, while for the

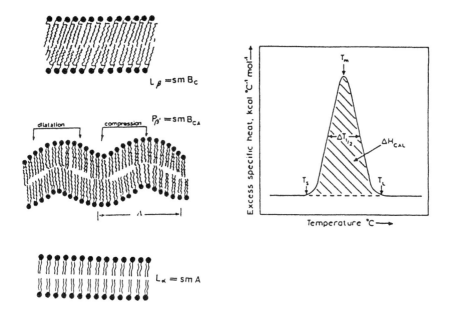

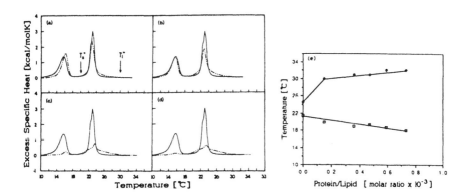

Figure 3. Above: Schematic thermogram of lipid chain melting and various stages of lipid orientation. Below: DSC-measurements of DMPC/DMPG with and without reconstituted talin. Note the shift of phase transitions with increasing protein/lipid ratios (a–d). In (e) the shift of solidus lines (■) and liquidus lines (♦) are a function of protein–lipid molar ratios. Taken from Heise et al., 1991, with permission.

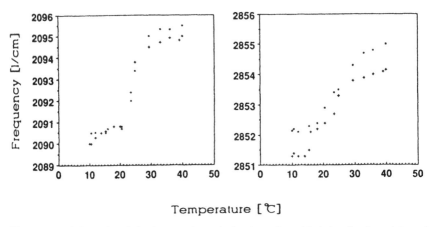

Temperature [℃]

Figure 4. FTIR study of the interaction of platelet talin with 1:1 mixed vesicles of chain-deuterated DMPC (left panel) and DMPG (right panel). The shift in frequency numbers plotted against temperature is significant for DMPG spectra. Taken from Heise et al., 1991, with permission.

charged component T_1 is shifted to a T_1^* of 33°C and the solidus temperature T_s^* is shifted to around 18°C (Figure 4). Both these shifts are in reasonable agreement with the calorimetric data. It is important to stress this point because only the combination of various techniques will allow us to obtain comprehensive answers to our questions. In line with this multimethodological approach is the application of photoactivatable lipid analogues for hydrophobic labeling of lipid-binding, membrane-associated proteins (Niggli et al., 1986; Goldmann et al., 1992; Brunner, 1993). [^{3}H]- PTPC/11, a photoactivatable PC derivative (Figure 5), has been used successfully to label talin and vinculin after reconstitution into lipid vesicles. Since these lipid probes selectively react with protein domains, which insert into the hydrophobic part of lipid membranes, it is concluded that talin and vinculin belong to this category of proteins (Niggli et al., 1986; Goldmann et al., 1992). A labeling efficiency of 0.004–0.01 mol of label/mol of protein suggests that only a minor portion of the talin molecule penetrates into the hydrophobic membrane core (see below).

Taking talin as a representative model protein for the studies of actin-binding protein–lipid interactions, the film balance technique remains to be discussed as a method to study the interaction of this protein with lipid monolayers (Dietrich et al., 1993). Lipid monolayers, spread on an air–water interface, are particularly suited to investigate the insertion behavior, surface pressure induction and pattern formation of proteins in two dimensions (Figure 6). Partitioning of talin was measured in mixed DPPC–DMPG lipid monolayers. For viewing with epifluorescence light microscopy, talin was labeled with NBD (Detmers et al., 1981). For

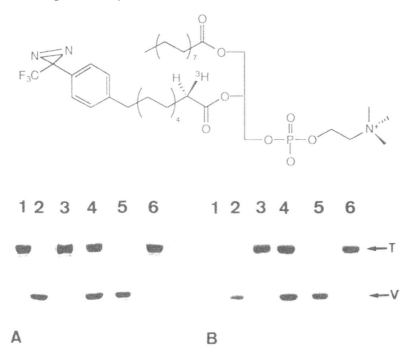

Figure 5. Above: Chemical formula of [^{3}H] PTPC/11, 1-palmitoyl-2-[11-[4-[3-(tri-fluoromethyl)diazirinyl] phenyl][2-^{3}H]undecanoyl]-*sn*-glycero-3-phosphocholine. Below: Hydrophobic photolabeling of talin upon incubation with phosphatidylserine liposomes (a) Coomassie blue stained gradient gel (b) the corresponding autora-diogram. (Lane 1) talin incubated with liposomes without photolysis; (Lane 2) vinculin incubated with liposomes in the presence of 130 mM KCl and 1 mM MgCl$_2$; (Lane 3) talin incubated with liposomes in the presence of 130 mM KCl and 1 mM MgCl$_2$; (Lane) talin and vinculin, incubated with liposomes in the absence of added salt; (Lane 5) vinculin incubated with liposommes in the absence of added salt; (Lane 6) talin incubated with liposomes in the absence of added salt. Arrows on the right indicate the position of talin (T) and vinculin (V). Taken from Goldmann et al., 1992, with permission.

lipid labeling, small amounts (~ 0.1%) of N-(texas-red-sulfonyl-dipalmitoyl-L-α-phospatidylethanolamine) were added to the lipid solution. Since binary lipid mixtures normally show a phase transition inducing partial separation of the two components (Frey et al., 1987), it is expected that dark domains (cf., Figure 6) are rich in components of a lower phase transition pressure (here, DPPC), while the more *fluid* regions are enriched in components with higher transition pressure (here, DMPG or DMPC). For DPPC–DMPG monolayers, which are composed of less charged, *crystalline* regions and negatively charged *fluid* regions, it was demonstrated that talin codistributes with lipids in the negatively charged region (cf., Figure 6 bottom row,

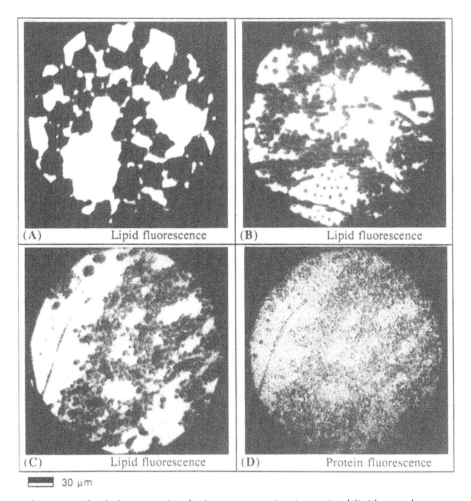

30 μm

Figure 6. Film balance study of talin incorporation into mixed lipid monolayers: DPPC/DMPG at a molar ratio of 6:4. Lipid fluorescence in the absence (a) and presence (b) of talin; lipid fluorescence (c) and protein fluorescence (d) 50 min after injection talin into the subphase. Taken from Dietrich et al., 1993, with permission.

where lipid and protein fluorescence are displayed in the same image). The film balance apparatus, when equipped with high resolution optics and data acquisition, is thus of great potential in supplying data complementary to DSC and FTIR.

The next step is to map the lipid interacting domain of talin. The primary sequence of talin is known (Rees et al., 1990), and it has been speculated from the apparent sequence homology with the membrane and actin-binding proteins 4.1 and ezrin that the lipid-binding site is localized in the smaller (47 kDa) calpain II cleavage product (Fox et al., 1985; Beckerle et al., 1987; Rees et al., 1990). We have recently confirmed (Niggli et al., 1994) that, out of a mixture of the 47-kDa and 190-kDa talin fragments, only the 47-kDa domain carrying the N terminus binds to lipid vesicles. We have additional reasons to believe that the actin-binding site is located in the 190-kDa C-terminal fragment. Hence, it is conceivable that both subunits can redistribute into focal contacts after microinjection (Nuckolls et al., 1990, 1992). So far, only the binding sites for vinculin (Burridge and Mangeat, 1984) have been identified on the tail portion of the talin molecule (Gilmore et al., 1993), and other binding sites, for instance, for actin (Muguruma et al., 1990), α-actinin (Muguruma et al., 1992) and integrin (Horwitz et al., 1986), await their precise characterization.

B. Vinculin

Vinculin was first identified and isolated by Geiger (1979) and soon discovered in a variety of cells and tissues (Geiger et al., 1987; Otto, 1990). The primary sequence of this 117-kDa protein (130 kDa on SDS gels) has been published for chicken (Price et al., 1987; Coutu and Craig, 1988), nematode (Barstead and Waterstone, 1989) and human species (Weller et al., 1990). The organization of the entire human vinculin gene, including its promotor sequence was reported recently (Moiseyeva et al., 1993). For many years vinculin was believed to be an actin-binding protein (Isenberg et al., 1982; Jockusch and Isenberg, 1982; Ruhnau and Wegner, 1988; Westmeyer et al., 1990), and indeed the vinculin sequence contains actin-binding sites that can be blocked by specific antibodies (Westmeyer et al., 1990). Binding domains for actin in talin have been mapped to residues 1–258 (Jones et al., 1989; Gilmore et al., 1992).

Although the molecular shape of vinculin can best be described by the "balloon-on-a-string" model (Eimer et al., 1993), the location of binding domains on the vinculin molecule has not been determined. Vinculin is a typical amphitropic protein (Burn, 1988), in that it exists both as a soluble cytoplasmic protein as well as a membrane-bound protein. Binding to negatively charged phosphlipids (PA, PI, PG) has been reported; however, neutral lipids (PC and PE) do not promote binding (Ito et al., 1983; Niggli et al., 1986). Vinculin was also reconstituted into lipid monolayers (Fringeli et al., 1986; Meyer, 1989). The latter study reports that the dissociation constant for vinculin–phospholipid interaction can vary dramatically (1.2×10^{-6} M–5.3×10^{-10} M) depending on temperature, surface pressure and different lipid composition and ratios. It is not clear if posttranslational modification of vinculin is important for lipid bilayer interactions because a minor fraction of vinculin has myristate (Kellie and Wigglesworth, 1987) and palmitate (Burn and Burger, 1987) covalently attached. The degree of posttranslational lipid modification may be related to the phosphorylation state: The phoshoplipid modified 10-fold increase of vinculin phosphorylation induced by the purified *src*-gene product (Ito et al., 1983) contrasts with a 3-fold lower level of palmitylated vinculin in Rous sarcoma virus-infected fibroblasts (Burn and Burger, 1987). Vinculin is one of the rare examples for which hydrophobic labeling by a lipid analogue (TID) may be applied *in vitro* (Niggli et al., 1986) and *in vivo* (Niggli et al., 1988).

The binding characteristics, which have been determined for two components do not necessarily remain unchanged in a more complex system: In a ternary complex of vinculin/α-actinin (Wachsstock et al., 1987) and phospholipids, photolabeling of α-actinin is markedly suppressed, whereas α-actinin alone (see below) is well labeled (Niggli and Gimona, 1993). On the other hand, ternary complex formation of vinculin–talin (Nuckolls et al., 1990) and phospholipids *in vitro* did not influence the insertion of talin into lipid membranes (Goldmann et al., 1992). Clearly, more *in vivo* labeling studies will be needed to establish whether all the lipid binding capacities reported *in vitro* are manifest in more complex structures like focal contacts.

C. α-Actinin

The biochemistry of α-Actinin (MW ~100 kDa) has been reviewed recently (Blanchard et al., 1989; Vandekerckhove, 1990). Agreement exists that α-actinin (i) is a homodimer with its subunits orientated in an antiparallel fashion (Wallraff et al., 1986; Schleicher et al., 1988), (ii) cross-links actin filaments into a three-dimensional network (Jockusch and Isenberg, 1981; Jockusch and Isenberg, 1982) and (iii) is involved in linking the cytoskeleton to the plasma membrane (Geiger et al., 1980). The influence of α-actin on actin polymerization is still controversial (Muguruma et al., 1992; Colombo et al., 1993) as is its regulation by Ca^{++} in muscle and nonmuscle tissues. Generally, it was believed that the nonmuscle isoforms bind to F actin in a Ca^{++}-sensitive manner whereas binding of muscle α-actinin is Ca^{++} insensitive (Condeelis and Vahay, 1982; Duhaiman and Bamburg, 1984). More recent data, however, (Pacaud and Harricane, 1993) clearly demonstrate that macrophage α-actinin binds to actin independent of regulation by Ca^{++}. Interaction of α-actinin with lipids has been reported by several laboratories (cf. Fritz et al., 1993). In skeletal muscle, a high ratio of PIP-2 (20–30 mol/mol protein) is endogenously bound to α-actinin, whereas smooth muscle α-actinin strongly binds exogenously added PIP-2 (Fukami et al., 1992). α-Actinin-induced gelation of F-actin *in vitro* is greatly enhanced by PIP-2 but not by PIP or PI, as long as the added inositolphosphate is below the critical micelle concentration. Lipid interaction of α-actinin *in vivo* is likely with the finding that in platelets, when physiologically activated, a 30-fold increase in lipid binding was detected in the nondetergent-lysed cytoskeleton (Burn et al., 1985). Furthermore, when immunoprecipitated from prelabeled, activated platelets, α-actinin was found to have PA and DAG bound in a molar ratio of 1:1 (Burn et al., 1985). Consistent with this result, Meyer et al. (1982) reported that out of a total lipid extract from yeast only two lipids, PA and DAG, formed a stable 1:1:1 complex with α-actinin. In summary, α-actinin has potential lipid-binding capacity but as mentioned above, however, it has not been confirmed if this is of particular cell biological importance.

D. MARCKS

The myristoylated alanine-rich C kinase substrate (MARCKS) is a protein belonging to a family of signal-transducing proteins (for review cf. Aderem, 1992). The protein, which is amphitropic in nature, can be isolated in a cytoplasmic and membrane-bound form from the same source (Manenti et al., 1992; 1993). Sequence analysis shows, the protein to have an actual mass of 31 kDa but when myristoylated on the N–terminus the molecular weight on SDS gels is about 70 kDa. Myristoylation is necessary for membrane binding and is thought to occur cotranslationally. The interaction of MARCKS with lipids involves hydrophobic and electrostatic components (Taniguchi and Manenti, 1993), the latter being phosphorylation dependent. Incorporation of negatively charged phosphate groups leads to dissociation from PS-containing lipid bilayers. Irrespective of its phosphorylation state, MARCKS binds to F actin with different affinity (Hartwig et al., 1992). Since MARCKS is regulated by two important chemotactic signals, it serves as a PKC and calcium–calmodulin-regulated transducer during cell stimulation.

VI. ACTIN-BINDING PROTEIN–LIPID INTERACTIONS IN THE LEADING EDGE OF MOVING CELLS

The mechanisms by which a cell moves forward is not completely understood. We still believe that the unidirectional polymerization of actin can be utilized for vectorial force production (Isenberg et al., 1978). This hypothesis received support by the finding that the majority of actin filaments in the extreme outer edge of advancing lamellipodia is polarized with the fast polymerizing end directed towards the growing front (Small et al., 1978). Under physiological conditions the barbed end growth is favored over the pointed end growth, due to the difference in critical concentrations at each end (Wegner and Isenberg, 1983), and polymerization primarily occurs in the desired direction unless the filaments are blocked (e.g., by capping proteins, see below). Theriot and Mitchison (1991) showed that the rate of actin polymerization directly correlates with the advancement of lamellipodia (Cramer and Mitchison, 1993). Hypotheses that developed included the notion that actin alone may be sufficient as a driving force for polymerization once the reaction is nucleated at the membrane interface (Heath and Holifield, 1991a; 1991b; Rinnerthaler et al., 1991; Sheetz et al., 1992).

It is noteworthy that monomeric actin is sequestered by both profilin (see below) and thymosin β_4, the latter a highly potent actin-sequestering protein in higher eukaryotic cells (Safer, 1992). One interesting hypothesis is that release of actin monomers from thymosin β_4 is regulated by the ratio of ATP/ADP in this region, rendering actin–ADP molecules an unfavorable but available source for polymerization (Carlier, 1993; Carlier et al., 1993). An actin-nucleating protein could have a tremendous impact on these events.

A. Talin

From microinjection studies using rhodamine-conjugated actin (Wang, 1984; 1985) and from high-resolution immunofluorescent studies coupled with interference reflection contrast (Izzard and Lochner, 1980; DePasquale and Izzard, 1987; 1991), it was established that stress fiber assembly is coupled with actin polymerization, starting at discrete foci, so-called talin-rich nodes, which represent precursor structures for subsequent developing focal adhesions (Izzard, 1988). Therefore, talin is not only a focal contact protein but also involved in actin rib formation and polymerization events at the leading membrane. Talin is a true nucleating protein for actin polymerization *in vitro*: Talin binds to G–actin (Muguruma et al., 1990; Goldmann and Isenberg, 1991), it overcomes the rate limiting steps in actin assembly by facilitating actin nuclei formation and it enhances actin polymerization by favoring an increase in filament number concentration over filament length (Kaufmann et al., 1991, 1992; Goldmann et al., 1992). Although talin nucleates actin filament growth, it does not restrict assembly of actin monomers at either end, because it is not a capping protein (see below). All these features match the requirements that predict the essentials of pseudopod formation during cell locomotion (Stossel, 1989; Condeelis et al., 1992). In addition, viscoelastic measurements (Ruddies et al., 1993) show that talin induces an increase in actin filament stiffness. Such a reduction in chain dynamics may avoid repulsion between filaments due to undulation forces and thus favor their parallel arrangement (Goldmann et al., 1993).

Further evidence indicates that talin can nucleate actin filament assembly at the lipid interface by (i) driving polymerization and (ii) anchoring the newly formed filaments into the lipid bilayer (Kaufmann et al., 1992; Figure 7). Purified talin, when reconstituted into lipid vesicles, has been directly visualized by video-enhanced microscopy to facilitate polymerization of actin filaments starting at the bilayer surface and proceeding into the surrounding medium. From these experimental findings, a model was presented to explain how talin, with a minimum of binding parameters, functions in (i) driving the assembly of actin filaments, (ii) anchoring the cytoskeleton to the lipid bilayer and (iii) transmitting signals to the extracellular space via integrins (Isenberg and Goldmann, 1992).

B. Ponticulin

Ponticulin is a membrane-spanning glycoprotein that mediates actin binding and nucleation (Wuesthube and Luna, 1987; Wuesthube et al., 1989; Luna et al., 1990; Shariff and Luna, 1990; Chia et al., 1991). The 17-kDa protein is present in the outer membrane of *Dictyostelium* cells and is particularly enriched in cell–cell adhesions and arched actin-rich membrane regions, reminiscent of presumptive stages of pseudopod formation. Consistent with morphological observations, pon-

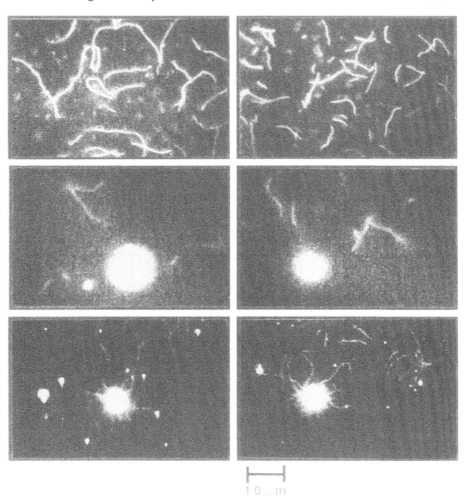

Figure 7. Fluorescent imaging of rhodamine-phalloidin-labeled actin filaments in the absence and presence of talin (top row; left to right). Polymerization of actin in the presence of lipid vesicles alone (middle row) and in the presence of lipid vesicles with reconstituted talin (bottom row). Taken from Kaufmann et al., 1992, with permission.

ticulin, like talin, triggers lateral association of actin filaments with the lipid bilayer, with both ends free for monomer assembly and disassembly (Chia et al., 1993). When ponticulin was added to commercially reconstituted lipid mixtures, it failed to nucleate actin polymerization (cf. talin, see above), but was active when incorporated into native *Dictyostelium* membrane vesicles (Chia et al., 1993). Thus, a specific lipid composition including DAG (Shariff and Luna, 1992) may be needed to effect actin nucleation by ponticulin. Alternatively, augmentation of nucleation may be achieved by dissociation of capping proteins or by activation of other

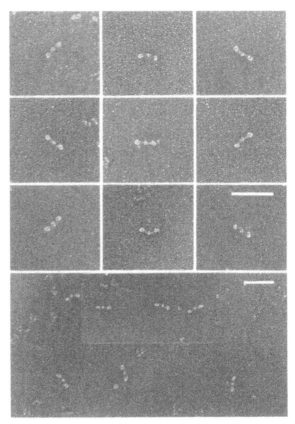

Figure 8. Glycerol-sprayed, rotary metal-shadowed human platelet talin in 50 mM Tris/HCl, 2 mM EDTA, 0.1 mM DTT, pH 8.0, at a final concentration of 0.1–0.3 mg/ml (in 30% glycerol). (**a**) Low magnification overview, and (**b**) gallery of selected talin molecules. Scale bars, 100 nm (a, b) For details see Goldman et al. (1994). Native Talin Is a Dumbbell-shaped Homodimer When it Interacts with Actin. J. Struct. Biol. 112, 3–10. Electron micrographs and montage courtesy of G. Isenberg, M. Häner and U. Aebi.

nucleating proteins. An analogue protein of ponticulin has been identified in the plasma membrane of polymorphonuclear leucocytes, which makes this protein an interesting candidate for triggering cell motility in general (Luna, 1990).

The protein hisactophilin (Scheel et al., 1989) is another protein from *Dictyostelium* involved in signal transduction by acting as a pH sensor. It is unusual that this low molecular mass (13.5 kDa) protein possesses 31 histidine residues out of 118 amino acids. Its structure was found to be similar to interleukin-1β and fibroblast growth factor (Habazettl et al., 1992). Though a lipid interaction of this actin nucleating protein is highly likely, this has yet to be demonstrated.

VII. CAPPING AND SEVERING PROTEINS

A.   Capping Proteins: Cap 32/34; Cap 100; gCap 39

The first actin filament capping protein ever identified was isolated from *Acanthamoeba* (Isenberg et al., 1980). The protein consists of two polypeptides (32 and 34 kDa). It is ubiquitous and was isolated from bovine brain (Kilimann and Isenberg, 1982), *Dictyostelium* (Schleicher et al., 1984) and skeletal muscle (Casella et al., 1986). The skeletal muscle analogue of this protein was later renamed CapZ (Casella et al., 1989; Heiss and Cooper, 1991). Capping proteins bind to the fast-growing end of actin filaments and inhibit polymerization. Since this is the membrane facing end in the outer edge of protruding cells (lamellipodia) capping proteins have been investigated with respect to their potential interaction with phospholipids. Membrane localization, amphitropic behavior or reconstitution into lipid model membranes has not been reported for any of the known capping proteins. However, binding to PIP-2 is common to most capping proteins from various sources and is inhibitory to their function. The heterodimeric capping proteins (Cap 32/34) (Heiss and Cooper, 1991; Haus et al., 1991), Cap-100, a novel capping but nonnucleating protein from *Dictyostelium* (Hofmann et al., 1992), and g-Cap 39, the Ca^{++}-regulated phosphoprotein (Yu et al., 1990; Onoda and Yin, 1993) belong to this group.

B.   Gelsolin

Gelsolin is an actin-binding protein that interacts with actin in several ways (Stossel et al., 1985; Yin, 1988; Stossel, 1990): (i) it binds to actin monomers and stimulates the formation of actin nuclei, (ii) it acts as a capping protein at the barbed ends and (iii) it severs actin filaments in a Ca^{++}-dependent manner. PIP-2, and to a less extent PIP, have been found to inhibit the Ca^{++}-dependent-severing activity of gelsolin specifically (Jamney and Stossel, 1987; Janmey et al., 1987). Moreover, it was shown that PIP-2 micelles dissociate the EGTA-resistent 1:1 gelsolin–actin complex and restore its severing activity (Janmey and Stossel, 1987; Yin et al., 1988). Since EGTA reacts with the Ca^{++}-sensitive actin-binding domain in the C-terminal half of gelsolin (Kwiatkowski et al., 1985; Chaponnier et al., 1986), it follows that the actin-binding site, which is inhibited by PIP-2, is distinguishable from the Ca^{++}-sensitive actin-binding domain. By analyzing proteolytic peptides in respect to their function (Yin et al., 1988) and by deletional mutagenesis, using COS cells and *Escherichia coli* to produce truncated plasma gelsolin after DNA transfection (Kwiatkowski et al., 1989; Way et al., 1989), the following model for the domain structure has emerged. Gelsolin potentially has three actin-binding sites (Yin et al., 1988; Bryan, 1988; Way et al., 1989). Ca^{++}-regulation of the intact gelsolin molecule as well as Ca^{++}-sensitive actin binding involved in nucleation, occurs in the C-terminal half which itself undergoes a Ca^{++}-induced conformational change. The two other actin-binding sites are located in the NH_2-terminal half of the molecule. CT 17 N (residues 1–149) contains a high-affinity binding site for actin monomers and filament ends, whereas CT 28 N (residues 150–406) contains the only PIP-2 inhibited binding site for actin molecules arranged in a filament (Kwiatkowski et al., 1986; Bryan, 1988; Kwiatkowski et al., 1989). Although severing by the NH_2-terminal half-fragment is inhibited by PIP-2 (Janmey et al., 1987), the PIP-2 binding fragment CT 28 N by itself does not have severing activity (Yin et al., 1988). Deletion of 79% from the C-terminal end yields a 160-amino-acid fragment (PG 160) that severs and is PIP-2 regulated. Hence by exclusion, the PIP-2 binding site must reside in a small sequence of about 11 amino acids (residues 150–160) (Bryan, 1988; Yin et al., 1988; Kwiatkowski et al., 1989). The recent published structure of gelsolin–segment-1-actin complex at 2.5 angstroms resolution (McLaughlin et al., 1993) supports this conclusion. This region contains the predicted hydrophobic sequence (Janmey and Stossel, 1987) capable of interacting with the acyl chains of the phospholipid as well as certain basic residues which have been suggested to bind to the negatively charged phosphate groups on PIP-2 (Kwiatkowski et al., 1989).

Like profilin, gelsolin binds to clusters of PIP-2 molecules (Janmey and Stossel, 1987). Studies by the same authors (1989) suggest that the physical state of phosphatidylinositolphosphates within the membrane is important for its inhibitory effects; even at low concentrations in mixed lipid vesicles (e.g., PC containing vesicles) PIP-2 fully inhibits gelsolin, provided that diffusion-limited clustering can occur and that the exposure of hydrophilic phosphatidylphosphoinisitol headgroups is facilitated (cf. talin–lipid interaction above). The latter can efficiently be hindered (i) by trapping into multilamellar lipid sheets, (ii) by the formation of hexagonally packed quasi-crystals of PIPs induced by divalent cations or neomycin and (iii) by specific high-affinity binding to other PIP-2 binding proteins, such as profilin. Quasi-elastic light scattering data (Janmey and Stossel, 1989) also suggest that PI incorporation directly influences the PC diffusion constant and hence affects the phospholipid arrangement and the overall membrane structure. It should, therefore, be intriguing to investigate the long-range effect of PIP-2 within the membrane on proteins that do not bind directly to phosphoinositolphosphates but interact in a hydrophobic or amphiphilic way with other lipid bilayer components.

C. Severin, Fragmin, Villin, Cofilin

Severin (40 kDa), fragmin (42 kDa) and villin (95 kDa) are F-actin fragmenting proteins with extensive sequence homologies to gelsolin (André et al., 1988). In structure and function severin is regarded as a prototype of gelsolin (Yin et al., 1990) since both proteins stem from an ancestral gene, from which gelsolin has been derived by duplication. Regulation by phosphoinositols *in vitro* is common to all three proteins. Severin has at least two PIP-2 binding sites since the activity of the two nonoverlapping severin fragments (domain 1 and 2+3) are strongly inhibited by PIP-2 (Eichinger and Schleicher, 1992). gCap39, which has 49% sequence homology with gelsolin and is not inhibited by PC, PE, PS, PI and IP$_3$ (Yu et al., 1990). In contrast Eichinger and Schleicher (1992) reported that PC and PC/PE vesicles as well as PC and PE/PS vesicles had no or only slightly an inhibitory effect on severing activity; however, vesicles composed of PC and PS surprisingly do have inhibitory effects! The same authors also reported a pH-dependence for severin and lipid interactions. Hence, the severin activity *in vivo* could be modulated by PIP-2, pH and membrane lipid composition.

The brush border protein villin has been reported to partition into the hydrophobic (Glenney and Glenney, 1984) and hydrophilic phases (Conzelman and Mooseker, 1986), depending on whether detergents are included in the isolation protocol. Cap-100 (Hofmann et al., 1992) is highly homologous to villin and appears to be a premature villin-type protein (protovillin) (Hofmann et at., 1993). Since Cap-100 binds to PIP-2, also villin might be regulated by phosphoinositolphosphates. Cofilin is also inhibited by PIP-2 (Yonezawa et al., 1990).

VIII. THE F-ACTIN CROSS-LINKING PROTEINS: SPECTRIN AND FILAMIN

The role of spectrin in red cell cytoskeleton–membrane interactions is reviewed in this Volume by Niggli. Evidence for electrostatic coupling of spectrin to charged phospholipids was obtained by analyzing the penetration into lipid monolayers and by evaluating the shifts of phase transitions and phase boundaries of lipid mixtures in the presence of spectrin (Mombers et al., 1980; Maksymiw et al., 1987). The selective binding of spectrin to negatively charged phospholipids (PS and PG) results from locally clustered positive charges along the folded spectrin polypeptide chain, although the whole protein has a net negative charge. Hydrophobic interactions with PC and PE are probable because hydrophobic ligands have been shown to act as strong quenchers for intrinsic protein fluorescence (Isenberg et al., 1981). *In situ* labeling of erythrocyte spectrin was achieved by application of the hydrophobic label phenylisothiocyanate (Sikorski and Kuczek, 1985). Interaction with lipid bilayers is facilitated by covalently bound palmitate; however, only a small fraction of spectrin was shown to be ^{3}H-palmitoylated (Mariani et al., 1993).

Filamin is a high molecular weight (250 kDA) actin-binding protein, the analogue of which in nonmuscle cells is known as actin binding protein (ABP) with a corresponding molecular weight of 280 kDa. The molecular design and function of filamin are summarized elsewhere (Hartwig and Kwiatkowsky, 1991; Small et al., 1992). In smooth muscle where filamin is an abundant protein, it is arranged in an alternative pattern with vinculin and dystrophin in regions closely aligned beneath the plasma membrane. Initial indications for filamin–lipid interactions were obtained by the work of Furuhashi et al. (1992). These authors reported inhibition of filamin–actin interaction and consequently an inhibition of gelation

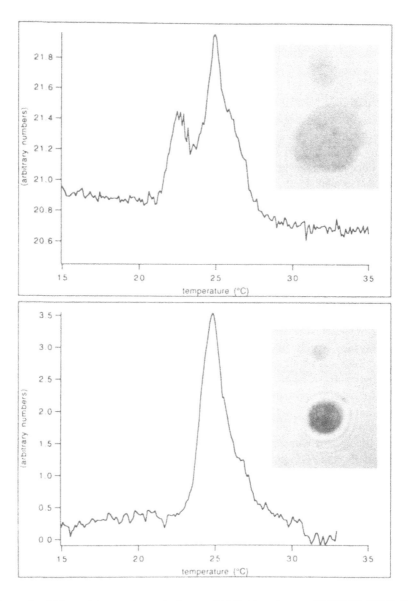

Figure 9. Differential scanning calorimetry (DSC). Top picture: DMPG/DMPC (1:1) in the absence of filamin. Bottom picture: DMPG/DMPC (1:1) in the presence of filamin. Inset: Corresponding phase contrast microscopy of vesicles (-/+) filamin. Taken from Goldmann et al., 1993, with permission.

upon incubation with phosphoinositols in the order PIP-2 > PIP > PI. Three mol PIP-2/mol filamin were sufficient to achieve complete inhibition of filamin cross-linking, which compares to the inhibition of capping proteins (see above). In addition, there is sound evidence from work in this laboratory that filamin not only binds to phosphoinositols in solution but also interacts directly with lipid membranes (Goldmann et al., 1993; Tempel et al., 1994). DSC measurements in combination with hydrophobic labeling and film balance studies show insertion of filamin into the lipid bilayer when reconstituted into vesicles and lipid monolayers. This lipid–filamin interaction is highly charge dependent as hydrophobic interactions diminish with rising salt concentrations. Filamin incubation with lipid vesicles (DMPG/DMPC, 1:1) leads to the formation of condensed rounded-up vesicles with a smooth surface when viewed under the light microscope. This observation agrees with similar effects described for PC and PE vesicles loading from the inside with actin–filamin mixtures (Cortese et al., 1989). The possibility that actin-binding proteins influence shape and deformability of lipid layers is significant for cellular shape changes and pseudopod formation.

IX. ACTIN-SEQUESTERING PROTEINS

A. Profilin

Some of the pioneer work concerning actin-binding proteins/lipid interactions was performed on profilin, an actin monomer sequestering protein (cf. Pollard and Cooper, 1986).

Profilin inhibits actin polymerization by binding to actin monomers and to a lesser extent to actin filament ends (Pollard and Cooper, 1986). Profilin and non-muscle actin monomers form a high affinity complex *in vitro* and *in vivo* (K_d ~10-400 nM). Lassing and Lindberg (1985) were the first to demonstrate that this high affinity complex can be dissociated by phospholipids. After preincubation with various lipids at low salt concentration, the effectiveness in dissociating the profilin-actin complex *in vitro* decreased in the following order of phospholipids: PIP-2 > PA > PS > PI > PG. Under physiological ionic conditions (80 mM KCl and Ca^{++}-concentrations below 10^{-5} M) only PIP-2 effectively dissociates the profil-actin complex. PIP, when reconstituted in lipid bilayers is much less active and PI, as well as the cationic phospholipids (PC and PE), is inactive (Lassing and Lindberg, 1988a; Goldschmidt-Clermont et al., 1990). From these results it was suggested that the onset of actin polymerization which is frequently observed to occur upon cell stimulation could be accounted for by increased production of PIP-2, a subsequent binding to the profilin-actin complex at the membrane interface and the liberation of G-actin (Lassing and Lindberg, 1988b).

More recently, the binding of profilin to PIP-2 has been studied in greater detail (Goldschmidt-Clermont et al., 1990). Large unilamellar vesicles (LUVETs) (Mayer et al., 1986) were used to determine the binding stoichiometry of profilin to PIP-2, which by nuclear magnetic resonance (Van Paridon et al., 1986) has been shown to partition between the two leaflets of such vesicles. According to this report (Goldschmidt-Clermont et al., 1990) profilin binds to reconstituted PIP-2 with a submicromolar affinity ($K_d < 0.1$ μM in pure PIP-2 micelles) and a stoichiometry of 1 : 7 or with a molar ratio of 1 profilin per 5 PIP-2 molecules and a $K_d < 1.0$ μM in vesicles containing PIP-2 and PC in a molar ratio of 1 : 5. Profilin probably does not interact hydrophobically with the acyl chains of the inositol glycerol backbone (Goldschmidt-Clermont et al., 1991) but interacts by a cluster of basic residues close to its COOH-terminal (Goldschmidt-Clermont et al., 1990). Since the actin-binding site is also localized within this sequence region, PIP-2 and actin can compete for profilin binding. However, with equal concentrations of monomeric actin and PIP-2 (at least in platelets; 140–240 mM) and an affinity, which is up to 10-fold higher for profilin than for actin, a large amount of the membrane integrated PIP-2 is probably bound to profilin (Goldschmidt-Clermont et al., 1991). In support of this notion is the demonstration using electronmicroscopy and immunolocalization that in platelets the membrane association of profilin reversibly increases upon activation (Hartwig et al., 1989). Most interestingly, the profilin-PIP-2 interaction is not only involved in actin regulation but also interferes with the cytosolic phospholipase C-catalyzed PIP-2 hydrolysis (Goldschmidt-Clermont et al., 1990) which normally leads to the generation of inositoltriphosphate (IP3) and diacyl-glycerol, a potent activator of protein kinase C. Profilin bound to PIP-2 is a negative regulator of phospholipase C (PLC) activity. Since the cycle of inositol turnover has to be complete there must also exist a mechanism which overcomes this inhibitory effect of profilin upon phospholipase C. Goldschmidt-Clermont et al. (1991) have found that PLC-γ, when phosphorylated by the EGF receptor can effec-tively compete with profilin for PIP-2 binding and thereby switches the system on. Hence, PIP-2 hydrolysis, dissociation of the profilin-PIP-2 complex and profilin-actin complex formation could be the reversal steps following cellular stimulation.

X. MOLECULAR MOTOR PROTEINS

A. The Myosin I Family

Myosin I, which was first identified in Acanthamoeba (Pollard and Korn, 1973), represents a new class of mechanoenzymes necessary for actin-based motility (Korn and Hammer, 1988, 1990; Adams and Pollard, 1989a). Myosin I isolated from Acanthamoeba (Adams and Pollard, 1989b; Miyata et al., 1989; Doberstein and Pollard, 1992; cf. Pollard et al., 1991) and brush border membranes (Hayden et al., 1990) has been shown to interact with membrane lipids. Three independent groups have reported that myosin I binds to salt-treated 'stripped' plasma mem-branes devoid of actin and myosin as well as to pure, negatively charged pho-

pholipid vesicles. In all cases myosin I follows saturation kinetics with an overall binding capacity which exceeds that of actin several fold. Apparent dissociation constants were found to range between 0.3–0.5×10^{-7} M for the binding of Acanthamoeba myosin I to KI stripped plasma membranes (Miyata et al., 1989) and 1.4–3.0×10^{-7} M for the binding of myosin I to NaOH stripped membranes and to phospholipid vesicles with selected lipid compositions (Adams and Pollard, 1989b; Hayden et al., 1990; this reflects the higher affinity of myosin I for lipids than for pure F-actin. No evidence was presented that myosin I binds to neutral phospholipids such as phosphatidylcholine (PC). Instead, myosin I associates with liposomes containing phosphatidylglycerol (PG), phosphatidylserine (PS), phosphatidylinositol (PI) or PIP-2 or mixtures of these anionic lipids with neutral ones at an estimated ratio of 4–5 pmol protein / nmol phospholipid; this is similar to that of other protein-lipid interactions.

Since a high ionic strength is required to solubilize myosin I from membranes (Miyata et al.,1989), it has been questioned if this salt treatment could lead to an artificial exposure of basic sequences which could in turn favor an electrostatic interaction of myosin I with membranes. However, a major fraction of the myosin I, including its associated kinase, is also linked to membranes *in situ* (Kulesza-Lipka et al.,1991; Baines et al., 1992). Interestingly, the kinase, which itself is activated by a phospholipid enhanced autophosphorylation *in vitro* (Brzeska et al., 1990, 1992), is no longer activated by phosphorylation when operating in a membrane-bound form, whereas lipids still stimulate myosin I phosphorylation through this kinase (Kulesza-Lipka et al., 1993). This example convincingly demonstrates that (i) the myosin-I-lipid interaction is not merely electrostatic and (ii) specific protein–lipid interactions exist and these may differ, depending on whether the purified components are mixed in solution, reconstituted into lipid-bilayers or react as constituents of purified plasma membranes.

XI. OTHERS

In the nervous system the vesicle specific phosphoprotein synapsin I is a lipid-and-actin binding protein. Synapsin I acts as a phophorylation-dependent, actin nucleating protein *in vitro* (Bähler and Greengard, 1987; Petrucci and Morrow, 1987; Valtorta et al., 1992; Benfenati et al., 1992). The proline-rich hydrophobic head region of the molecule inserts into the hydrophobic core of lipid membranes and simultaneously reacts with acidic phospholipids (Benfenati et al., 1989 a,b; Südhof et al., 1989; Benfenati, personal communication).

Hydrophobic and electrostatic interactions with phospholipids (PS) have also been measured for caldesmon (Vorotnikov and Gusev, 1990; Czurylo et al., 1993), an actin-binding regulatory protein in smooth muscle contraction. The near future will show whether the cytoskeleton and membrane associated members of the rho-family (small GTP-binding proteins) and the connected activating proteins (e.g. GAP) interfere with membrane lipids (Ridley & Hall, 1992).

Finally, enzymes like 5'-nucleotidase and proteins from the erythrocyte membrane skeleton such as protein 4.1 appear to interact with both actin and membrane components (see Niggli, this volume and previous reviews: Isenberg, 1991; Luna & Hitt, 1992).

XII. CONCLUSIONS

It is surprising how many of the known actin-binding proteins can bind to lipids (see Table I). It also appears likely that some of these actin-binding proteins whose lipid-binding properties have not been investigated the ezrin-radixin-moesin family, ERM; Algrain et al., 1993, zyxin; Sadler et al., 1992, or the membrane associated protein CD43; Yonemura et al., 1993, may have to lipid-binding properties.

Table 1. Actin and Lipid Binding Proteins

Protein	Molecular weight (kDa)	Origin	Lipid
Talin	269	vertebrates	PS;PG;PC
Vinculin	116	vertebrates	PS;PG;PI;PA
α-Actinin	200	higher and lower organisms	PA;DG;PIP-2
MARCKS	31 (70)	vertebrates	PS;PC
Ponticulin	17	*Dictyostelium*	DAG
Cap 32/34	32/34	*Acanthamoeba, Dictyostelium*, brain, skeletal muscle etc.	PIs
Cap 100	100	*Dictyostelium*	PIP-2
gCap 39	39	macrophages	PIP-2
Gelsolin	90	vertebrate cells	PIP-2;PIP
Severin (Fragmin)	42	lower eukaryotic cells	PIP-2;PIP;PC/PS
Villin	95	vertebrates	PIs;acidic phospholipids
Cofilin	15–20	vertebrates	PIs
Spectrin	240 (2x)	erythrocytes	PS;PG;PE
Filamin	250–270	smooth muscle	PS;PG
Profilin	12–15	higher and lower organisms	PIP-2;PIP
Myosin I	110–140	higher and lower organisms	PG;PS;PI;PIP-2
Protein 4.1	78	erythrocytes	PS;PIP-2
Synapsin	80/86	brain	acidic phospholipids
Caldesmon	70–80	smooth muscle, non-muscle cells	PS
5'-Nucleotidase	71	eurkaryotic cells	GPI

Note: *Abbreviations*: PC, phosphatidylcholine; PG, phosphatidylglycerol; PS, phosphatidylserine; PE, phosphatidylethanolamine; PA, palmitic acid; DAG, diacylglycerol; PI, phosphatidylinositol; PIP-2, phosphatidylinositol-4,5-biphosphate; PPI, phosphatidylphosphoinositol; IP3, inositoltriphosphate; GPI, glycosylphosphoinositol.

Claims have been made that most of the lipid binding capacity is artefactual and of no biological relevance. As we have tried to point out, lipid interactions, however, are not as trivial as they appear at first glance. Clearly, investigation of lipid-binding in solution is of limited interest since one has to expect that a charged protein with hydrophobic pockets will readily interact with polarized lipid molecules.

On the other hand, one has to realize that a group of actin-binding proteins primarily binds phosphoinositolphosphates and is regulated by these lipids in its functions. Hence, there exists *specificity*. We have pointed out the possibility that lipid interactions can be modulated by the presence of additional proteins. This stresses the factor of *competition* and *selective binding*. We know that even a reconstituted bilayer system may not be sufficient enough to mirror the *in vivo* situation in biological membranes. Many other factors including physical factors such as diffusion rates, curvature, lateral pressure and microviscosities will certainly be important in determining the biological role of actin-binding protein/lipid interactions inside a cell.

ACKNOWLEDGMENTS

We would like to thank Dr. V. Niggli for stimulating discussions, Ms. Liz Nicholson (MA) and Dr. S. Kaufmann for careful reading of this manuscript. This work was supported by the following grants: DFG 25/7-2 and SFB 266/C5, go 598/3-1 and Nato CRG 940 666 to W.H.G.

REFERENCES

Adams, R.J. & Pollard, T.D. (1989b). Binding of myosin I to membrane lipids. Nature 340, 565–568.

Adams, R.J. & Pollard, T.D. (1989a). Membrane-bound myosin I provides new mechanisms in cell motility. Cell. Motil. Cytoskel. 14, 178–182.

Aderem, A. (1992). The MARCKS brothers: A family of protein kinase C substrates. Cell 71, 713–716.

Algrain, M., Turunen, O., Vaheri, A., Louvard, D., & Arpin, M. (1993). Ezrin contains cytoskeleton and membrane binding domains accounting for its proposed role as a membrane-cytoskeletal linker. J. Cell Biol. 120, 129–139.

Anderson, R.A. & Marchesi, V.T. (1985). Regulation of the association of membrane skeletal protein 4.1 with glycophorin by a polyphosphoinositide. Nature 318, 295–298.

André, E., Lottspeich, F., Schleicher, M., & Noegel, A. (1988). Severin, gelsolin and villin share a homologous sequence in regions presumed to contain F-actin severing domains. J. Biol. Chem. 263, 722–727.

Bähler, M. & Greengard, P. (1987). Synapsin I bundles F-actin in a phosphorylation-dependent manner. Nature 326, 704–707.

Baines, I.C., Brzeska, H., & Korn, E.D. (1992). Differential localization of *Acanthamoeba* myosin I isoforms. J. Cell Biol. 119, 1193–1203.

Barstead, R.J. & Waterstone, R.H. (1989). The basal component of the nematode dense-body is vinculin. J. Biol. Chem. 264, 10177–10185.

Beckerle, M.C., Burridge, K., Demartino, G.N., & Croall, D.E. (1987). Colocalization of calcium-dependent protease II and one of its substrates at sites of cell adhesions. Cell 51, 569–577.

Beckerle, M.C. & Yeh, R.K. (1990). Talin: Role at sites of cell-substratum adhesion. Cell Motil. Cytoskel. 16, 7–13.

Benfenati, F., Bähler, M., Jahn, R., & Greengard, P. (1989a). Interactions of synapsin I with small synaptic vesicles: Distinct sites in synapsin I bind to vesicle phospholipids and vesicle proteins. J. Cell Biol. 108, 1863–1872.

Benfenati, F., Greengard, P., Brunner, J., & Bähler, M. (1989b). Electrostatic and hydrophobic interactions of synapsin I and synapsin I fragments with phospholipid bilayers. J. Cell Biol. 108, 1851–1862.

Benfenati, F., Valtorta, F., Chieregatti, E., & Greengard, P. (1992). Interaction of free and synaptic vesicle-bound synapsin I with F-actin. Neuron 8, 377–386.

Bennett, V. & Stenbuck, P.J. (1979). Identification and partial purification of ankyrin, the high affinity membrane attachment site for human erythrocyte spectrin. J. Biol. Chem. 254, 2533–2541.

Blanchard, A., Ohanian, V., & Critchley, D.R. (1989). The structure and function of alpha-actinin. J. Muscle. Res. Cell Motil. 10, 280–289.

Brunner, J. (1993). New photolabeling and crosslinking methods. Ann. Rev. Biochem. 62, 483–514.

Bryan, J. (1988). Gelsolin has three actin binding sites. J. Cell Biol. 106, 1553–1562.

Brzeska, H., Lynch, T.J., & Korn, E.D. (1990). Ancanthamoeba myosin I heavy chain kinase is activated by phosphatidylserine-enhanced phosphorylation. J. Biol. Chem. 265, 3591–3594.

Brzeska, H., Lynch, T.J., Martin, B., Corigliano-Murphy, A., & Korn, E.D. (1992). Substrate specificity of Acanthamoeba myosin I heavy chain kinase as determined by synthetic peptides. J. Biol. Chem. 265, 16138–16144.

Burn, P. (1988). Amphitropic proteins: A new class of membrane proteins. Trends in Biochem. 13, 79–84.

Burn, P. & Burger, M.M. (1987). The cytoskeletal protein vinculin contains transformation sensitive, covalently bound lipid. Science 235, 476–479.

Burn, P., Rotman, A., Meyer, R.K., & Burger, M.M. (1985). Diacylglycerol in large alpha-actinin/actin complexes and in the cytoskeleton of activated platelets. Nature 314, 469–472.

Burridge, K. & Connell, L. (1983a). A new protein of adhesion plaques and ruffling membranes. J. Cell. Biol. 97, 359–367.

Burridge, K. & Connell, L. (1983b). Talin: A cytoskeletal component concentrated in adhesion plaques and other sites of actin-membrane interaction. Cell Motil. 3, 405–417.

Burridge, K., Faith, K., Kelly, T., Nuckolls, G., & Turner, C. (1988). Focal contacts: Transmembrane links between the extracellular matrix and the cytoskeleton. Ann. Rev. Cell Biol. 4, 487–525.

Burridge, K. & Mangeat, P. (1984). An interaction between vinculin and talin. Nature 308, 744–746.

Burridge, K., Nuckolls, G., Otey, C., Pavalko, F., Simon, K.O., & Turner, C. (1990). Actin-membrane interaction in focal adhesions. Cell Diff. Dev. 32, 337–342.

Carlier, M.F. (1993). Dynamic actin. Curr. Biol. 3, 321–323.

Carlier, M.F., Jean, C., Rieger, K.J., Lenfant, M., & Pantaloni, D. (1993). Modulation of the interaction between G-actin and thymosin β_4 by the ATP/ADP ratio: Possible implication in the regulation of actin dynamics. Proc. Natl. Acad. Sci. USA 90, 5034–5038.

Casella, J.F., Casella, S.J., Hollands, J.A., Caldwell, J.E., & Cooper, J.A. (1989). Isolation and characterization of cDNA encoding the alpha subunit of CapZ (32/36) an actin capping protein from the Z-line of skeletal muscle. Proc. Natl. Acad. Sci. USA 86, 5800–5804.

Casella, J.F., Maack, D.J., & Lin, S. (1986). Purification and initial characterization of a protein from a skeletal muscle that caps the barbed ends of actin filaments. J. Biol. Chem. 261, 10915–10921.

Chaponnier, C., Janmey, P.A., & Yin, H.L. (1986). The actin filament severing domain of plasma gelsolin. J. Cell Biol. 103, 1473–1481.

Chia, C.P., Hitt, A.L., & Luna, E.J. (1991). Direct binding of F-actin to ponticulin, an integral plasma membrane glycoprotein. Cell Motil. Cytoskel. 18, 164–179.

Chia, C.P., Shariff, A., Savage, S.A., & Luna, E.J. (1993) The integral membrane protein, ponticulin, acts as a monomer in nucleating actin assembly. J. Cell Biol. 120, 909–922.

Cohen, C.M. & Foley, S.F. (1982). The role of band 4.1 in the association of actin with erythrocyte membranes. Biochim. Biophys. Acta 688, 691–701.

Colombo, R., Dalle Donne, I., & Milzani, A. (1993). Alpha-actinin increases actin filament end concentration by inhibiting annealing. J. Mol. Biol. 230, 1151–1158.

Condeelis, J. (1992). Are all pseudopods created equal?. Cell Motil. Cytoskel. 22, 1–6.

Condeelis, J.S. & Vahey, M. (1982). A calcium and pH regulated protein from *Dictyostelium discoideum* that crosslinks actin filaments. J. Cell Biol. 94, 466–471.

Conzelman, K.A. & Mooseker, M.S. (1986). Re-evaluation of the hydrophobic nature of the 110 kD calmodulin-, actin-, and membrane binding protein of the intestinal microvillus. J. Cell Biochem. 30, 271–279.

Cortese, J.D., Schwab III, B., Frieden, C., & Elson, E. L. (1989). Actin polymerization induces a shape change in actin-containing vesicles. Proc. Natl. Acad. Sci USA 86, 5773–5777.

Coutu, M.D. & Craig, S. (1988). cDNA-derived sequence of chicken embryo vinculin. Proc. Natl. Acad. Sci. USA 85, 8535–8539.

Cramer, L. & Mitchison, T.J. (1993). Moving and stationary actin filaments are involved in spreading of postmitotic PtK2 cells. J. Cell Biol. 122, 833–843.

Czurylo, E.A., Zobrowski, J., & Dabrowska, R. (1993). Interaction of caldesmon with phospholipids. Biochem. J. 291, 403–408.

DePasquale, J.A. & Izzard, C.S. (1987). Evidence for an actin-containing cytoplasmic precursor of the focal contact and the timing of incorporation of vinculin at the focal contact. J. Cell Biol. 105, 2803–2810.

DePasquale, J.A. & Izzard, C.S. (1991). Accumulation of talin in nodes at the edge of lamellipodium and separate incorporation into adhesion plaques at focal contacts in fibroblasts. J. Cell Biol. 113, 1351–1359.

Detmers, P., Weber, A., Elzinga, M., & Stephens, R.E. (1981). 7-chloro-4-nitrobenzeno-2-oxa-1,3-diazole actin as a probe for actin polymerization. J. Biol. Chem. 256, 99–104.

Dietrich, D., Goldmann, W.H., Sackmann, E., & Isenberg, G. (1993). Interaction of NBD-talin with lipid monolayers: A film balance study. FEBS Lett. 324, 37–40.

Dill, K. A. (1990). Dominant forces in protein folding. Biochemistry 29, 7133–7155.

Doberstein, S.K. & Pollard, T.D. (1992). Localization and specificity of the phospholipid and actin binding sites on the tail of *Acanthamoeba* myosin I$_C$. J. Cell Biol. 117, 1241–1249.

Duhaiman, A.S. & Bamburg, J.R. (1984). Isolation of brain alpha-actinin. Its characterization and a comparison of its properties with those of muscle alpha-actinins. Biochemistry 23, 1600–1608.

Dunlevy, J.R. & Couchman, J.R. (1993). Controlled induction of focal adhesion disassembly and migration in primary fibroblasts. J. Cell Sci. 105, 489–500.

Egger, M., Ohneoorge, F., Weisenhorn, A., Heyn, S.P., Drake, B., Prater, C.B., Gould, S.A.C., Gaub, H.E., & Hansma, P. (1990). Wet lipid-protein membranes at submolecular resolution by atomic force microscopy. J. Struct. Biol. 103, 89–99.

Eichinger, L. & Schleicher, M. (1992). Characterization of actin- and lipid binding domains in severin, a Ca^{++}-dependent F-actin fragmenting protein. Biochemistry 31, 4779–4787.

Eimer, W., Niermann, M., Eppe, M.A., & Jockusch, B.M. (1993). Molecular shape of vinculin in aqueous solution. J. Mol. Biol. 229, 146–152.

Fox, J.E.B., Goll, D.E., Reynolds, C.C., & Philips, D.R. (1985). Identifications of two proteins (actin-binding-protein and P$_{235}$) that are hydrolized by endogenous Ca^{++}-dependent protease during platelet aggregation. J. Biol. Chem. 260, 1060–1066.

Frey, W., Schneider, J., Ringsdorf, H., & Sackmann, E. (1987). Preparation, microstructure and thermodynamic properties of homogeneous and heterogeneous compound monolayers of polymerized and monomeric surfactants on the air/water interface and on solid substrates. Macromolecules 20, 1312–1321.

Fringeli, U.P., Leutert, P., Thurnhofer, H., Fringeli, M., & Burger, M.M. (1986). Structure-activity relationship in vinculin: An IR/attenuated total reflection spectroscopic and film balance study. Proc. Natl. Acad. Sci. USA 83, 1315–1319.

Fritz, M., Zimmermann, R.M., Bärmann, M., & H:E. Gaub (1993). Actin binding to lipid-inserted α-actinin. Biophys. J. 65, 1–8.

Fukami, K., Furuhashi, K., Inagaki, M., Endo, T., Hatano, S., & Takenawa. T. (1992). Requirement of phosphatidylinositol 4,5-bisphosphate for alpha-actinin function. Nature 359, 150–152.

Furuhashi, K., Inagaki, M., Hatano, S., Fukami, K., & Takenawa, T. (1992). Inositol phospholipid-induced suppression of F-actin-gelating activity of smooth muscle filamin. Biochem. Biophys. Res. Comm. 184, 1261–1265.

Geiger, B. (1979). A 130K protein from chicken gizzard: Its localization at the termini of microfilament bundles in cultured chicken cells. Cell 18, 193–197.

Geiger, B., Tokuyasu, K.T., Dutton, A.H., & Singer, S.J. (1980). Vinculin, an intracellular protein localized at specialized sites where microfilament bundles terminate at cell membranes. Proc. Natl. Acad. Sci. USA 77, 4127–4131.

Geiger, B., Volk, T., Volberg, T., & Bendori, R. (1987). Molecular interactions in adherens-type contacts. J. Cell Sci. (Suppl) 8, 251–272.

Gilmore, A.P., Jackson, P., Waites, G.T., & Critchley, D.R. (1992). Further characterization of the talin binding site in the cytoskeletal protein vinculin. J. Cell Sci. 103, 719–731.

Gilmore, A.P., Wood, C., Ohanian, V., Jackson, P., Patel, B., Rees, D.J.G., Hynes, R.O., & Critchley, D.R. (1993). The cytoskeletal protein talin contains at least two distinct vinculin binding domains. J. Cell Biol. 122, 337–347.

Glenney, J.R. & Glenney, P. (1984). The microvillus 110K cytoskeletal protein is an integral membrane protein. Cell 37, 743–751.

Goldmann, W.H. & Isenberg, G. (1991). Kinetic determination of talin-actin binding. Biochem. Biophys. Res. Comm. 178, 718–723.

Goldmann, W.H., Käs, J., Sackmann, E., & Isenberg, G. (1993). Direct visualization of lipid vesicle changes on addition of filamin. Bioch. Soc. Trans. 21, 133S.

Goldmann, W.H., Niggli, V., Kaufmann, S., & Isenberg, G. (1992). Probing actin and liposome interaction of talin and talin-vinculin complexes: A kinetic, thermodynamic and lipid labelling study. Biochemistry 31, 7665–7671.

Goldmann, W.H., Käs, J., & Isenberg, G. (1993b). Talin decreases the bending elasticity of actin filaments. Biochem. Soc. Trans. 22, 46S.

Goldmann, W.H., Bremer, A., Häner, M., Aebi, U., & Isenberg, G. (1994). Native talin is a dumbbell-shaped homodimer when it interacts with actin. J. Struct. Biol. 112, 3–10.

Goldschmidt-Clermont, P.J., Machesky, L.M., Baldassare, J.J., & Pollard, T.D. (1990). The actin binding protein profilin binds to PIP-2 and inhibits its hydrolysis by phospholipase C. Science 247, 1575–1578.

Goldschmidt-Clermont, P.J., Kim, J.W., Machesky, L.M., Rhee, S.G., & Pollard, T.D. (1991). Regulation of phospholipase C-γ1 by profilin and tyrosine phosphorylation. Science 251, 1231–1233.

Habazettl, J., Gondol, D., Wiltscheck, R., Otlewski, J., Schleicher, M., & Holak, T.A. (1992). Structure of histactophilin is similar to interleukin-1β and fibroblast growth factor. Nature 359, 855–858.

Hartwig, J.H., Chambers, K.A., Hopcia, K.L., & Kwiatkowski, D.J. (1989). Association of profilin with filament-free regions of human leukocyte and platelet membranes and reversible membrane binding during platelet activation. J. Cell Biol. 109, 1571–1579.

Hartwig, J.H., & Kwiatkowski, D.J. (1991). Actin-binding proteins. Curr. Opinion Cell Biol. 3, 87–97.

Hartwig, J.H., Thelen, M., Rosen, A., Janmey, P.A., Nairn, A.C., & Aderem, A. (1992). MARCKS is an actin filament crosslinking protein regulated by protein kinase C and calcium-calmodulin. Nature 356, 618–622.

Haus, U., Hartmann, H., Trommler, P., Noegel, A., & Schleicher, M. (1991). F-actin capping by cap 32/34 requires heterodimeric conformation and can be inhibited with PIP-2. Biochem. Biophys. Res. Comm. 181, 833–839.

Hayden, S.M., Wolenski, J.S., & Mooseker, M.S. (1990). Binding of brush border myosin I to phosphlipid vesicles. J. Cell Biol. 111, 443–451.

Heath, J.P. & Holifield, B.F. (1991a). Actin alone in lamellipodia. Nature 352, 107–108.

Heath, J.P. & Holifield, B.F. (1991b). Cell locomotion: New research tests old ideas on membrane and cytoskeletal flow. Cell Motil. Cytoskel. 18, 245–257.

Heise, H., Bayerl, T., Isenberg, G., & Sackmann, E. (1991). Human platelet P-235, a talin like actin binding protein, binds selectively to mixed lipid bilayers. Biochim. Biophys. Acta 1061, 121–131.

Heiss, S.G. & Cooper, J.A. (1991). Regulation of capZ, an actin capping protein of chicken muscle, by anionic phospholipids. Biochemistry 30, 8573–8578.

Heyn, S.P., Tillmann, R.W., Egger, M., & Gaub, H.E. (1991). A miniaturized computer controlled micro fluorescence film balance for protein-lipid monolayers. J. Biochem. Biophys. Meth. 22, 145–158.

Hock, R.S., Sanger, J.M., & Sanger, J.W. (1989). Talin dynamics in living microinjected non-muscle cells. Cell Motil. Cytoskel. 14, 271–287.

Hofmann, A., Eichinger, L., André, E., Rieger, D., & Schleicher, M. (1992). Cap100, a novel phosphatidylinositol 4,5-bisphosphate-regulated protein that caps actin filaments but does not nucleate actin assembly. Cell Motil. Cytoskel. 23, 133–144.

Hofmann, A., Noegel, A.A., Bomblies, L., Lottspeich, F., & Schleicher, M. (1993). The 100 kDa F-actin capping protein of Dictyostelium amoebae is a villin prototype('protovillin') FEBS Lett. 328, 71–76.

Horwitz, A., Duggan, K., Buck, C., Beckerle, M.C., & Burridge, K. (1986). Interaction of plasma membrane fibronectin receptor with talin—a transmembrane linkage. Nature 320, 531–533.

Hörber, J.K.H., Lang, C.A., Hänsch, T. W., Heckl, W.M., & Möhwald, H. (1988). Scanning tunneling microcopy of lipid films and embedded biomolecules. Chem. Phys. Lett. 145, 151–158.

Hui, S.W. & Huang, C. (1986). X-ray diffraction evidence for fully interdigitated bilayers of 1-stearoyl-phosphatidylcholine. Biochemistry 25, 1330–1335.

Isenberg, G. (1991). Actin binding proteins - lipid interactions. J. Muscle Res. Cell Mot. 12, 136–144.

Isenberg, G., Aebi, U., & Pollard, T.D. (1980). An actin-binding protein from Acanthamoeba regulates actin filament polymerization and interactions. Nature 288, 455–459.

Isenberg, G. & Goldmann, W.H. (1992). Actin-membrane coupling: A role for talin. J. Muscle Res. Cell Mot. 13, 587–589.

Isenberg, G., Hülsmann, N., Rathke, P.C., Franke, W.W., & Wohlfarth-Bottermann, K.E. (1976). Cytoplasmic actomyosin fibrils in tissue culture cells: Direct proof of contractility by laser microbeam dissection. Cell and Tissue Res. 166, 427–443.

Isenberg, G., Small, J.V., & Kreutzberg, G.W. (1978). Correlation between actin polymerization and surface receptor segregation in neuroblastoma cells treated with concanavalin A. J. Neurocytol. 7, 649–661.

Isenberg, H., Kenna, J.G., Green, N.M., & Gratzer, W.B. (1981). Binding of hydrophobic ligands to spectrin. FEBS Lett. 129, 109–112.

Ito, S., Werth, D.K., Richert, N.D., & Pastan, I. (1983). Vinculin phosphorylation by the *src*-kinase. J. Biol. Chem. 258, 14626–14631.

Izzard, C.S. (1988). A precursor of the focal contact in cultured fibroblasts. Cell Motil. Cytoskel. 10, 137–142.

Izzard, C.S. & Lochner, L.R. (1980). Formation of cell-to-substrate contacts during fibroblast motility: An interference reflection study. J. Cell Sci. 42, 81–85.

Janmey, P.A. & Stossel, T.P. (1987). Modulation of gelsolin function by phosphatidylinositol 4,5-biphosphate. Nature 325, 362–364.

Janmey, P.A. & Stossel, T.P. (1989). Gelsolin-polyphosphoinositide interaction. J. Biol. Chem. 264, 4825–4831.

Janmey, P.A., Iida, K., Yin, H.L., & Stossel, T.P. (1987). Polyphosphoinositide micelles and polyphosphoinositide-containing vesicles dissociate endogenous gelsolin-actin complexes and promote actin assembly form the fast growing end of actin filaments blocked by gelsolin. J. Biol. Chem. 262, 12228–12236.

Jockusch, B.M. & Füchtbauer, A. (1983). Organization and function of structural elements in focal contacts of tissue culture cells. Cell Motil. Cytoskel. 3, 391–397.

Jockusch, B.M. & Isenberg, G. (1981). Interaction of alpha-actinin and vinculin with actin: Opposite effects on filament network formation. Proc. Natl. Acad. Sci. USA 78, 3005–3009.

Jockusch, B.M. & Isenberg, G. (1982). Vinculin and alpha-actinin interaction with actin and effect on microfilament network formation. Cold Spring Harbor Symp. Quant. Biol. 46, 613–623.

Jones, P., Jackson, P., Price, G.J., Patel, B., Ohanion, V., Lear, A.L., & Critchley, D.R. (1989). Identification of a talin binding site in the cytoskeletal protein vinculin. J. Cell Biol. 109, 2917–2927.

Kaufmann, S., Pieckenbrock, T., Goldmann, W.H., Bärmann, M., & Isenberg, G. (1991). Talin binds to actin and promotes filament nucleation. FEBS Lett. 284, 187–191.

Kaufmann, S., Käs, J., Goldmann, W.H., Sackmann, E., & Isenberg, G. (1992). Talin anchors and nucleates actin filaments at lipid membranes: A direct demonstration. FEBS Lett. 314, 203–205.

Keenan, Th., Heid, H.W., Stadler, J., Jarasch, E.D., & Franke, W.W. (1982). Tight attachment of fatty acids to proteins associated with milk lipid globule membrane. Eur. J. Cell Biol. 26, 270–276.

Kellie, S. & Wigglesworth, N.M. (1987). The cytoskeletal protein vinculin is acylated by myristic acid. FEBS Lett. 213, 428–432.

Kilimann, M.W. & Isenberg, G. (1982). Actin filament capping protein from bovine brain. EMBO J. 1, 889–894.

Kimelberg, H. & Papahadjopoulos (1971). Phoslipid-protein interactions: Membrane permeability correlated with monolayer penetration. Biochim. Biophys. Acta 233, 805–809.

Korn, E.D. & Hammer, J.A. (1988). Myosins of nonmuscle cells. Ann. Rev. Biophys. Chem. 17, 23–45.

Korn, E.D. & Hammer, J.A. (1990). Myosin I. Curr. Opin. Cell Biol. 2, 57–61.

Kulesza-Lipka, D., Baines, I.C., Brzeska, H., & Korn, E.D. (1991). Immunolocalization of myosin I heavy chain kinase in Acanthamoeba castellanii and binding of purified kinase to isolated plasma membranes. J. Cell Biol. 115, 109–119.

Kulesza-Lipka, D., Brzeska, H., Baines, I.C., & Korn, E.D. (1993). Auto-phosphorylation-independent activation of Acanthamoeba myosin I heavy chain kinase by plasma membranes. J. Biol. Chem. 268, 17995–18001.

Kwiatkowski, D.J., Janmey, P.A., Mole, J.E., & Yin, H.L. (1985). Isolation and properties of two actin-binding domains in gelsolin. J. Biol. Chem. 260, 15232–15238.

Kwiatkowski, D., Janmey, P.A., & Yin., H.L. (1989). Identification of critical functional and regulatory domains in gelsolin. J. Cell Biol. 108, 1717–1726.

Kwiatkowski, D.J., Stossel, T.P., Orkin, S.H., Mole, J.E., Colten, H.R., & Yin, H.L. (1986). Plasma and cytoplasmic gelsolins are encoded by a single gene and contain a duplicated actin-binding domain. Nature 323, 455–458.

Lassing, I. & Lindberg, U. (1985). Specific interaction between phosphatidylinositol 4,5-bisphosphate and profilactin. Nature 314, 472–474.

Lassing, I. & Lindberg, U. (1988a). Evidence that the phosphatidylinositol cycle is linked to cell motility. Exp. Cell Res. 174, 1–15.

Lassing, I. & Lindberg, U. (1988b). Specificity of the interaction between phosphatidylinositol 4,5-biphosphate and the profilin:actin complex. J. Cell. Biochem. 37, 255–267.

Lazarides, E. & Moon, R.T. (1984). Assembly and topogenesis of the spectrin-based membrane skeleton in erythroid development. Cell 37, 354–356.

Luna, E.J. & Hitt, A.L. (1992). Cytoskeleton-plasma membrane interactions. Science 258, 955–964.

Luna, E.J., Wuesthube, L.J., Chia, C.P., Shariff, A., Hitt, A.L., & Ingalls, H.N. (1990) Ponticulin, a developmentally-regulated plasma membrane glycoprotein, mediates actin binding and nucleation. Dev. Genetics 11, 354–361.

Maksymiw, R., Sui, S., Gaub, H. & Sackmann, E. (1987). Electrostatic coupling of spectrin dimers to phosphatidylserine containing lipid lamellae. Biochemistry 26, 2983–2990.

Manenti, S., Sorokine, O., Van Dorsselaer, A., & Taniguchi, H. (1992). Affinity purification and characterization of myristoylated alanine-rich protein kinase C substrate (MARCKS) from bovine brain. J. Biol. Chem. 267, 22310–22315.

Manenti, S., Sorokine, O., Van Dorsselaer, A., & Taniguchi, H. (1993). Isolation of the non-myristoylated form of a major substrate of protein kinase C (MARCKS) from bovine brain. J. Biol. Chem. 268, 6878–6881.

Mariani, M., Maretzki, D., & Lutz, H.U. (1993). A tightly membrane-associated subpopulation of spectrin is ^{3}H-palmitoylated. J. Biol. Chem. 268, 12996–13001.

Marsh, D. (1990). Lipid-protein interactions in membranes. FEBS Lett. 268, 371–375.

Mason, J.T., Huang, C., & Biltonen, R.L. (1981). Calorimetric investigations of saturated mixed-chain phosphatidylcholine bilayer dispersions. Biochemistry 20, 6086–6092.

Mayer, L.D., Hope, M.J., & Cullis, P.R. (1986). Vesicles of variable sizes produced by a rapid extrusion procedure. Biochim. Biophys. Acta 858, 161–168.

McLaughlin, P.J., Gooch, J.T., Mannherz, H.G., & Weeds, A.G. (1993). Structure of gelsolin segment 1-actin complex and the mechanism of filament severing. Nature 364, 685–692.

Meyer, R.K. (1989). Vinculin-lipid monolayer interactions: A model for focal contact formation. Eur. J. Cell Biol. 50, 491–499.

Meyer, R.K., Schindler, H., & Burger, M.M. (1982). Alpha-actinin interacts specifically with model membranes containing glycerides and fatty acids. Proc. Natl. Acad. Sci. USA 79, 4280–4284.

Miyata, H., Bowers, B., & Korn, E.D. (1989). Plasma membrane association of *Acanthamoeba* myosin I. J. Cell Biol. 109, 1519–1528.

Möhwald, H. (1990). Phospholipid and phospholipid-protein monolayers at the air/water interface. Ann. Rev. Phys. Chem. 41, 441–476.

Moiseyeva, E.P., Weller, P.A., Zhidkova, N.I., Corben, E.B., Patel, B., Jasinska, I., Koteliansky, V.E., & Critchley, D.R. (1993). Organization of the human gene encoding the cytoskeletal protein vinculin and the sequence of the vinculin promotor. J. Biol. Chem. 268, 4318–4325.

Mombers, C., De Gier, J., Demel, R.A., & Van Deenen, L.L.M. (1980). Spectrin phospholipid interaction - A monolayer study. Biochim. Biophys. Acta 603, 52–62.

Muguruma, M., Matsumura, S., & Fukazawa, T. (1990). Direct interactions between talin and actin. Biochem. Biophys. Res. Comm. 171, 1217–1223.

Muguruma, M., Matsumura, S., & Fukazawa, T. (1992). Augmentation of alpha-actinin-induced gelation of actin by talin. J. Biol. Chem. 267, 5621–5624.

New, R.R.C. (Ed.) Liposomes: A Practical Approach. (1990). IRL Press, Oxford University. UK.

Niggli, V. (1993). Lipid-cytoskeleton interactions. Nature 361, 214.

Niggli, V. & Burger, M.M. (1987). Interaction of the cytoskeleton with the plasma membrane. J. Membrane Biol. 100, 97–121.

Niggli, V., Dimitrov, D.P., Brunner, J., & Burger, M.M. (1986). Interaction of the cytoskeletal component vinculin with bilayer structures analyzed with a photoactivable phospholipid. J. Biol. Chem. 261, 6912–6918.

Niggli, V. & Gimona, M. (1993). Evidence for a ternary interaction between α-actinin, (meta)vinculin and acidic phospholid bilayers. Eur. J. Biochem. 213, 1009–1015.

Niggli, V., Kaufmann, S., Goldmann, W.H., Weber, T., & Isenberg, G. (1994). Identification of functional domains in the cytoskeletal protein talin. Eur. J. Biochem. 224, 951–957.

Niggli, V., Sommer, L., Brunner, J., & Burger, M.M. (1988). Interaction of cytoskeletal protein vinculin with membranes in intact cells. In: Structure And Function Of The Cytoskeleton (Rousset, B., Ed. Colloque INSERM/John Libbey Eurotext Ltd., Paris Vol. 171, pp. 121–126.

Nuckolls, G.H., Romer, L.H., & Burridge, K. (1992). Microinjection of antibodies against talin inhibits the spreading and migration of fibroblasts. J. Cell Sci. 102, 753–762.

Nuckolls, G.H., Turner, C.E., & Burridge, K. (1990). Functional studies of the domains of talin. J. Cell Biol. 110, 1635–1644.

Ohnesorge, F., Egger, M., Heyn, S.P., Gaub, H.E., Weisenhorn, A., Drake, B., Prater, C.B., Gould, S.A.C., & Hansma, P. (1990). Immobilized proteins in buffer imaged at molecular resolution by atomic force microscopy. Biophys. J. 58, 1251–1257.

Onoda, K. & Yin, H.L. (1993). gCap39 is phosphorylated. J. Biol. Chem. 268, 4106–4112.

O'Leary, T.J. & Levin, I. W. (1984). Raman spectroscopic study of an interdigitated lipid bilayer dipalmitoylphosphatidylcholine dispersed in glycerol. Biochim. Biophys. Acta 776, 185–189.

Otto, J.J. (1990). Vinculin. Cell Motil. Cytoskel. 16, 1–6.

Pacaud, M. & Harricane, M.C. (1993). Macrophage alpha-actinin is not a calcium-modulated actin-binding protein. Biochemistry 32, 363–374.

Petrucci, T.C. & Morrow, J.S. (1987). Synapsin I: An actin bundling protein under phosphorylation control. J. Cell Biol. 105, 1355–1363.

Pollard, T.D. & Cooper, J.A. (1986). Actin and actin binding proteins. A critical evaluation of mechanisms and functions. Ann. Rev. Biochem. 55, 987–1035.

Pollard, T.D., Doberstein, S.K., & Zot, H.G. (1991). Myosin I. Ann. Rev. Physiol. 53, 653–681.

Pollard, T.D. & Korn, E.D. (1973). Acanthamoeba myosin I. Isolation from Acanthamoeba castellanii of an enzyme similar to muscle myosin. J. Biol. Chem. 248, 4682–4690.

Price, G.J., Jones, P., Davison, M.D., Patel, B., Eperon, I.C., & Critchley, D.R. (1987). Isolation and characterization of a vinculin cDNA from chick embryo fibroblasts. Biochem. J. 245, 595–603.

Rees, D.J.G., Ades, S.E., Singer, S.J., & Hynes, R.O. (1990). Sequence and domain structure of talin. Nature 247, 685–689.

Ridley, A.J. & Hall, A.A. (1992). Distinct patterns of actin organization regulated by the small GTP-binding proteins rac and rho. Cold Spring Harb. Symp. Quant. Biol. 57, 661–671.

Rinnerthaler, G., Herzog, M., Klappacher, M., Kunka, H., & Small, J.V. (1991). Leading edge movement and ultrastructure in mouse macrophages. J. Struct. Biol. 106, 1–16.

Rotman, A., Heldman, J., & Linder, S. (1982). Association of membrane and cytoplasmic proteins with the cytoskeleton in blood platelets. Biochemistry 21, 1713–1719.

Ruddies, R., Goldmann, W.H., Isenberg, G., & Sackmann, E. (1993). The viscoelasticity of entangled actin networks: Influence of defects and modulation by talin and vinculin. Eur. Biophys. J. 22, 309–321.

Ruhnau, K. & Wegner, A. (1988). Evidence for direct binding of vinculin to actin filaments. FEBS Lett. 228, 105–108.

Ruocco, M.J., Makriyannis, A., & Siminovitch, D. (1985a). Deuterium NMR investigation of ether- and ester-linked phophatidylcholine bilayers. Biochemistry 24, 4844–4851.

Ruocco, M.J., Siminovitch, D.G., & Grifin, R.G. (1985b). Comparative study of the gel phases of ether- and ester-linked phosphatidylcholines. Biochemistry 24, 2406–2411.

Sadler, I., Crawford, A.W., Michelsen, J.W., & Beckerle, M.C. (1992). Zyxin and cCRP: Two interactive LIM domain proteins associated with the cytoskeleton. J. Cell Biol. 119, 1573–1587.

Safer, D. (1992). The interaction of actin with thymosin β_4. J. Muscle Res. Cell Motil. 13, 269–271.

Samuelsson, S.J., Luther, P.W., Pumplin, D.W., & Bloch, R.J. (1993). Structures linking microfilament bundles to the membrane at focal contacts. J. Cell Biol. 122, 485–496.

Scheel, J., Ziegelbauer, K., Kupke, T., Humberl, B.M., Noegel, A., Gerisch, G., & Schleicher, M. (1989). Histoactophilin, a histidine-rich actin binding protein from Dictyostelium discoideum. J. Biol. Chem. 264, 2832–2839.

Schleicher, M., Gerisch, G., & Isenberg, G. (1984). New actin binding proteins from Dictyostelium discoideum. EMBO J. 3, 2095–2100.

Schleicher, M., Noegel, A., Schwarz, T., Wallraff, E., Brink, M., Faix, J., Gerisch, G., & Isenberg, G. (1988). A Dictyostelium mutant with severe defects in alpha-actinin: Its characterization using cDNA probes and monoclonal antibodies. J. Cell Sci. 90, 59–71.

Shariff, A. & Luna, E.J. (1990). *Dictyostelium discoideum* plasma membranes contain an actin-nucleating activity that requires ponticulin, an integral membrane glycoprotein. J. Cell Biol. 110, 681–692.

Shariff, A. & Luna, E.J. (1992). Diacylglycerol-stimulated formation of actin nucleation sites at plasma membranes. Science 256, 245–247.

Sheetz, M.P. (1993). Glycoprotein motility and dynamic domains in fluid plasma membranes. Ann. Rev. Biophys. Biomolec. Struct. 22, 417–431.

Sheetz, M.P., Wayne, D.B, & Pearlman, A.L. (1992). Extension of filapodia by motor-dependent actin assembly. Cell Motil. Cytoskel. 22, 160–169.

Sikorski, A.F.& Kuczek, M. (1985). Labelling of erythrocyte spectrin in situ with phenylisothiocyanate. Biochim. Biophys. Acta 820, 147–153.

Siminovitch, D.J., Wong, P.T., & Mantsch, H.H. (1987). High pressure infrared spectroscopy of ether- and ether-linked phosphatidylcholine aqueous dispersions. Biophys. J. 51, 465–473.

Small, J.V., Fürst, D.O., & Thornell, L.E. (1992). The cytoskeletal lattice of muscle cells. Eur. J. Biochem. 208, 559–572.

Small, J.V., Isenberg, G., & Celis, J.E. (1978). Polarity of actin at the leading edge of cultured cells. Nature 272, 638–639.

Stossel, T.P. (1989). From signal to pseudopod. J. Biol. Chem. 264, 18261–18264.

Stossel, T.P. (1990). Actin-membrane interactions in eukaryotic mammalian cells. In: Curr. Topics In Membranes And Transport (Hoffman J.F. & Giebisch, G., Eds.), Vol. 36, pp. 97-107, Academic Press, New York.

Stossel, T.P., Chaponnier, C., Ezzell, R.M., Hartwig, J.H., Janmey, P.A., Kwiatkowski, D.J., & Lind, S.E. (1985). Nonmuscle actin binding proteins. Ann. Rev. Biol. 1, 353–402.

Südhof, T.C., Czernik, A.J., Kao, H.T., Takei, H., Johnston, P.A., Horiuchi, A., Kanazir, S.D., Wagner, M.A., Perin, M.S., DeCamilli, P., & Greengard, P. (1989). Synapsins: Mosaics of shared and individual domains in a family of synaptic vesicle phosphoproteins. Science 245, 1474–1480.

Taniguchi, H. & Manenti, S. (1993). Interaction of myristoylated alanine-rich protein kinase C substrate (MARCKS) with membrane phospholipids. J. Biol. Chem. 268, 9960–9963.

Tempel, M., Goldmann, W.H., Dietrich, C., Niggli, V., Weber, T., Sackmann, E., & Isenberg, G. (1994). Insertion of filamin into lipid membranes examined by calorimetry, the film balance technique, and lipid photolabeling. Biochemistry 33, 12565–12572.

Tendian, S.W. & Lentz, B.R. (1990). Evaluation of membrane phase behaviour as a tool to detect extrinsic protein induced domain formation: Binding of prothrombin to phosphatidylserine/phosphatidylcholine vesicles. Biochemistry 29, 6720–6729.

Theriot, J.A. & Mitchison, T. (1991). Actin microfilament dynamics in locomoting cells. Nature 352, 126–131.

Valtorta, F, Greengard, P,, Fesce, R., Chieregatti, E., & Benfenati, F. (1992). Effects of the neuronal phophoprotein synapsin I on actin polymerization. J. Biol. Chem. 267, 11281–11288.

Vandekerckhove, J. (1990). Actin-binding proteins. Curr. Opinion Cell Biol. 2, 41–50.

Van Paridon, P.A., De Kruiff, B., Ouwerkerk, K., & Wirtz, W.A. (1986). Polyphosphoinositides undergoe charge neutralization in the physiological pH range: a ^{31}P-NMR study. Biochim Biophys. Acta 877, 216–219.

Vorotnikov, A. & Gusev, N. B. (1990). Interaction of smooth muscle caldesmon with phospholipids. FEBS Lett. 277, 134-136.

Wachsstock, D.H., Wilkins, J.A., & Lin, S. (1987). Specific interaction of vinculin with alpha-actinin. Biochem. Biophys. Res. Comm. 146, 554–560.

Wallraff, E., Schleicher, M., Modersitzki, M., Rieger, D., Isenberg, G., & Gerisch, G. (1986). Selection of Dictyostelium mutants defective in cytoskeletal proteins: Use of an antibody that binds to the ends of alpha-actinin rods. EMBO J. 5, 61–65.

Wang, Y.L. (1984). Reorganization of actin filament bundles in living fibroblasts. J. Cell Biol. 99, 1478–1484.

Wang, Y.L. (1985). Exchange of actin subunits at the leading edge of living fibroblasts: Possible role of treadmilling. J. Cell Biol. 101, 597–602.

Watson, P.A. (1991). Function follows form: Generation of intracellular signals by cell deformation. FASEB J. 5, 2013–2019.

Way, M., Gooch, J., Pope, B., & Weeds, A.G. (1989). Expression of human plasma gelsolin in *Escherichia coli* and dissection of actin binding sited by segmental deletion mutagenesis. J. Cell Biol. 109, 593–605.

Wegner, A. & Isenberg, G. (1983). 12-fold difference between the critical monomer concentrations of the two ends of actin filaments in physiological salt conditions. Proc. Natl. Acad. Sci. USA 80, 4922–4925.

Weisenhorn, A., Egger, M., Ohnesorge, F., Gould, S.A.C., Heyn, S.P., Hansma, H.G., Sinsheimer, R.L., Gaub, H.E., & Hansma, P. (1991). Molecular resolution images of Langmuir Blodgett films and DNA by atomic force microcopy. Langmuir 7, 8–12.

Weller, P.A., Ogryzko, E.P., Corben, E.B., Zhidkova, N.I., Patel, B., Price, G.J., Spurr, N.K., Koteliansky, V.E., & Critchley, D.R. (1990). Complete sequence of human vinculin and assignment of the gene to chromosome 10. Proc. Natl. Acad. Sci. USA 87, 5667–5671.

Westmeyer, A., Ruhnau, K., Wegner, A., & Jockusch, B.M. (1990). Antibody mapping of functional domains in vinculin. EMBO J. 9, 2071–2078.

Wuesthube, L.J., Chia, C.P., & Luna, E.J. (1989). Immunofluorescence localization of ponticulin in motile cells. Cell Motil. Cytoskel. 13, 245–263.

Wuesthube, L.J. & Luna, E.J. (1987). F-actin binds to the cytoplasmic surface of ponticulin, a 17 kDa integral glycoprotein from *Dictyostelium discoideum* plasma membranes. J. Cell Biol. 105, 1741–1751.

Yin, H.L. (1988). Gelsolin: Calcium- and polyphosphoinositide-regulated actin-modulating protein. Bio Essays, 7, 176–179.

Yin, H.L., Iida, K., & Janmey, P.A. (1988). Identification of a polyphosphoinositide-modulated domain in gelsolin which binds to the sides of actin filaments. J. Cell Biol. 106, 805–812.

Yin, H.L., Janmey, P.A., & Schleicher, M. (1990). Severin is a gelsolin prototype. FEBS Lett. 264, 78–80.

Yonemura, S., Nagafuchi, A., Sato, N., & Tsukita, S. (1993). Concentration of an integral membrane protein, CD43 (leukosialin, sialophorin), in the cleavage furrow through the interaction of its cytoplasmic domain with actin-based cytoskeletons. J. Cell Biol. 120, 437–449.

Yonezawa, N., Nishida, E., Iida, K., Yahara, I., & Sakai, H. (1990). Inhibition of the interactions of cofilin, destrin and deoxyribonuclease I with actin by phosphoinositides. J. Biol. Chem. 265, 8382–8386.

Yu, F.X., Johnston, P.A., Südhof, T.C., & Yin, H. (1990). gCap39, a calcium ion-and polyphosphoinositide-regulated actin capping protein. Science 250, 1413–1415.

Zachowski, A. (1993). Phospholipids in animal eukaryotic membranes: Transverse asymmetry and movement. Biochem. J. 294, 1–14.

Zhang, F., Lee, G.M., & Jacobson, K. (1993). Protein lateral mobility as a reflection of membrane microstructure. BioEssays (in press).

THE PROTEINS OF INTERMEDIATE FILAMENT SYSTEMS

Robert L. Shoeman and Peter Traub

The Cytoskeleton, Volume 1
Structure and Assembly, pages 205–255.
Copyright © 1995 by JAI Press Inc.
All rights of reproduction in any form reserved.
ISBN: 1-55938-687-8

I. INTRODUCTION

The cytoskeleton of a typical eukaryotic cell consists of three filament systems: microfilaments, intermediate filaments (IFs) and microtubules, as well as a number of filament-associated proteins that cross-link these systems to each other and to cell organelles. At least two filament systems are found within many or most nuclei: the nuclear lamina, which is composed of one or more subunit proteins of one class of IF proteins (the nuclear lamins), and the nuclear matrix, whose precise molecular composition is currently being defined. One component of the nuclear matrix is the 10-nm filament whose morphology and axial repeat pattern are highly reminiscent of the cytoplasmic IFs (Jackson and Cook, 1988; He et al., 1990) and which may very well be similar to them in subunit composition (Wang and Traub, 1991). At the periphery of a cell in a tissue, IFs are linked to the elements of the extracellular matrix and, within the interior, the cytoplasmic and nuclear IFs form a continuum that extends from the plasma membrane into the interior of the nucleus (French et al., 1989; Figure 1). While no universally accepted model has been presented to explain how IFs physically traverse the nuclear membrane, it is quite clear that this continuum is real in a functional sense and may play a role in the transduction of cellular signals. For example, perturbation of cytoplasmic IFs affects chromatin organization (Hay and De Boni, 1991) and interferes with conformational changes induced in the nuclear lamina by exogenously applied phorbol ester (Collard and Raymond, 1992). A recent study based on immunoelectron microscopy has described the presence of a transcellular IF network extending from the plasma membrane into the interior of the nucleus of cardiac myocytes (Lockard and Bloom,

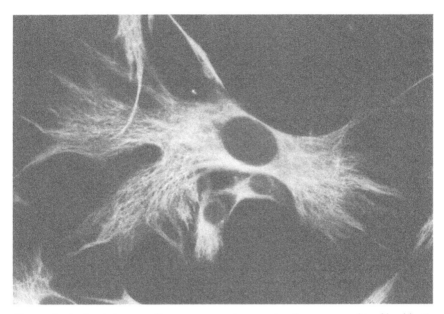

Figure 1. Indirect immunofluorescence micrograph of a mouse skin fibroblast's vimentin IF network. Vimentin IFs were decorated with a primary goat antivimentin antibody, which was visualized with a fluorescent-labeled rabbit anti-goat Ig antibody. The IFs traverse the cell, from one end to the other and from the plasma membrane to the nucleus. No vimentin IFs are found within the nucleus, which is seen as the large dark circle in the center of the cell. The few IFs in this region actually lie within the thin zones of cytoplasm above and below the nucleus. In some regions of the cell, particularly adjacent to the nucleus, the IFs appear diffuse. This artifact is due to the thickness of the cell exceeding the limited depth of field of the conventional optical system. This micrograph was kindly provided by Dr. A. Janetzko.

1993). The major theme of this contribution is the cytoplasmic IFs, their subunit proteins as well as their interactions with organelles and other components of the cytoskeleton. Given that IFs form a continuum throughout the cell, some discussion of the nuclear IFs and the nuclear lamina will be included. However, more information on the nucleoskeleton and the other filament systems of the cytoskeleton can be found in other chapters of this book.

IFs, with few exceptions, are universal components of eukaryotic cells (Lazarides, 1980; Steinert and Roop, 1988), although their unequivocal demonstration in most plant cells is still lacking (see Menzel, 1993). Analogues of the nuclear lamins have been reported to occur in yeast (Georgatos et al., 1989) and *Physarum* (Lang and Loidl, 1993). IFs received their name since their characteristic diameter of 8 to 12 nm (hereafter referred to as 10 nm) falls between that of the previously

Table 1. Classes and characteristics of vertebrate IF proteins. The molecular masses are based on the amino acid sequences and may vary dramatically from the apparent molecular weights or mobilities as a result of posttranslational modifications (particularly for the neurofilament triplet proteins) and/or anomalous electrophoretic behavior.

Class	Subunit Protein	Occurrence	Mol. Mass (kDa)
type I	acidic keratins	epithelia	40–60
type II	basic keratins	epithelia	50–70
type III	vimentin	mesenchyme, most tissue culture cells	53
	desmin	muscle	53
	GFAP	glial cells and astrocytes	50
	peripherin	neurons	53
	plasticin	neurons	53
type IV	neurofilament	neurons	NFP-L: 62
	triplet proteins		NFP-M: 96
			NFP-H: 116
	internexin	neurons	55
type V	lamins A, B, C	nucleated cells	72, 67, 65
type VI	nestin	neuroepithelial stem cells, developing myoblasts	200

identified microfilaments (6–7 nm) and that of the thick myosin filaments and microtubules (25 nm). Unlike microfilaments, which are composed of actin subunits and microtubules, which are polymerized from tubulins, IFs may be formed from one or more than 40 different subunit proteins. Many of these proteins have been identified by immunoreaction with the monoclonal antibody α-IFA (Pruss et al., 1981), which reacts with a highly conserved domain (see below). The vertebrate IF proteins have been grouped into six classes on the basis of their gene structure, their developmental and tissue-specific expression (see van de Klundert et al., 1993 for a recent review), their polymerization characteristics and their cellular localization (Table I).

As more diverse organisms are studied, it is likely that additional individual as well as new classes of IF proteins will be discovered and this grouping will have to be updated or revised. For example, several studies have demonstrated IF proteins in a variety of invertebrates (Zackroff and Goldman, 1980; Bartnik and Weber, 1989; Dodemont et al., 1990; see Riemer et al., 1991) and some of these await assignment to a new or seventh class of IF proteins. One important result of these studies was the identification of an amino acid residue essential for the α-IFA epitope and the observation that the lack of reactivity of a tissue or cell with the monoclonal antibody α-IFA does not preclude the presence of IFs (Riemer et al., 1991). This effectively laid to rest the dogma that all classes of IF share the α-IFA epitope (Pruss et al., 1981). Two additional IF proteins from vertebrate lens cells, filensin and phakinin, do not possess the α-IFA epitope and, while they share some properties of keratins and copolymerize into 10-nm filaments (Masaki and Watanabe, 1992; Gounari et al., 1993; Hess et al., 1993; Merdes et al., 1993; Remington, 1993), they do not fit neatly into the classification scheme of Table 1.

II. INTERMEDIATE FILAMENT PROTEIN STRUCTURE

All of the IF proteins share a common secondary structure consisting of three distinct domains (Figure 2A). Non-α-helical head and tail domains of variable size and amino acid sequence flank a central rod domain that consists of about 310 amino acid residues (about 350 for the nuclear lamins and invertebrate cytoplasmic IF proteins) and has a high propensity for α-helix formation (Parry and Fraser, 1985;

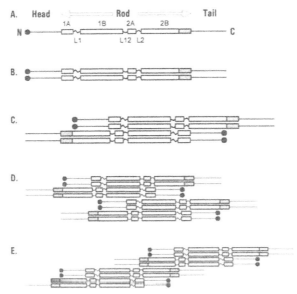

Figure 2. Schematic representation of IF protein monomer (A), dimer (B), tetramer (C) and octamer structure (D,E). The α-helical regions are represented by boxes, the non-α-helical linkers by wavy lines and the head and tail domains by straight lines (the head domain has a circle at its amino terminus for identification). The reversal of the α-helix towards the end of coil 2B is indicated by the shaded box. The length of the rod domain is about 47 nm. The length of the head and tail domains is variable and little is known about their conformation. The model of tetramer structure in (C) and potential modes of interaction of tetramers (D,E) have been proposed by Geisler et al. (1992) on the basis of cross-linking experiments with the type III IF protein desmin. The length of the head domain in these drawings has been scaled to agree with peptide binding results that suggest the head domain can bind to the carboxy-terminal end of the rod domain (Traub et al., 1992b).

see Traub, 1985, for more references). The amino acid sequence of the central rod domain is highly conserved, especially at its ends. Throughout the rod domain the sequence homology is quite high (70% or higher) among members of the same class of IF proteins and remains at 30% to 50% between classes (Conway and Parry, 1988; Albers and Fuchs, 1992). The homologies at the beginning and end of the rod domain are much higher, supporting experimental observations that these regions play an important and invariant or universal role in interactions of IF proteins. As a general rule, the head domain is rich in arginine residues and positively charged, whereas the tail domain is acidic. The non-α-helical head and tail domains are primarily responsible for the unique reactivities or properties of the IF proteins, while the rod domain is primarily responsible for the one property all of these diverse proteins share: the ability to polymerize into 10-nm filaments.

III. TISSUE-SPECIFIC DISTRIBUTION

As summarized in Table I, the IF protein complement of cells in an adult vertebrate organism has a tissue-specific distribution. This pattern results from an orderly cascade of expression of IF proteins during development. For example, of the more than 30 different keratin subunit proteins, the keratins K8 and K18 are among the first IF proteins to be synthesized in the developing vertebrate embryo (Jackson et al., 1980). As development and organogenesis proceed, a general trend develops: Epithelia express characteristic subsets of keratins; keratin expression is down-regulated in cells differentiating to mesenchyme, myogenic and neuronal tissues, while vimentin synthesis increases. Later, desmin replaces vimentin within muscle tissue, as do the neurofilament triplet proteins (NFPs) in neurons, etc. Some of the subunit proteins listed in Table I, such as plasticin, α-internexin and nestin, are expressed transiently prior to the expression of NFPs in development or during regeneration of the central nervous system following wounding (Glasgow et al., 1992; reviewed by Liem, 1993) and concomitantly with vimentin and desmin during early skeletal muscle development (Sejersen and Lendahl, 1993).

This tissue-specific distribution of IF proteins and the continued development of mono-specific monoclonal anti-IF antibodies has a practical application in human medicine in the sensitive and rapid typing of tumor tissues (Osborn and Weber, 1983). After transformation, cells maintain their cell-lineage-specific IF protein(s) expression permitting their precise identification. Many tumor cells and most tissue culture cells additionally express vimentin, suggesting that the transformed pheno-type represents a "dedifferentiation," at least with respect to IF protein expression. As new anti-IF antibodies are developed and detailed studies are performed, new insights into the IF phenotype correlating with invasive or proliferating activities are being made (for example, Schüssler et al., 1992 and Schaafsma et al., 1993). Thus, IF typing, particularly that of keratin IFs since most human tumors originate from epithelia (Lane and Alexander, 1990), will be of increasing importance in making clinical decisions concerning treatment regimes for human cancers. As an aside, it should be mentioned that there have been a number of reports describing the exposure of IF proteins on the surface of a variety of cultured cell lines, although, given the lack of conventional membrane-spanning domains in IF proteins, these data have been difficult to integrate into a coherent concept of IF localization and function. A recent report has presented data that suggest these cell-surface exposed IF proteins actually represent intracellular IF proteins exposed as a result of cell membrane damage during the labeling procedure (Riopel et al., 1993).

IV. ASSEMBLY OF INTERMEDIATE FILAMENTS

A. General Properties

Historically, the assembly of IFs has been studied *in vitro* using highly purified components. While most of the results have been confirmed *in vivo* by recent experiments, there have been several unexpected and, as yet, unexplained differences between the seemingly straightforward, orderly polymerization of IFs *in vitro* and their dynamic behavior *in vivo* (Georgatos, 1993; Stewart, 1993). Accordingly, the discussion of assembly of IFs will be discussed in two parts and an attempt will be made to resolve some of the discrepancies and point out avenues for future research.

Coiled-Coil Dimer Formation

A schematic diagram of IF protein structure and polymerization is presented in Figure 2. The rod domain of all IF proteins consists of many imperfect repeats of a heptad sequence with a high propensity for α-helix formation, interrupted by short stretches of non-α-helical structure. The rod domains of two IF protein molecules wrap around each other to form a coiled-coil superhelical structure. This motif is common to a variety of proteins, including myosins, tropomyosins and the hard keratins found in hoof and hair. Within the heptads, whose sequence may be represented by the letters a-b-c-d-e-f-g, the a and d positions are usually apolar amino acids, e and g are generally charged residues and b, c and f are generally polar amino acids (Conway and Parry, 1988). The pitch of an α-helix is such that the equivalent residues, in this case the hydrophobic a and d amino acids, are rotated relative to each other about the long axis of the helix, producing a hydrophobic seam that winds around the rod domain. It follows that the rod domains of the individual monomers must twist around each other to permit the interaction of these residues and, furthermore, to permit the more polar and charged groups to be arrayed primarily on the surface of the molecules. In addition to the hydrophobic backbone, there are alternating and repeating clusters of charged and uncharged residues along the helices of the rod domain (Geisler et al., 1982). A conserved feature of the rod domain of IF proteins is the reversal of the direction of the pitch of the α-helix about in the middle of the helix 2B (Figure 2). This probably serves to limit the register of the coiled-coil dimer to a parallel, in-register orientation, which would maximize the number of hydrophobic interactions between the two monomers. Another conserved feature of the rod domain is the location and length, but not the sequence of the non-α-helical linker regions (Figure 2). These linker regions probably function as swivels or hinges about which the α-helices of the rod domain rotate or flex to achieve a maximum interaction between subunit proteins during dimer and tetramer formation. This has been demonstrated *in vivo* by the normal assembly of modified keratin K19 molecules, having long inserts (up to 24 residues) within the L1 region, into the existing keratin network of kangaroo rat kidney epithelial cells (Rorke et al., 1992).

Higher Order Interactions

At low ionic strength (see following discussion), IF-protein dimers spontaneously associate *in vitro* into tetramers (also referred to as protofilaments) with a multitude of orientations, parallel/antiparallel, staggered/unstaggered, etc., only a few of which are productive for higher order polymerization leading to the formation of 10-nm filaments (Aebi et al., 1988; Steinert, 1991). The assembly of IFs from tetramers is dependent on a variety of factors including tetramer concentration, ionic strength (cation type and concentration) and temperature. A key component in models of IF structure are the protofibrils, essentially overlapping end-to-end polymers of octamers. Protofibrils are often observed during disruption or depolymerization of IFs (Aebi et al., 1983) or are sometimes seen to protrude from the ends of incompletely polymerized IFs (Franke et al., 1982a). Furthermore, cross-sections of IFs *in vivo* appear to be a cylindrical arrangement of four protofibrils (octamers; Small and Squire, 1972; Eriksson and Thornell, 1979). A variety of studies have shown that the mass-per-unit length of IFs would suggest that on average, in cross-section, an IF is composed of about 32 subunit proteins (i.e., 16 dimers, 8 tetramers or 4 octamers). There is some variation in this number, particularly in images obtained from IFs *in vivo*, so that an IF may be made up of 12 to 16 dimers in cross-section (Steven et al., 1983; Engel et al., 1985). The initial steps in IF assembly (dimer to tetramer to octamer) appear to be rate limiting (Steinert, 1991), whereas the polymerization from tetramer to 10-nm filament proceeds rapidly under permissive conditions (see below). This supports the concept that all of the information necessary to construct an IF is present within the tetramer. While these properties are primarily a function of the rod domain, the non-α-helical head and tail domains play an important, modulating role in IF formation. For example, deletion of a large part of the desmin head domain gives rise to a molecule that can form tetramers, but not higher order polymers (Kaufmann et al., 1985). Likewise, the addition of short peptides whose sequence is identical to regions of the head domain of vimentin or the tail domain of desmin to *in vitro* assembly reactions results in the formation of long polymers with an unravelled or loosely braided structure (Traub et al., 1992c; Birkenberger and Ip, 1990).

Ion Effects

Unlike the assembly of microfilaments or microtubules, the polymerization of IF proteins is not obligatorily dependent on either accessory (protein) cofactors or exogenous energy sources (such as ATP), nor is there a fastidious requirement for a particular cation. This does not mean that the assembly process is indifferent to cations and indeed a number of physiological (i.e., Ca^{2+}, Mg^{2+}, Na^+, K^+) and nonphysiological cations (i.e., Cu^{2+} or Ni^{2+}) may support IF polymerization, albeit with different kinetics (i.e., Stromer et al., 1987). In general, divalent cations support IF polymerization *in vitro* at much lower concentrations than monovalent cations (i.e., low mM *vs.* 80–160 mM; Yang and Babitch, 1988; Hofmann et al., 1991). At a concentration higher (~5 mM) than that (1–2 mM) needed for normal assembly of bovine vimentin, calcium induces the formation of shorter and thicker IFs, with pronounced axial repeats (Hofmann et al., 1991). While the significance of these results for *in vivo* IF assembly is not immediately clear, since free divalent cation concentrations rarely or never reach low mM levels, it is likely that the IFs formed *in vivo* have a variety of mono- and divalent cations, as well as polycations, bound to them. In addition to its ability to support IF assembly, calcium (or other divalent cations) may play a regulatory role in modulating IF function since GFAP (and other IF proteins by homology) possess an *EF-hand-like* calcium binding site (Yang et al., 1988). Certain ions, such as PO_4^{3-}, exert a disrupting effect on IF structure (Aebi et al., 1983) and others, like Al^{3+}, induce bundling of NFPs and may be of relevance in disease development (Leterrier et al., 1992).

Individual, purified IF proteins or matching pairs of type I and type II keratins spontaneously polymerize *in vitro* into long, smooth filaments. With the exception of the nuclear lamins, all of the IF proteins form characteristic 10-nm filaments when incubated above a critical concentration, and this reaction is dependent upon several variables including temperature, pH and the specific cations present (Chou et al., 1990). However, the details of polymerization for the individual classes of IF proteins differ. For example, *in vitro* keratin subunits can assemble into normal IFs when dialyzed from urea-containing solutions against 5 mM Tris, 1 mM dithiothreitol, pH 7.5 (Eichner et al., 1986). At pH 8.5, vimentin exists as a tetramer at a concentration of 10–20 mM NaCl, eight of these tetramers assemble at 50 mM NaCl into a short, full-width (10 nm) intermediate structure (Ip et al., 1985). At higher NaCl concentrations (170 mM), vimentin polymerizes into authentic IFs. At low to physiological salt concentrations, purified nuclear lamins tend to aggregate, often forming paracrystals that become increasingly soluble with increasing ionic strength (Moir et al., 1991). IFs formed *in vitro* are remarkably stable and do not disassemble upon protein dilution. In general, the various types of IFs are sensitive to a lowering of the ionic strength, becoming increasingly soluble at low ionic strength. Among those best studied *in vitro*, IFs formed from the type III proteins (i.e., vimentin and desmin) are most labile in buffers of intermediate salt concentration (i.e., 50–80 mM) and dissociate into tetramers at low salt concentration (10 mM or below). Keratin IFs, on the other hand, are much less sensitive to ionic strength and some combinations of keratins yield IFs that are stable in distilled water. The tetramer is the smallest polymer that can be isolated without the use of chaotropic agents, such as urea or guanidinium-HCl (in whose presence one can produce and isolate dimers and monomers), or detergents, such as sodium dodecylsulfate (with which one can produce monomers). One remarkable feature of the IF proteins is that they may refold and regain (apparently) all of their functions following denaturation with these reagents upon return to the appropriate buffers. This has not only permitted their purification but also enabled the production of mixed tetramers (Traub et al., 1993). As will be discussed later, posttranslational modifications of IF proteins (particularly of the head domain) affect IF assembly and stability and are one mechanism by which IF structures are modulated during cytokinesis and differentiation (Goldman et al., 1991; Georgatos, 1993; Moir and Goldman, 1993).

B. Polymerization Details

Three Groups of Intermediate Filament Proteins

Molecular genetics has provided the most detailed data for the subdivision of the IF proteins into six classes. In a recent review on IF structure and assembly, Stewart (1993) has made a more logical proposal dividing these proteins into three groups based on their polymerization characteristics. Group A consists of the type I (acidic) and type II (basic) keratins that polymerize only as heterodimers containing one of each subunit type. Group B includes the type III, IV and VI proteins, all of which polymerize into IFs from homodimers. When two or more group B proteins are expressed simultaneously in a cell or the proteins are mixed together *in vitro*, homodimers of the respective proteins may polymerize to give rise to heterotypic IFs. That is, the tetramer structure shown in Figure 2C would be composed of two homodimers (Figure 2B) of different IF proteins, such as vimentin and desmin. For most of the group B proteins, with the exception of the NFPs *in vivo* (Ching and Liem, 1993; Lee et al., 1993), there is no absolute requirement for this copolymerization as is the case for the keratins. Group C consists of the nuclear lamin proteins, whose central rod domain is both longer and, as will be discussed, simpler in its protein–protein interactions during polymerization. While by and large the polymerization pathway of these three groups of proteins is quite similar, giving rise to the common IF, each group possesses some unique characteristics which, presumably, are the reasons why two (or more) individual subunit proteins from different classes do not copolymerize to give heterotypic IFs.

Two different models of IF polymerization have recently been published, one for the group B, type III desmin (Geisler et al., 1992), and the other for the group A keratins (Steinert et al., 1993a; Steinert, 1993).

Group B (Desmin)

The diagram of IF polymerization of group B IF proteins into tetramers and the organization of tetramers in IFs presented in Figure 2C and Figure 2D is based on the results obtained by analysis of cross-links of desmin polymers (Geisler et al., 1992). One structure, a tetramer composed of anti-parallel dimers staggered by about 15–20 nm (Figure 2C), was found to be common to rod domains, protofilaments and filaments. Another cross-link obtained in high yield from filaments was compatible with an overlap of protofilaments by either 1/3 or 2/3 (Figure 2D, E), which implies that one or both orientations exist within filaments. Both of these overlaps bring the ends of the rod domains of neighboring protofilaments in close apposition within the IFs and result in the positioning of the head domain of one protofilament close to the end of coil 2B of the rod domain of another protofilament (for the overlap in Figure 2D) or close to the end of coil 1A of the rod domain of a second protofilament (Figure 2E). These overlaps are explicitly mentioned since they involve regions of the rod domain that are most highly conserved in sequence (Conway and Parry, 1988), and the head domain has been shown to bind to peptides containing these conserved sequences (Hofmann and Herrmann, 1992; Traub et al., 1992c). Furthermore, synthetic peptides with a sequence identical to that surrounding the highly conserved arginine dipeptide present close to the amino-terminal end of the head domain of type III IF proteins not only bind to the beginning of the rod domain, but also interfere with IF assembly (Hofmann and Herrmann, 1992; Herrmann et al., 1992; see also Raats et al., 1992). Two additional types of observations support the proposal of Geisler et al. (1992) that the interaction of the ends of the rod domain play a role in IF assembly: Single amino acid substitutions within these regions alter or abolish polymerization and synthetic peptides with sequences identical to those of the conserved end regions inhibit IF assembly and affect IF stability (Hatzfeld and Weber, 1991; 1992; Geisler et al., 1993).

Group A (Keratins)

As mentioned, the solubility properties of the group A proteins, the keratins, differ from those of the other IF proteins in that the subunit proteins polymerize to IFs at much lower ionic strength and the resulting IFs are less sensitive to changes in ionic composition of reconstitution. The composition of keratin IFs differs from that of the other groups of IFs in that they are assembled from obligate heterodimers of a type I (acidic) and a type II (neutral or basic) subunit protein. Thus, charge neutralization within the dimer is an important characteristic of keratin IF polymerization and explains the propensity for keratins to occur *in vivo* in pairs: Generally

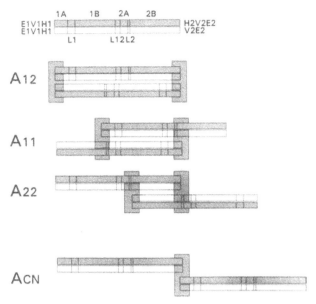

Figure 3. Four modes of keratin dimer alignments adduced from cross-linking studies (Steinert et al., 1993a). The top drawing represents a heterodimer of a type I (keratin 10) and a type II (keratin 1) chain. The N- and C-terminal subdomains are indicated by the letters to the left and right (respectively) of the box representing the central rod domain (open for the type I and diagonally striped for the type II). The α-helical subdomains of the rod are labeled above the box and the non-α-helical linker segments are labeled below the box. The H1 and H2 subdomains are unique to the type II keratins. The dispositions of the N- and C-terminal domains are not known and are not shown in the following alignments. The 4 modes of molecule alignment correspond to A_{12}, antiparallel, approximately in register; A_{11}, antiparallel, approximately half-staggered with segment 1B overlap; A_{22}, antiparallel, approximately half-staggered with segment 2B overlap and A_{CN}, end-to-end overlap of 10 to 11 residues of similarly directed molecules. The five important sequence regions that overlap frequently within a keratin IF are boxed (diagonal stripes): H1; H2; beginning of 1A and end of 2B (A_{12}, A_{CN}); H1 with L12 region (A_{11}) and H2 with L12 region (A_{22}). Redrawn from the data of Steinert et al. (1993a).

only one set of (or at most a few) individual keratins can productively dimerize, although not all possible combinations that can pair *in vitro* have been observed to occur *in vivo* (Hatzfeld and Franke, 1985). Quantitative mass measurements have shown that keratin IFs may be quite polymorphic, containing from 12 to 24 dimers in cross-section (Steven et al., 1984). The keratins are also special among the IF proteins in that they contain highly conserved subdomains flanking the central rod domain (Figure 3), and these play important roles in assembly (Steinert and Parry, 1993; Steinert, 1993).

Cross-linking studies have provided information on the molecular alignments in keratin IFs formed from keratins K1 and K10 (Figure 3; Steinert et al., 1993a) and thus provide insight into how keratin IFs assemble (Figure 4; Steinert et al., 1993a; Steinert, 1993). Two of these alignments within tetramers, the A11 mode (in which the heterodimers are antiparallel and staggered with an overlap of segment 1B) and the A22 mode (in which the heterodimers are antiparallel and staggered with an overlap of segment 2B), are very similar to those observed with desmin (Geisler et al., 1992; Figure 2). The other two alignments have only been observed with keratin IFs (although this does not rule out their existence for the other classes and groups of IFs). These alignments have been termed A12 (in which the heterodimers are antiparallel and approximately in register) and ACN (in which the heterodimers of the same polarity overlap end to end by about 10 residues) (Figure 3). These observed overlaps, together with results of previous model building (Fraser et al., 1990) and use of synthetic peptides to affect keratin IF assembly, support the model of surface lattice arrangement of keratin heterodimers shown in Figure 4 (Steinert et al., 1993a). One of the features of this arrangement is the dislocation of the lattice when the pattern is rolled or folded into a cylinder, producing a seam or discontinuity. Recent data (Sarria et al., 1990; Vikstrom et al., 1992) have shown that IFs are dynamic structures *in vivo* and oligomeric subunits may exchange or incorporate along the length of preexisting IFs. Steinert et al. (1993a) propose that the seam, or discontinuity, in their surface lattice model (Figure 4) may be the site of this exchange or incorporation.

Cross-linking experiments with keratin IFs formed from keratins K5 and K14 yielded axial molecular alignments and dimensions essentially identical to those found for keratins K1 and K10 IFs (Steinert et al., 1993a, 1993b). This suggests that the replacement of subunit proteins within preexisting IFs (for example, the replacement of keratins K5 and K14 by K1 and K10 during terminal differentiation in the epidermis) may take place simply because these molecules have the same linear dimensions and axial configurations as those in the IFs (Steinert et al., 1993b). This same logic would apply to the other classes and groups of IF proteins and would also explain why the different groups of IF proteins fail to copolymerize. While the general scheme of subunit organization with the various types of IFs is similar (i.e., the models in Figures 2 and 3 are not radically different), the interactions between different groups of IF proteins fail to support polymerization into authentic 10-nm filaments. Thus, cells expressing both keratins (group A IF proteins) and vimentin (a group B IF protein), such as keratinocytes in tissue culture, have two distinct and separate IF networks, whereas cells expressing both desmin and vimentin (both are group B IF proteins), such as myocytes in tissue culture, have a single IF network.

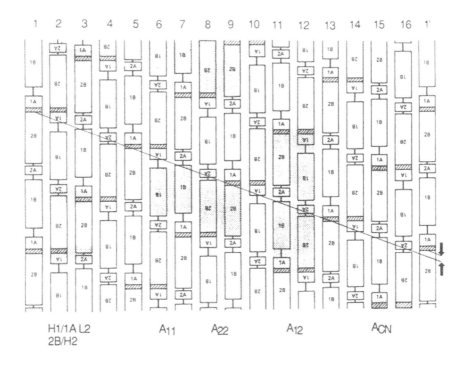

Figure 4. A possible surface lattice model of keratin IF dimer organization based on chemical cross-linking studies (Steinert et al., 1993a). In this model, 8 A_{11}-A_{22} polymers related by A_{12} interactions are shown, leading to 16 dimers in cross-section (numbered above the columns of dimers). A 17th column corresponding to the first in the next axial row is shown at the far right (number 1'). Examples of A_{11}, A_{22} and A_{12} alignments are stippled. The A_{CN} head-to-tail overlap is hatched throughout the diagram (and is stippled in the column above the A_{CN} label). Additional modes of interaction are shown for columns 2 and 3 (stippled). There is a close axial juxtaposition of segment L2 with the head-to-tail overlap region containing the first 10 to 11 residues of segment 1A and its preceding H1 subdomain sequences, and the last 10 to 11 residues of segment 2B with its preceding H2 subdomain sequences. The diagonal line arbitrarily connecting the beginning of one set of dimers across the surface lattice has a dislocation since the pattern does not align when wrapped around a cylinder. The dislocation of molecules 16 and 1' is shown by the arrows at the right. Steinert et al. (1993a) speculate that this dislocation may serve as a seam along which dynamic exchanges or net insertions of additional molecules may take place. This surface lattice model has a 44.4-nm axial repeat of structure with diagonal lines of equivalent points at axial intervals of 22.2 nm, which are interrupted at the dislocation. Used with permission from Steinert et al. (1993a)

Why 10-nm Filaments?

The structures represented in Figures 2 and 3 are symmetrical and provide no direct clues as to why the higher order polymerization should be limited to the 10-nm filament. A variety of studies (see Traub, 1985 for references) suggest that the protofibrils are helically wound about each other, with the result that the individual components are helically displayed on the surface of the IF with a pitch of about 23 nm and a spacing of about 45–50 nm along the long axis. This has been shown, for example with the head domain, by the decoration of IFs by a gold-labeled oligomer of deoxyguanosine (25-mer) with a periodicity of 23 nm (Traub et al., 1992b); the head is the only domain of vimentin that binds to nucleic acids (Shoeman et al., 1988). Electron microscopy (Small and Squire, 1972; Eriksson and Thornell, 1979), model building (Fraser et al., 1990) and protease accessibility experiments (Perides et al., 1987b) suggest that an IF is not a smooth cylinder, but rather is a symmetrical arrangement of 4 octamers (in cross-section) whose surfaces are involuted so that most or all of the non-α-helical end domains are exposed. The propensity to form filaments of 10-nm diameter may be a simple consequence of the length and helical pitch of the tetramers and the presence of a limited number of types of lateral and longitudinal binding sites such that the 10-nm filament is the most stable and energetically favorable configuration (Shoeman and Traub, 1993) that is compatible with the requirement of surface exposure of the head and tail domains. This concept is supported by data obtained from experiments employing recombinant DNA technology to perform domain swapping between two different IF proteins (McCormick et al., 1991). For example, filaments formed from wild-type keratin K5 and a modified K14, possessing the head and helix 1A domain of vimentin (i.e., regions known to be important in modulating IF assembly), have a larger diameter (18 nm), more numerous branch points and extensive intertwining compared to either keratin or vimentin IFs.

In vivo, cells expressing both keratins and vimentin possess two independent IF networks (Tölle et al., 1985). Thus, while the head and rod end domains of both group A (keratins) and group B (vimentin) IF proteins have the same function in IF assembly, they are not physically interchangeable. This also implies that a *de facto* function of the head and tail domains is to limit the maximum diameter of an IF polymer to about 10 nm by their requirement for surface or solvent access: Larger diameter structures composed of more tetramers in cross-section would require these domains to be buried within the IF, a process for which there is no good experimental evidence. The 10-nm filament can therefore be regarded as the assembly state of IF proteins that represents a balance between the strong interactions between tetramers (driving the lateral and, through the helical pitch, longitudinal associations) and the necessity for the end domains to remain displayed on the surface of the assembly (and thereby limiting the ultimate diameter). Support for this latter conjecture is demonstrated by the tendency of IF proteins with truncated or modified head and/or tail domains to polymerize into paracrystals or large aggregates (Quinlan et al., 1989; Herrmann et al., 1992).

Nuclear Lamins

The nuclear lamin proteins differ from the other IF proteins in that they possess an additional six heptads in the 1B segment of the central rod domain and, since all of the linker regions (L1, L12 and L2 in Figure 2A) are predicted to have an α-helical structure, it has been proposed that the entire rod domain is α-helical (Conway and Parry, 1988). The nuclear lamin proteins are stabilized within the coiled-coil dimer by a much larger number of interchain ionic interactions relative to the other IF proteins (see Conway and Parry, 1988 for references). Interestingly, the nuclear lamins have a strong propensity to assemble into a variety of paracrys-talline and lattice polymorphs (Parry et al., 1987). These observations led Conway and Parry (1988) to propose that the broad spectrum of nuclear lamin assemblies may be a consequence of their simpler, more regular structure (relative to other IF proteins) and, furthermore, that the nuclear lamins may be the most closely related (of the known IF proteins) to the evolutionary precursor of the IF protein family. As discussed in a following section, more recent studies on the evolution of IF proteins have supported this latter proposal (Weber et al., 1991). The nuclear lamins possess a nuclear localization signal, which is responsible for their segregation from their cytoplasmic relatives. While the nuclear lamina has traditionally been re-garded as a karyoskeletal or structural component of the periphery of the eukaryotic cell nucleus, it is becoming quite clear that the nuclear lamina may indeed penetrate into the interior of the nucleus and, furthermore, is a dynamic structure, particularly during mitosis (Moir and Goldman, 1993). More detailed information on the nuclear lamin proteins are presented in another chapter of this treatise (see Stick and Weber, this volume).

V. SPECIAL ASPECTS OF INTERMEDIATE FILAMENT ASSEMBLY

A.   Pathological States

In addition to these conventional modes of polymerization of IF proteins, several variations on this theme are seen in pathological situations or in extreme speciali-zations. Inclusion bodies of IFs (and other cytoskeletal elements) are characteristic of several major diseases (Yen et al., 1986). As in differentiation, the type of IF protein found is characteristic of the cell type: Keratins are found in Mallory bodies in alcoholic liver disease, desmin in cytoplasmic bodies in cytoplasmic body myopathy, glial fibrillary acidic protein in Rosenthal fibers of neoplasms derived from astrocytes, NFPs in both Pick bodies of Pick's disease (a form of dementia correlated with altered cortical neurons) and Lewy bodies in Parkinson's disease. All of the IF proteins within these inclusion bodies have been shown to be ubiquinated (Lowe et al., 1988; 1989). Ubiquitin is a 76-residue protein that, when attached to other proteins, provides a signal for the specific degradation of the complexes. The formation of ubiquitin–protein conjugates is an ATP-dependent process and is generally elevated in response to stress. It is not known whether the ubiquination of these IF proteins in these disease states represents an authentic defect in the IF proteins themselves or in the regulation of the ubiquitination pathway. In this regard, it is of interest to note that newly synthesized keratins are rapidly incorporated into preexisting Mallory bodies in hepatocytes of griseofulvin-fed mice, suggesting that the Mallory bodies exhibit a dynamic flux like that of more normal IF polymers (Kachi et al., 1993). Others have shown that the proteins of Mallory bodies possess a great number of ε-(γ-glutamyl) lysine cross-links, bonds which are formed by the Ca^{2+}-dependent transglutaminases found in a variety of tissues (Zatloukal et al., 1992). Transglutaminase-catalyzed cross-linking of keratin IFs is a hallmark of terminal differentiation of epidermal keratinocytes, and elevated levels of these enzymes correlate with hyperkeratosis in disease states such as psoriasis (Schroeder et al., 1992).

B. Hagfish Thread Cell Intermediate Filaments

The most fantastic IF polymer is found in the hagfish slime gland thread cell (Downing et al., 1984). These thread cells synthesize IF proteins that have a high degree of homology to the keratins based on their immunoreactivity and polymerization properties (Spitzer et al., 1984; 1988). To quote from Downing et al. (1984): "Thread cell differentiation is remarkable in that the life history of the cell is largely dedicated to the production of a single, tapered, cylindrical, highly coiled and precisely packaged cytoplasmic thread that may attain lengths of 60 cm and diameters approaching 1.5 μm. Each tapered thread, in turn, is comprised almost entirely of large numbers of intermediate filaments (IFs) bundled in parallel." In contrast to all of the other known IFs, the gland cell threads are secreted (following stress of the animal) and function extracellularly, together with the contents of the gland mucus cells, to form a viscous mucus. These threads play an important role in facilitating hydration of the mucus mass and modulate its viscoelastic and cohesive properties (Koch et al., 1991). It remains to be seen what unique sequences these keratin-like IF proteins possess that enable them to assemble into such long and large polymers.

VI. INTERMEDIATE FILAMENT ASSEMBLY AND DYNAMICS *IN VIVO*

One of the dogmas in the field of IFs that has come under fire as a result of studies of IF assembly *in vivo* is that which presents the IFs as static structures. This dogma arose from the early studies of IF structure and distribution *in vivo*: Most studies were performed with interphase cells and most if not all IF proteins were found to be associated with the nonionic detergent-resistant cytoskeletons (see Traub, 1985 or Goldman et al., 1986 for discussion and references). However, the study of dividing cells has shown that many cells disassemble their IF networks as they enter mitosis and reassemble them after mitosis is complete. For example, in fibroblasts, the vimentin IFs retract during late prophase into perinuclear caps, followed at prometaphase by formation of many discrete aggregates of protofilamentous vimentin (Rosevear et al., 1990). During anaphase–telophase, this process reverses: Short vimentin fibrils form and true IFs form as the aggregates disappear (Rosevear et al., 1990). In other cell types, for example keratinocytes, the keratin IF network remains largely intact throughout mitosis, with only a local disassembly of the IFs along the cleavage furrow (Jones et al., 1985). These cell-cycle-related dynamics of IF networks are a result of specific and reversible posttranslational modifications, such as phosphorylation (discussed in another section). The differential response of the different IF networks in various cells represents the sum of the susceptibility of the individual IF proteins to modification by a specific enzyme, as well as to the distribution of the various enzyme systems. For example, vimentin, NFPs and nuclear lamin proteins are phosphorylated by the $p34^{cdc2}$ kinase, whereas keratins are not (Goldman et al., 1991). Since keratins are phosphorylated in HeLa cells during mitosis (Bravo et al., 1982), there must be additional kinases that play a role in modulating IF structure *in vivo*. In contrast to these mitosis-related dynamics, the IF network of interphase was classically regarded as a somewhat static, rigid structure.

In this light, it was somewhat unexpected that biotinylated keratin, when microinjected, was rapidly incorporated throughout the preexisting keratin IF network of cultured epithelial cells in interphase (Miller et al., 1991). Likewise, fluorescently labeled vimentin was incorporated uniformly into the endogenous vimentin IF network of interphase 3T3 cells and, following photobleaching, rapidly repopulated the IFs with no evidence of polarity (Vikstrom et al., 1992). This has led to the suggestions that IFs are highly dynamic entities (Skalli and Goldman, 1991) and that the IF structure *in vivo* is more like a disordered braid, rather than a rigid, crystalline lattice (Vikstrom et al., 1992; Skalli et al., 1992), which is in agreement with the observation of variable numbers of subunit proteins per unit length and the prediction of a seam or discontinuity on the long axis of the IF along which subunits could be exchanged or incorporated (see previous text for references). Point mutations, either naturally occurring in humans or engineered via recombinant DNA technology in transgenic mice, in one or both members of the keratin pairs of skin basal cells (keratins K5 and K14) or suprabasal cells (keratins K1 and K10) have been correlated with abnormal keratin IF networks, the development of the blistering skin diseases epidermolysis bullosa simplex and epidermolytic hyperkeratosis in humans and diseases of similar appearance in the transgenic mouse systems (see Fuchs and Coulombe, 1992; Coulombe, 1993 or Steinert and Bale, 1993 for references). Surprisingly, the correct assembly of keratins K1 and K10 in fibroblasts required the presence of an established keratin IF network containing keratins K5 and K14. In the absence of a keratin K5 and K14 network, the newly expressed keratins K1 and K10 formed dense bundles or aggregates (Kartasova et al., 1993). While retrospectively this may seem to make sense, since basal cells (which express keratins K5 and K14) are the precursors for the suprabasal cells (which down regulate the synthesis of keratins K5 and K14 and express keratins K1 and K10), it is difficult to reconcile with the observation that keratins K1 and K10 readily polymerize into 10-nm filaments *in vitro* and, in fact, represent one of the best studied model systems for IF assembly (see previous discussion of Steinert and collaborators' studies).

A recent commentary has pointed out some of the difficulties associated with establishing an all-encompassing model of IF assembly that adequately explains the results obtained *in vitro* and *in vivo* (Georgatos, 1993). For example, IFs do not disassemble following dilution of an *in vitro* assembly mixture or cell extract, yet new subunits are rapidly incorporated into them both *in vitro* and *in vivo*. Thus, it is difficult to imagine that a simple equilibrium exists between the soluble and polymeric forms of IF proteins. Furthermore, most IF proteins spontaneously assemble *in vitro* when present at more than some low critical concentration, while expression of these same proteins at high levels *in vivo* often results in the collapse or aggregation of the IFs, or even in the failure to assemble unless a partner protein or network is present in the cell, even though such partners are not required for *in vitro* assembly (i.e., the keratin K1/K10 and K5/K14 system mentioned above (Kartasova et al., 1993) or NFPs (Ching and Liem, 1993; Lee et al., 1993)). Another enigmatic observation has been made in cells expressing dominant mutant forms of IF proteins: The more potent of these are capable of disrupting the entire cytoplasmic IF networks even when present at a level of only 1% to 2% of the endogenous, wild-type (and assembly–competent) IF proteins (Gill et al., 1990; Wong and Cleveland, 1990).

Clearly, additional cellular factors, such as other proteins or the exact ionic composition, must play important or even dominant roles in IF assembly *in vivo*. In this context, there has been discussion of the possible role of molecular chaperones in modulating the assembly of IFs *in vivo* (Quinlan, 1993), although their precise role is unclear given that IF assembly proceeds well in their absence *in vitro*. There has even been the suggestion that the tail domain of the type III IF proteins may itself function as a molecular chaperone during IF assembly (McCormick et al., 1993). Although a detailed discussion is beyond the scope of this article, it is appropriate to mention that IF proteins are capable of binding a wide variety of biological molecules including, but not restricted to, negatively charged phospholipids and nucleic acids; furthermore, these compounds exert an influence on IF assembly and stability, primarily through binding to the head domain and thereby perturbing IF structures (see Traub and Shoeman, 1994 for references). To put these reactions into perspective, it is interesting to note that phosphatidylinositol 4-phosphate and phosphatidylinositol 4,5-diphosphate exert a 40-fold stronger influence on vimentin IF formation and stability than the ionic detergent sodium dodecylsulfate (SDS) (Perides et al., 1986). Thus, the assembly of IFs *in vivo* is not simply a function of the IF proteins, but is rather a complex composite of the properties of both the IF proteins and the total cellular milieu.

VII. POSTTRANSLATIONAL MODIFICATIONS OF INTERMEDIATE FILAMENT PROTEINS

IF proteins are subjected to a number of fundamentally different reversible and irreversible modifications. Some of the posttranslational modifications, which give rise to anomalies in assembly and are associated with pathogenic states, were mentioned in a previous section. The spectrum of reversible modifications of IF proteins includes phosphorylation (Inagaki et al., 1990; Yano et al., 1991), O-glycosylation (Haltiwanger et al., 1992; Chou and Omary, 1993; Dong et al., 1993) and ADP-ribosylation (Huang et al., 1993). These modifications are largely restricted to the head domain of the IF proteins (Robson, 1989; see Stewart, 1993 for more references) and appear to be, to a certain extent, mutually exclusive events because they occur on similar residues. That is, glycosylation or ADP-ribosylation apparently blocks the phosphorylation of the head domain and may represent modifications of general importance in modulating the structure and function of IFs (Chou and Omary, 1993). Phosphorylation of a single site in the head domain of GFAP is sufficient to cause disassembly of GFAP IFs (Nakamura et al., 1992), and limited site-specific phosphorylation of vimentin and other IF proteins causes similar effects (Geisler et al., 1989; Hisanaga et al., 1990). The phosphorylation state of the IF proteins is highly dynamic throughout the cell cycle (Goldman et al., 1991; Skalli and Goldman, 1991), resulting in the formation of a variety of nonfilamentous structures (Franke et al., 1982b), and is a complex function of the action of a multitude of protein kinases and protein phosphatases (Eriksson et al., 1992a; 1992b). This partial destabilization of IF structure may play a role in subunit exchange in IFs *in vivo*, in addition to the irregularity or seam that is thought to exist along the long axis of the IF (Figure 4).

In general, the IF proteins are excellent substrates for the ubiquitous calpains (calcium-activated, neutral thiol proteases; reviewed in Shoeman and Traub, 1990a), and the IFs themselves become labile and fall apart after treatment with these proteases, primarily as a result of truncation of the head domain. One of the earliest observations of this phenomenon was in Wallerian degeneration of central nervous system neurons, a term applied to the characteristic changes that occur in nerve fibers distal to a nerve transection, during which NFPs are degraded by a calcium-activated protease (Schlaepfer and Micko, 1978). Calpains not only degrades NFPs, but may also remove small fragments resulting in modified NFPs that are apparently important in morphogenesis or modulation of IF structure or activity (Nixon et al., 1983; Schlaepfer et al., 1985).

A cell-cycle-specific degradation of vimentin (presumably calpain-modulated as the cleavage products are the same as those obtained after treatment of vimentin with calpain *in vitro*) has been observed in analysis of cellular protein patterns in two-dimensional gel electrophoresis (Bravo and Celis, 1980; Bravo et al., 1982). The degradation of vimentin is highest in the G_2 phase and is barely detectable in G_1 and S phases and has been proposed to be an indicator of some as yet unknown but important event in the process of cell division. A general scheme has been put forth, the "activation of the membrane" theory (Suzuki and Ohno, 1990), which proposes the activation of IFs and IFAPs by the unusually high concentrations of Ca^{2+}-activated protease around and in the nucleus and other membrane systems as soon as optimal concentrations of free Ca^{2+} are attained at the border regions.

Recent studies have shown that the end domains of IF proteins are readily released from IFs both *in vitro* and *in vivo* by viral proteases (Shoeman et al., 1990; Höner et al., 1991; Chen et al., 1993; Lindhofer et al., 1993), a process we believe to be important in the pathogenesis of, and the development of cancer in, human immunodeficiency virus type 1 (HIV-1) infection (Shoeman et al., 1992). Other viruses also interact with, and perturb, the cytoskeleton in general (Luftig, 1982) and IFs in particular: for example, adenovirus (Belin and Boulanger, 1987; White and Cipriani, 1989), frog virus 3 (Murti et al., 1988), human papillomavirus 16 (Doorbar et al., 1991), human respiratory syncytial virus (Garcia-Barreno et al., 1988), Moloney mouse sarcoma virus (Singh and Arlinghaus, 1989; Bai et al., 1993) and simian virus 40 (SV 40; Ben-Ze'ev, 1984).

As mentioned in a previous section, IF proteins are extensively modified by the addition of ubiquitin residues or by the Ca^{2+}-dependent, transglutaminase-catalyzed formation of glutamine–lysine cross-links in various disease states; these same modifications presumably occur to lesser extents in normal physiological states. Another Ca^{2+}-dependent modification of IF proteins is the deimination of arginine to citrulline (Inagaki et al., 1989). This modification is restricted to residues located in the head domain and results in a disassembly of the IFs. Thus, calcium plays an important role in regulating a wide variety of modifications of IF proteins, which is reflected by the intimate associations of IFs and Ca^{2+}-dependent enzyme systems (e.g., calpain) with membranes. This close association permits the rapid modification and resultant redistribution of IFs due to loosening of their anchorage to membranes and other structures in response to intra- and extracellular signals. Together with other posttranslational events like phophorylation, the Ca^{2+}-mediated modifications probably serve to activate the IF proteins by permitting their release from the IFs in response to physiological signals.

These posttranslational modifications are not restricted to the cytoplasmic IF proteins. As discussed in detail in another chapter of this treatise (see also Nigg, 1992a; 1992b), the nuclear lamins are also subjected to isoprenylation, proteolysis and carboxymethylation concomitant with their import into the nucleus, and to cyclic phosphorylation/dephosphorylation causing disassembly/reassembly of the nuclear lamina throughout the cell cycle. Interestingly, phosphorylation of lamin B in response to phorbol ester treatment of cells prevents its import into the nucleus, demonstrating that the nuclear transport process is dependent on the activity of cytoplasmic kinases and is responsive to extracellular signals (Hennekes et al., 1993).

VIII. EVOLUTION

Analysis of the structure and evolution of IF protein genes has proceeded rapidly (Weber et al., 1991) as a result of the cloning and sequencing of genes coding for a nuclear lamin protein (Döring and Stick, 1990) and an invertebrate cytoplasmic IF protein (Dodemont et al., 1990); this has enabled the filling-in of gaps in previous evolutionary schemes (Steinert and Roop, 1990). The localization and number of introns within the open reading frames of the IF protein genes has been especially useful in delineating the family tree of these IF protein genes. All IF protein genes are thought to have their roots in a single, common progenitor that code for an IF protein with a nuclear localization. The present day nuclear lamin B-type genes are proposed to be directly derived from and, of the known IF genes, to be most closely related (i.e., show the least divergence) to this primordial IF protein gene (Stick, 1992); the lamin A-type genes are probably derived from the B-type genes by exon shuffling (Stick, 1992). The dichotomy between the nuclear and cytoplasmic IF protein genes occurred early in evolution, as indicated by the high degree of similarity between the nuclear lamin genes and the genes coding for invertebrate, nonneuronal IF proteins: Both sets of genes code for a six heptad insert in coil 1 B of the rod domain that is missing from the other IF proteins and the positions of 8 of the 10 introns in these genes are identical. The archetypical cytoplasmic IF protein gene probably arose from the laminlike progenitor by the loss of two nuclear localization signals as a result of both acquisition of a new intron (intron 7) and loss of intron 10 as well as substitution of a new, last exon (Weber et al., 1991). The vertebrate cytoplasmic IF protein genes, with the exception of the NFP and nestin genes, have a similar organization, and the locations of most introns are identical; the most obvious differences are within the regions coding for the end domains, as might be expected from the homologies at the protein level. The vertebrate type IV and type VI protein genes (coding for NFPs and nestin) differ from all of the other IF protein genes by having only two or three introns, prompting the proposal that these genes arose as a result of a mRNA transposition event (Lewis and Cowan, 1986) and thus represent one branch of the IF family tree (Dahlstrand et al., 1992).

The squid NFP gene differs from its vertebrate counterparts in that low and high molecular weight NFPs are produced as a result of alternative splicing of a single gene product (Szaro et al., 1991; Way et al., 1992). Likewise, a *Drosophila* gene for a nuclear lamin has an unusually small number of introns and is an exception to the general scheme outlined here (see Lin and Worman, 1993).

IX. INTERMEDIATE FILAMENT-ASSOCIATED PROTEINS (IFAPs)

A. Cellular Proteins

The three major cytoplasmic filament systems of eukaryotic cells are cross-linked within and among themselves, as well as to various organelles, via associated proteins (Foisner and Wiche, 1991). The best studied of these are the microtubule-associated proteins (MAPs), which perform a variety of functions in modulating microtubule structure and activities. Furthermore, antibodies to MAPs have been used to identify certain IFAPs: The monoclonal antibody MA-01 recognizes an IFAP in 3T3 cells that shares an epitope with, but is distinct from, MAP2, and microinjection of this antibody causes a collapse of vimentin IFs (Dráberová and Dráber, 1993). IFAPs have been identified primarily on the basis of their coisolation or colocalization with IFs, with the net result being that very little is known about the function of most of them. The list of IFAPs (see Steinert and Roop, 1988; Yang et al., 1990 for references) includes IFAP-300k, filaggrin, plectin, synemin, para-nemin, epinemin, p50, NAPA-73 and p230. Other IFAPs, originally identified through their association with other cellular components, include nonerythrocyte spectrin, ankyrin, MAP-2, filamin (also identified as gyronemin, Brown and Binder, 1992) and desmoplakins.

Desmoplakins, which are components of desmosomes, apparently associate with IFs via their C-terminal domain (Stappenbeck and Green, 1992) and may thus be one point of anchorage of IFs with desmosomes, along with a 140-kDa protein that is immunologically related to lamin B (Cartaud et al., 1990). Interestingly, the carboxy-terminal domain of desmoplakin I interacts differentially with vimentin and keratin IFs, binding more tightly to the latter (Stappenbeck et al., 1993).

Filaggrin, a histidine-rich basic protein named for its ability to bundle IFs *in vitro*, is the only IFAP for which detailed information on its interaction with IFs is known (Dale et al., 1990). A large precursor containing many filaggrin repeats, profilag-grin, is synthesized and extensively phosphorylated in the initial stages of devel-opment of keratinizing epithelia. The conversion to filaggrin occurs as a result of proteolysis after dephosphorylation of profilaggrin. *In vitro*, filaggrin, but not profilaggrin, is capable of bundling keratin IFs into large aggregates or macrofibrils reminiscent of those seen in the terminally differentiated keratinocytes. This binding probably occurs as a result of simple ionic and hydrogen bond formation between β-turns of filaggrin and the rod domain of keratins, resulting in an ionic zipper (Mack et al., 1993). *In vivo*, filaggrin interacts with keratin IFs during normal epidermal differentiation (Manabe et al., 1991). Filaggrin is therefore regarded as the IF-bundling protein of keratinocytes.

Likewise, plectin, which interacts with both the nuclear lamin and cytoplasmic IF proteins, is thought to be a cytoplasmic element cross-linker and to function in the formation of cell junctions (Wiche, 1989); in contrast to filaggrin, plectin has a wide cell and tissue distribution (Wiche et al., 1983). Deletion experiments have shown that the plectin carboxy-terminal domain, which contains a sixfold tandem repeat, is responsible for the association of plectin with IFs *in vivo* (Wiche et al., 1993). Like the situation seen with IFs, more than one type of IFAP may be found within a given cell, but the exact complement of IFAPs present seems to be cell or tissue specific. IFAPs are thought to modulate the interactions of IFs with themselves and other cellular components, although the hows and whys of these processes have not been addressed experimentally.

Several enzymes have recently been shown to be associated with IFs and may therefore be regarded as IFAPs: protein kinase C (Murti et al., 1992), cAMP-dependent protein kinase (Dosemeci and Pant, 1992) and the microtubule-associated motor protein kinesin (Gyoeva and Gelfand, 1991). The implications of an IF-bound protein kinase are clear: The close association should enable the rapid modification of IF structure and function upon activation of the kinase, in response to intra- or extracellular signals. In many cell types IFs co-align with microtubules, and many treatments, such as application of colchicine, that cause disassembly of microtubules also effect the redistribution or reorganization of IFs (Forry-Schaudies et al., 1986). Kinesin has been shown to be responsible for the directed movement of membranous organelles along microtubules. Surprisingly, the microinjection of antikinesin antibodies into human fibroblasts caused a collapse of the IF network without affecting the microtubules, which suggests that kinesin is also intimately involved in cross-linking IFs and microtubules (Gyoeva and Gelfand, 1991).

B. Viral Proteins

A few viral proteins have been reported to be associated with IFs (or the IF proteins) and thus make up a special class of IFAPs. These include the adenovirus E1B 19K protein, which plays a role in cell transformation and causes the collapse of vimentin and lamin networks (White and Cipriani, 1989; 1990), the Epstein Barr virus latent membrane protein, which perturbs keratin expression and disrupts keratinocyte differentiation (Liebowitz et al., 1987; Dawson et al., 1990; Fahraeus et al., 1990), the human papillomavirus (HPV)-16 E1-E4 protein, which causes a collapse of the keratin networks in keratinocytes (Doorbar et al., 1991) and the Moloney murine sarcoma virus v-Mos protein kinase, which can not only phosphorylate vimentin but also associate with vimentin in mitotic cells (Bai et al., 1993).

The HPV-16 E1-E4 protein is remarkable in that it is highly divergent among viral types; it is not virally associated, but rather accumulates in the host cell cytoplasm, and it causes the collapse of keratin IFs without affecting either the other IF networks (i.e., the nuclear lamina and/or the vimentin IFs of other, nonepithelial cells) or the microtubules or microfilament portions of the keratinocyte cytoskeleton (Doorbar et al., 1991). Undoubtedly, future research will uncover additional viral proteins that are IFAPs whose functions will include either disruption of IFs to permit viral budding to take place or subversion of the IFs' ability to anchor or organize structures, such as those seen in the assembly centers of frog virus 3 (Murti et al., 1988).

X. INTERACTIONS WITH OTHER CELLULAR SYSTEMS AND ORGANELLES

A. Microtubules, Membranes and Lipids

In addition to the interconnections of the IFs with the microtubules, stress fibers and organelles through the IFAPs, IFs in axons are directly associated with microtubules via their phosphorylated tail domains (Miyasaka et al., 1993), and in most cell types IFs are attached to various membranous structures including the nucleus, mitochondria, the Golgi apparatus and the plasma membrane and its intrinsic structures such as desmosomes, etc. (see Carmo-Fonseca and David-Ferreira, 1990 and Goldman et al., 1990 for references). IFs have been shown to serve as transient docking sites for inner nuclear membrane vesicles during mitosis (Maison et al., 1993). IFs are often found attached to the plasma membrane at sites of cell surface attachment of extracellular matrix components, raising the possibility that there is a direct connection between these elements (Green and Goldman, 1986). This supposition is further supported by the observation of a direct interaction between IFs and the tail (cytoplasmic) domain of desmosomal cadherins (Troyanovsky et al., 1993).

The interaction of IFs with the plasma membrane seems to be partly via the head domain of the IF proteins, which interacts both with membrane-associated proteins such as ankyrin and spectrin (Georgatos et al., 1987; Langley and Cohen, 1986) and directly with lipids; this is not surprising given its amphipathic nature, resemblance to mitochondrial signal sequences (Ouellet et al., 1988) and ability to bind to and perturb negatively charged lipid membranes (Perides et al., 1986; 1987a; Horkovics-Kovats and Traub, 1990). Additionally, the rod domain of IF proteins binds neutral lipids, including long chain fatty acid esters and diglycerides (Traub et al., 1987a), a property that is probably important in the interaction of IFs with membranes and also significant in the regulation of IF–membrane-associated enzymes in conjunction with their affinity for phosphorylated phophatidylinositols. The tail domain of IF proteins is responsible for the binding of IFs to the nuclear envelope, either directly to lamin B or to an intermediate receptor protein (Georgatos and Blobel, 1987a; 1987b). These data were originally interpreted as supporting the hypothesis that IF assembly and organization is vectorial and polar, beginning with the association of the tail domains of the nascent filament with the nuclear surface and ending with the association of the head domains of the IF with components of the plasma membrane. While this hypothesis has not been directly refuted, it has fallen somewhat by the wayside as a result of the aforementioned *in vivo* experiments that have shown that the assembly is not vectorial and by the predictions of model building (Figures 2–4) that suggest an individual IF is not polar, since it is composed of antiparallel protofilament components. Given the association of the IF proteins with various membrane-bounded organelles, it is not surprising that anomalies in membrane structures are a hallmark of infection of cells with HIV-1 (discussed in Shoeman et al., 1992), since the IF protein vimentin is cleaved by the HIV-1 protease *in vivo* during infection of lymphocytes (Lindhofer et al., 1993).

A trans-cellular desmin–lamin B IF network has been described in cardiac myocytes, resulting in a connection of the sarcolemma and Z-disks with the nuclear surface and nuclear lamina (Lockard and Bloom, 1993). In this study, the cytoplasmic and nuclear IFs were observed to contact each other at or near the nuclear pores, although it was not possible to determine if the anchorage was direct or mediated by other proteins.

During differentiation of pre-adipocytes in cell culture, vimentin IFs are reorganized and form a cage surrounding the membrane-free lipid droplets (Franke et al., 1987 and references therein). There have also been reports of a colocalization of glycosphingolipids with IFs in a variety of cell types (Gillard et al., 1991; 1992) and of the association of cholesterol ester-containing lipid droplets with vimentin IFs (Almahbobi and Hall, 1990; Almahbobi et al., 1991). In all of these cases, the association of the lipid and IFs seems to be direct; that is, the lipid droplets have no detectable membrane.

B. Proteins Destined for Autophagy

It has been shown that a variety of proteins, such as glycolytic enzymes and Sendai virus envelope proteins, associates with IFs following their microinjection or introduction via membrane-vesicle fusion into cells (Doherty et al., 1987; Earl et al., 1987). The association of these proteins is rapid, taking from one to several hours, and precedes their degradation in lysosomes, which is a slow process with a $t_{1/2}$ of several days. The transfer of these proteins could also be blocked by the protease inhibitor leupeptin, so that the proteins accumulated on and remained associated with the IFs. It is likely that this system of protein transport via IFs to autophagic vesicles occurs with endogenous proteins during normal cell metabolism.

C. Nucleic Acids

As mentioned previously, the IF proteins are multidomain proteins and each domain possesses a unique structure with unique biochemical properties. One outstanding property of the IF proteins is the ability of the arginine-rich head domain to bind nucleic acids. While the significance of this property has baffled many researchers, it has been widely employed to isolate the IF proteins via affinity chromatography (Nelson et al., 1982). Vimentin IFs have been demonstrated to bind ribosomal subunits *in vitro* via their RNA (Traub et al., 1992b), making it likely that this also occurs *in vivo* and thereby providing an understandable mechanism for the observed association of components of the protein-synthesizing machinery with the cytoskeleton. While this association has been best documented to occur via microfilaments (see Hesketh and Pryme, 1991), there are specific instances where the cotranslational assembly of IF proteins has been observed (see Fulton and L'Ecuyer, 1993). For example, GFAP mRNA was localized to GFAP IFs alone or to the IFs associated with ribosomes and polyribosomes (Erickson et al., 1992), and vimentin has been observed to undergo cotranslational assembly (Isaacs et al., 1989). Other RNA-containing complexes, such as prosomes (Olink-Coux et al., 1992), and RNP particles containing the Ro antigen (Carmo-Fonseca and David-Ferreira, 1990) are also bound to cytoplasmic IFs. Indirect immunofluorescence has shown ribosomes and eukaryotic elongation factor 2 to be associated with IFs, especially in G_0-arrested cells (Shestakova et al., 1993).

XI. MODULATION OF INTERMEDIATE FILAMENT NETWORKS BY EXTRACELLULAR STIMULI

A. UV Light and Thermal Stress

In addition to the changes in IF networks that accompany progression through the cell cycle (see Goldman et al., 1991), a variety of external stimuli can modulate IF structure and/or function. For example, exposure of cultured human keratino cytes to polychromatic UV or UVC radiation results in a disruption of the keratin IF network that is reminiscent of the sunlight-induced changes characteristic of sunburn cells *in vivo* (Zamansky and Chou, 1990). This treatment also inhibits the reorganization of keratin IFs that is an important marker for the calcium-induced differentiation of cultured keratinocytes. Elevated temperature, which induces the heat shock response, causes a variety of changes in the cytoskeleton of mammalian cells. Heat shock causes the collapse of the vimentin-IF network in fibroblasts (Welch and Suhan, 1985), and in mammary epithelial cells the induced collapse of keratin IFs is accompanied by an aggregation of organelles and inhibition of protein synthesis (Shyy et al., 1989). It was observed that during the initiation of and recovery from heat shock, the redistribution of the keratin IFs coincided with the inhibition and resumption of protein synthesis, the disassembly and reassembly of polyribosomes and the translocation of organelles (Shyy et al., 1989). Thermotolerance in both protein synthesis and the keratin IF system was induced by an initial heat stress and recovery period; somewhat surprisingly, protein synthesis was not required for the acquisition of thermotolerance in the keratin IFs (Shyy et al., 1989). This suggests that the integrity of the IF network may play a role in modulating protein synthesis, which is also supported by the observation that heat-resistant mutant cell lines derived from Chinese hamster ovary (CHO) cells contain elevated levels of vimentin IFs that do not collapse at elevated temperatures (Lee et al., 1992). Interestingly, thermal stress caused by chilling also results in a collapse and aggregation of keratin IFs *in vivo*, which is reversible upon warming (Schliwa and Euteneuer, 1979). *Xenopus* wild-type vimentin has been shown to exhibit temperature-sensitive assembly *in vitro* and *in vivo*, forming IFs at 28 °C and aggregates above 34 °C (Herrmann et al., 1993). While the exact biochemical basis for this remains obscure, it was observed that various chimeric vimentin molecules, containing either half of mammalian vimentin in place of *Xenopus* vimentin, possessed the ability to assemble into normal IFs at 37 °C. This suggests that vimentin genes that code for thermotolerant proteins have evolved concomitantly with the acquisition of a regulated, higher (37–39 °C) body temperature in mammals.

B.   Extracellular Stimuli Probably Modulate Known Pathways of Protein Modification

The molecular basis for the effect(s) of UV radiation and elevated temperature on IFs is not well defined; in the latter case, it is likely that the activity of a variety of the cellular pathways that modify IF proteins (for example, protein kinases) is affected, resulting in a net change in the modification state of the IF proteins and thus in an alteration of the IF network. Striking examples of this type of effect are seen when exogenous agents are added to cultured cells: Acrylamide causes a collapse of keratin IFs and a net reduction in the phosphorylation level of the keratins (Eckert and Yeagle, 1988; 1990), aluminum causes aggregation and abnormal phosphorylation of NFPs (Gilbert et al., 1992 and references therein) and cAMP (or agents that raise the intracellular cAMP concentrations) applied to malignant CHO cells results in phosphorylation of vimentin and reorganization of IFs early in the reverse transformation process (Chan et al., 1989).

XII.   FUNCTIONS OF INTERMEDIATE FILAMENTS

A.   Mechanical Integrators of Cellular Space

IFs have been classically regarded as structural elements involved in organizing and stabilizing the various cellular compartments (Lazarides, 1980; see Traub, 1985 and Skalli and Goldman, 1991 for references). For example, the recently identified novel IF proteins of the vertebrate eye lens cells, filensin and phakinin, form a cortical IF network attached to the plasma membrane and, through the interaction of filensin with vimentin, anchor the vimentin IF network to the plasma membrane (Merdes et al., 1991; 1993). The identification of this anchorage system provides an explanation for the previously observed end-on association of the vimentin IFs with the plasma membrane in these cells (Ramaekers et al., 1982).

NFPs are thought to impart structural integrity to neurons, in part because of their cross-linking via the long tails of NFP-H and their associations with microtubules. NFPs are also involved in axonal extension and regeneration and, together with microtubules, participate in the transfer of metabolites and molecules from the cell soma to the synapses (Angelides et al., 1989; Hollenbeck, 1989). NFPs have been proposed to play a role in the determination of the diameter of axons, since there is a good correlation between the number of NFP IFs and the axonal caliber of myelinated axons (Hoffmann and Cleveland, 1988 and references therein). Further evidence of this hypothesis has been provided by a study (Zhao et al., 1993) on the large myelinated fibers of the peripheral nervous system of a Japanese quail mutant deficient in NFPs (Ohara et al., 1993). These quail have a smaller number of large and a larger number of small myelinated fibers relative to the normal, control quail (Zhao et al., 1993). The net significance of these observations is unclear since these mutant quail, except for some generalized quivering, seem to be more or less normal in their behavior and morphology, despite a complete lack of NFP IFs (Ohara et al., 1993; see following discussion). Overexpression of NFP proteins *in vivo* results in aberrations of NFP IF organization as well as pathological and behavioral symptoms reminiscent of neurological disease (Xu et al., 1993; Côte et al., 1993).

In keeping with their proposed role as mechanical organizers of intracellular space, desmin IFs have been shown to play a role in the early events of human skeletal muscle cell differentiation by binding to and translocating titin (van der Ven et al., 1993), one of the important organizing molecules of the sarcomere (Fulton and Isaacs, 1991). However, since myogenesis still occurs in the absence of desmin IFs (Schultheiss et al., 1991), this organizing role of desmin IFs in myogenesis is not absolute, but rather is redundant to or operates in concert with other myofibril-linking networks (discussed in van der Ven et al., 1993).

B. Keratins and Skin Diseases

The cytoskeletal view of IF function has been strengthened by the recent demonstrations of a correlation between the fidelity of keratin IFs and blistering skin diseases (Fuchs and Coulombe, 1992). While there is no question that the conclusions derived from these studies are correct and that the IFs in general function as cytoskeletal elements, a point of contention among researchers in this field has been whether this is the "only" function of IFs. Given the multidomain structure and diverse reactivities of the head, rod and tail domains, we believe it likely that IFs function in multiple roles in many cell types. Thus, the keratin IF networks of terminally (in the truest sense of the word, since the ultimate fate of these cells is to dehydrate, die and thereby form a strong and resistant covering) differentiated keratinocytes *in vivo* fulfill a role that is unlikely to be the same as that of the other IF systems in nonepithelial cells. It must also be remembered that whatever functions IFs fulfill in nonepithelial cells, they must be of a fine-tuning nature or be redundant, since it is clear that several cell types do not possess detectable IFs and are viable nonetheless; in addition, it has been shown that IF networks collapse after microinjection of anti-IF antibodies (Klymkowsky, 1981; Klymkowsky et al., 1983; see Klymkowsky et al., 1989) and the treated cells are viable and demonstrate no detectable defects (reviewed in Sarria and Evans, 1989). If nonepithelial IFs are not absolutely required, yet are ubiquitously distributed among eukaryotic cells, what possible function(s) might they possess?

C. Transport of Cholesterol

A specific function of vimentin IFs has clearly been demonstrated (Sarria et al., 1992) in human adrenal cortex tumor cells lacking endogenous vimentin and in cell lines stably transfected with vimentin cDNA and expressing varying amounts of vimentin. In these cells, it could be demonstrated that the ability to transport low-density lipoprotein-derived cholesterol from the lysosome to the site of cholesterol esterification was a function of the vimentin IF network (Sarria et al., 1992). This is in keeping with the previously discussed observations of a close physical association of lipids and lipid droplets with the cytoplasmic IF networks, as well as with the concept that IFs transport proteins to the lysosomes as a normal link in the process of autophagy (Doherty et al., 1987; Earl et al., 1987).

D. Modulation of Mucus Properties: Special Reactions of Special Polymers

The hagfish gland thread cell IFs are unique among the known IFs in that they are secreted and function extracellularly (Koch et al., 1991). They have been shown to facilitate the hydration and modulate the viscoelastic and cohesive properties of the mucus that forms following the release and rupture of the gland thread cells and gland mucus cells from the epidermal slime glands (Koch et al., 1991). The mucus glycoproteins are bound to the surface of the IF threads, which uncoil but do not depolymerize, remaining as threads many centimeters long (Koch et al., 1991). It remains to be seen what special primary and secondary structures the hagfish keratin-like IF proteins possess that enable them to engage in these unique activities. As pointed out by Koch et al. (1991), it is possible that specialized IFs are hitherto unrecognized components of mucus in a variety of organisms and locations.

E. Naturally Occurring Mutants and Dominant-Negative Mutants

A confusing body of literature has accumulated concerning the description of natural mutations and attempts to employ recombinant DNA technology to elucidate the *in vivo* function of various IF proteins. In some of these experiments, the operating premise has been that the IF proteins function only as filaments and no special care was taken to completely eliminate IF protein gene expression or to consider the effect of expression of truncated IF proteins. Several examples of the seemingly discordant results of the various approaches will serve to illustrate the difficulty in both interpreting these experiments and trying to define a *single* function for all IF proteins.

The expression of dominant-negative mutants of various IF proteins causes collapse of the IF network in these cells (for example, Schultheiss et al., 1991). To date, with the exception of the keratin IFs and skin diseases, no overt phenotype has been associated with this phenomenon, which has unfortunately given rise to the dogma that IFs (or their proteins) are not involved in other cellular processes. We believe caution is necessary in the formulation of such dogma, since the dominant-negative mutant experiments formally demonstrate that IFs as such are not necessary for differentiation, but do not address the question of whether IF subunit proteins, either free or in another aggregation state, exert important functions. Likewise, a naturally occurring mutant strain of Japanese quail, which does not express detectable levels of NFPs as a result of a nonsense mutation within the 1A region of the rod domain of the NFP-L protein, is (with the exception of some generalized quivering), completely normal and viable, demonstrating that NFP IFs are not absolutely needed for CNS development or function (Ohara et al., 1993). In these quail mutants, one would expect to find a truncated NFP-L protein which, while it is not capable of forming IFs, should still exert functions associated with the head domain of IFs (i.e., nucleic acid binding; see below). Another experimental approach is to employ knockout mutations of IF protein genes in embryonic stem cells and to regenerate transgenic animals. Targeted inactivation of both keratin K8 alleles in mice results in a midgestational lethality in 94% of the embryos; 6% survive and develop into apparently healthy, fertile animals (Baribault et al , 1993). Thus, aside from a critical point in embryogenesis, keratin IFs are not necessary in those epithelia where K8 is normally expressed. This includes the epithelial cells of the digestive tract (which normally express K8 and K 18). Since these cells are subjected to extremes of mechanical stress and yet function adequately in the K 8 knockout mice, it is unlikely that the only function of all keratin IFs is to provide mechanical stability to a cell or tissue.

On the other hand, a complete inhibition of desmin expression was found to block subsequent differentiation in myoblasts (Choudhary and Capetanaki, 1990). There has been one report demonstrating an inhibition of differentiation of astrocytic processes in an astrocytoma cell line following expression of RNA antisense to GFAP mRNA (Weinstein et al., 1991) and another demonstrating a transient requirement for vimentin in neuritogenesis *in vitro* revealed by intracellular application of antivimentin antibodies and antisense oligonucleotides (Shea et al., 1993).

To summarize a variety of seemingly discordant observations on the expression of mutant IF proteins *in vivo*: Alterations in keratin IFs may reduce the ability of some (but not all) epithelia to resist mechanical forces resulting in lethal damage to embryos or blistering skin diseases. Alterations in structure and intracellular distribution of filaments from the type III or IV proteins are generally associated with no apparent phenotype, whereas the complete absence of these IF subunit proteins may have drastic and demonstrable effects on cell structure and function, particularly in development. One consistent exception to this rule is the ectopic expression or overexpression of endogenous IF proteins in the lens of transgenic animals, which results in cataract formation (Capetanaki et al., 1989; Pieper et al., 1989; Dunia et al., 1990; Monteiro et al., 1990); most likely, this is due to a perturbation in the enucleation process of the lens fiber cells rather than to a specific effect of the IF proteins (Dunia et al., 1990). Our conclusion is that the nonepithelial IFs as such are not essential for cell functions, but rather that the IF subunit proteins are. That is, they can still function when present in non-IF aggregates or when truncated forms of the proteins are expressed, as in the quivering quail system.

F. Binding of Nucleic Acids

The nonepithelial IF proteins bind nucleic acids via their non-α-helical, N-terminal polypeptides. The nucleic acid binding site of the vimentin head domain has a high degree of structural homology to the DNA-binding regions of several prokaryotic ssDNA-binding proteins (Traub et al., 1992a and references therein). Vimentin filaments, protofilaments and even the isolated head domain peptide are all capable of binding nucleic acids. Vimentin IFs preferentially bind dsDNA in an apparently cooperative manner. The binding propensities of the various forms of vimentin are similar: There is a sequence specificity and a preference for G-rich sequences (Shoeman et al., 1988; Traub and Shoeman, 1994), and vimentin IFs are able to selectively bind G-rich repeat sequences of mouse genomic DNA (Wang and Traub, unpublished results) that are homologous to G-rich repeats that occur within regulatory regions of active genes (Vogt, 1990). The additional ability to bind a variety of DNA repeat elements that share homology to G-rich telomere repeats and, surprisingly, AT-rich centromere repeats (Shoeman et al., 1988; Shoeman and Traub, 1990b; Wang and Traub, unpublished results) suggests that the nucleus-associated, cytoplasmic IFs may play a role in fixing the location and thereby determining the exposure and activity of chromatin domains in the nucleus. While a complete discussion of this topic is beyond the scope of this chapter (see Traub and Shoeman, 1994 for further details), these suggestions are compatible with observations that chromosome telomeres are attached to the nuclear matrix via their G-rich repeats (De Lange, 1992) and that telomeres as well as centromeres are often localized adjacent to the nuclear envelope (Manuelidis and Borden, 1988; Gilson et al., 1993).

G. Regulation of Gene Expression

A Hypothesis

We have proposed that the IF proteins play a role in the regulation of gene expression and may participate in DNA replication (Traub et al., 1987b; Traub and Shoeman, 1994). Puck and coworkers have shown that vimentin plays an important role in the reverse transformation reaction (Chan et al., 1989) and have put forth a model in which the cytoskeleton (including IFs) participates in the global regulation of gene expression (Puck and Krystosek, 1992). In their proposal, gene activation occurs via a cytoskeleton-dependent anchorage and exposure of repetitive DNA sequence elements in the nuclear periphery. In our hypothesis, the head domain of cytoplasmic IF proteins is responsible for the binding of nucleic acids and the rod domain may play a role in the transient binding of core histones, resulting in IF-mediated anchorage and activation (or exposure) of chromosomal domains or genes in the nuclear periphery. Since the head domains of the IF proteins each have a unique primary sequence, they presumably have unique DNA-binding properties and may each be responsible for activating, in concert with other transcriptional factors, large regions of the genome that code for the variety of differentiation-specific proteins necessary for a particular type of global gene expression; that is, vimentin participates in a cascade of reactions that lead to the exposure and transcription of genes responsible for the differentiation of mesenchymal cell lineages, desmin for muscle cells, NFPs for neurons, etc. In this respect, it is important to point out that the nonepithelial IF proteins share a number of structural and functional features with known regulatory DNA-binding proteins, including a dimerization motif adjacent to a positively charged DNA-binding domain (Lamb and McKnight, 1991) and considerable homologies within the DNA-binding domain amino acid sequences (Capetanaki et al., 1990; Traub and Shoeman, 1994).

One function of the filament form of IF proteins may be to organize their aggregation state, permitting multisite and cooperative interactions with repetitive DNA sequence elements, as has been shown to be important for other proteins involved in the global regulation of gene expression, such as the *Drosophila zeste* protein (Chen and Pirrotta, 1993; see Traub and Shoeman, 1994). Another function of IFs may be to simply sequester the highly reactive head domains of their constituent subunit proteins and therefore regulate their activity. All of the various posttranslational modifications that lead to reorganization of the IF network would then, in this context, be mechanisms by which the nucleic acid binding or gene regulation would be activated! It should be emphasized that in this model the IF proteins are postulated to play a "fine-tuning" role and must function together in a concerted fashion with a battery of other proteins to permit regulation of gene expression. That is, IFs are not required for cell viability, but rather play an important role in regulating differentiation. This would explain both the existence of cell lines with no detectable cytoplasmic IF protein expression (these cells are characteristically undifferentiated and when induced to differentiate express IF proteins) (see Giese and Traub, 1986) and the specific cascade of IF protein gene expression observed during embryonic development.

Feedback Regulation of Gene Expression by Intermediate Filament Proteins

Support for this hypothesis is also provided by the observations of a feedback regulation at the level of IF protein expression, as well as an apparent regulation of the coordinate expression of pairs or partners of IF proteins. In transfected mouse cells, expression of human keratin K 18, the homologue of mouse Endo B, resulted in a reduction of expression of the mouse keratin Endo B (and a degradation of the excess human protein produced) so that the amount of the type I keratin (human keratin K 18 plus the mouse Endo B) equalled that of the type II keratin (mouse Endo A; Kulesh and Oshima, 1988). This effect was noted to be due at least in part to a reduction in the rate of synthesis of the mouse Endo B, although it was not determined if it occurred at the level of transcription or posttranscriptionally. In this same system, suppression of mouse Endo B expression by antisense RNA inhibited the expression of its polymerization partner Endo A by lowering the level of stable Endo A mRNA (Trevor et al., 1987), suggesting that Endo B either represses Endo A transcription or exerts an influence on Endo A mRNA stability. In the previously mentioned Japanese quail NFP-L mutant (Ohara et al., 1993), the amount of the NFP-L mRNA was reduced by ~95% and the levels of the NFP-M and NFP-H mRNAs were also reduced. It is attractive to speculate that the truncated NFP-L protein binds to the G-rich CpG islands that occur in the 5'-flanking region of the NFP protein genes (Bruce et al., 1993) and thereby regulates NFP gene expression. All of these results are compatible with a direct interaction of the IF proteins with the transcriptional machinery that controls the levels of the IF proteins in a feedback mechanism.

The Problem of Compartmentalization

One major reason why investigators in this field have been reluctant to seriously consider this concept is the fact that cytoplasmic IFs and nuclear DNA are, by definition, compartmentalized and separated by the nuclear membrane. While IFs are tightly attached to the surface of the nucleus, they possess no known nuclear localization signal and there is no evidence that they directly penetrate the double nuclear membrane as authentic IFs or that they go through the nuclear pores. So, if the preceding hypothesis is correct, how can cytoplasmic IF proteins gain access to the interior of the nucleus? One straightforward mechanism entails the liberation of the head domain from IFs by proteolysis, after which the free head domain could freely diffuse into the nucleus. In the quail mutant, this liberation is not necessary since the mutant protein is polymerization-incompetent. Another mechanism of activation might entail posttranslational modification, such as phosphorylation, glycosylation, ADP-ribosylation or Ca^{2+}-dependent proteolysis, allowing the "piggy-back" transport of modified IF proteins (after their release from filaments) by karyophilic carrier proteins during interphase or their incorporation into the nuclear membrane or nuclear matrix during telophase. Candidates for such a carrier protein include the nuclear lamin B protein, which interacts stably and specifically with the cytoplasmic IF proteins vimentin, desmin, peripherin and cytokeratin D (Georgatos and Blobel, 1987a; 1987b; Georgatos et al., 1987; Djabali et al., 1991; Bastos et al., 1992) and a nuclear matrix protein that associates with cytoplasmic IFs during mitosis (Marugg, 1992). A last possibility is that IFs penetrate the nuclear membrane directly, due to their amphiphilic head domains and lipophilic rod domains, in a manner similar to that described for mitochondrial protein import (Roise et al., 1986; Maduke and Roise, 1993), but are not recognizable on a morphological or immunological basis due to alterations in the protein structure as a result of interaction with the membrane.

Electron micrographs suggestive of such a penetration have been published (Franke, 1971), and the interaction between the nucleus and IFs withstands treatment with nonionic detergents and high salt (Fey et al., 1984; French et al., 1989). In these preparations, the IFs are interwoven with the lamina filaments of the residual nuclei. While they provide no information as to how the IF proteins get there, there have been numerous reports of the detection of IF protein epitopes within the nucleus, for example, the presence of a vimentin epitope in chromatin structures and chromosomes (Fidlerová et al., 1992). IF proteins have been found to be cross-linked to DNA *in vivo* in normal, untreated cells (Cress and Kurath, 1988) and also in cells exposed to UV light or heavy metals (Wedrychowski et al., 1986a, b; Galcheva-Gargova and Dessev, 1987). Furthermore, the *in vivo* expression of modified IF proteins, such as C terminally truncated vimentin (Eckelt et al., 1992), NFP-L (Gill et al., 1990) or keratins (Bader et al., 1991), as well as the ectopic expression of keratin K 1 in pancreatic islet cells of transgenic mice (Blessing et al., 1993), leads to the accumulation of fibers or granules of the IF proteins within the nucleus! At present, it is completely unclear how and why this happens, but nonetheless shows that modified IF proteins may indeed enter the nucleus.

H. Interactions with Nucleic Acids and Nucleoprotein Structures in the Cytoplasm

As a result of observations of the binding of rRNAs and ribosomal subunits to vimentin IFs (Traub et al., 1992b) and of prosomes, subcomplexes of untranslated, nonpolyribosomal mRNP particles (Olink-Coux et al., 1992), and RNP particles containing the Ro antigen (Carmo-Fonseca and David-Ferreira, 1990) to cytoplasmic IFs, it has been suggested that the cytoplasmic IFs may play an important role in the organization and transport of ribosomes and mRNPs in the cell (Traub et al., 1992b). The moderate binding affinity of these components for IFs would result in their continual association/disassociation, effectively limiting their movement within the cytoplasm to a path extending along the network of the IFs. Active movements of these components might be possible in conjunction with other cytoskeletal elements, such as microtubules and their associated motor proteins like kinesin, which have been shown to be interconnected with vimentin IFs (Gyoeva and Gelfand, 1991). The joint reactivities of the binding of IFs to the nuclear envelope and to nucleic acids may not be fortuitous, especially when one realizes that IF proteins also bind core histones (Traub et al., 1986), but may be a mechanism in the fine-tuning of the regulation of gene expression (Traub and Shoeman, 1994). The replication of the RNA virus, frog virus 3, takes place in cytoplasmic assembly centers that are surrounded by, and organized by, cytoplasmic IFs (Murti et al., 1988). These various observations suggest that there may be a general requirement for a solid-state matrix for the transport, metabolism and activity of nucleic acids and nucleic acid containing structures in the cytoplasm, like that proposed for the nucleus, and that the IFs contribute to this matrix.

XIII. FUTURE RESEARCH DIRECTIONS

Much of the published data on IFs have been purely anecdotal or descriptive and based on the appearance of cells decorated with anti-IF antibodies. With the widespread application of molecular biology techniques, new methods of identification of IF proteins are being employed. These include oligonucleotide-based hybridization and polymerase chain reaction techniques for the isolation and sequencing of new IF protein cDNAs and genes. Whereas the major problem in the more classical immunological identification of IF proteins and structures was demonstrating the specificity of the antibody reaction, the major problem with the sequence-based identification of new IF proteins lies in the still-limited ability of secondary structure prediction algorithms to accurately predict the structure and interactions of large proteins. Thus, simply isolating a gene, whose deduced sequence codes for a protein related to another IF protein or that has many IF proteinlike heptad repeats, is not enough evidence to label such a gene as coding for a new IF protein. It is still necessary to do the appropriate biochemistry and cell biology. Progress in this field will increasingly require a balanced application of molecular genetic and cell biology techniques. One good example of such an approach is the discovery and characterization of the novel IF network of the eye lens cell, made up of new members of the IF protein family (Gounari et al., 1993; Merdes et al., 1993), which connects the cytoplasmic vimentin IFs to the plasma membrane, providing an explanation for these interactions that had been observed years ago (Ramaekers et al., 1982). We believe this will not be the last example of new and novel IFs exhibiting new and novel activities.

The major challenge for the future remains the elucidation of the function of the IFs. Why are so many different subunit proteins employed to assemble a common structure, the 10-nm filament, in a variety of cells? The next few years will see additional applications of molecular genetics in attempts to define the functions of IF proteins. This will be done, not just because of the relative ease with which such experiments can be performed as a result of the commercial availability of almost all of the "tools" necessary, but because *proper* application of this technology will permit either the complete knockout or the regulated expression of specific IF protein genes in combination with other gene products. A severe handicap in this process has been the lack of extensive knowledge of and ability to manipulate the genetics of the vertebrate systems commonly employed to study IF proteins. In this respect, it has been disappointing that so little is known of IF or IF-like proteins in organisms with "user-friendly" genetics, such as yeast or *Drosophila*. Thus, a logical predication would be the intensification of research efforts on yeast and *Drosophila*. In contrast to the wealth of information on the structure and assembly properties of the various IF proteins, little or nothing is really known about their function(s) *in vivo*.

Another area for future research is in applied biotechnology for the development of unique biopolymers. As the individual sequences, structures and biochemical properties of the various IF proteins (especially the non-α-helical end domains) become known, it is likely that some of these principles may be bioengineered into novel polymers with as yet unforeseen applications (such as tissue replacements for skin grafts, the development of biosensors, etc.). One good (but completely unrelated) example of a potential useful biopolymer is the mouse mast cell secretory granule matrix. In response to various electrical stimuli, this matrix either swells unexpectedly rapidly, producing a pressure of several atmospheres, or condenses (Nanavati and Fernandez, 1993), properties that undoubtedly will be applied to the solution of unique problems. This topic and one example are mentioned here with respect to future studies on IF proteins because all of the IF proteins possess, in addition to their unique properties, a number of characteristics important for the application of a chemical strategy for the synthesis of nanostructures (Whitesides et al., 1991), especially self-organization and molecular self-assembly. Thus, IF proteins may one day play an important role in the bioengineering and production of very large, yet very ordered biopolymers.

ACKNOWLEDGMENT

We would like to thank Dr. A. Janetzko for providing Figure 1 and Ms. A. Gawenda for the photos.

REFERENCES

Aebi, U., Fowler, W.E., Rew, P., & Sun, T.-T. (1983). The fibrillar substructure of keratin filaments unraveled. J. Cell Biol. 97, 1131–1143.

Aebi, U., Häner, M., Troncoso, J., Eichner, R., & Engel, A. (1988). Unifying principles in intermediate filament (IF) structure and assembly. Protoplasma 145, 73–81.

Albers, K. & Fuchs, E. (1992). The molecular biology of intermediate filaments. Int. Rev. Cytol. 134, 243–279.

Almahbobi, G. & Hall, P.F. (1990). The role of intermediate filaments in adrenal steroidogenesis. J. Cell Sci. 97, 679–687.

Almahbobi, G., Williams, L.J., & Hall, P.F. (1991). Attachment of steroidogenic lipid droplets to intermediate filaments in adrenal cells. J. Cell Sci. 101, 383–393.

Angelides, K.J., Smith, K.E., & Takeda, M. (1989). Assembly and exchange of intermediate filament proteins of neurons: Neurofilaments are dynamic structures. J. Cell Biol. 108, 1495–1506.

Bader, B.L., Magin, T.M., Freudenmann, M., Stumpp, S., & Franke, W.W. (1991). Intermediate filaments formed de novo from tail-less cytokeratins in the cytoplasm and in the nucleus. J. Cell Biol. 115, 1293–1307.

Bai, W., Arlinghaus, R.B., & Singh, B. (1993). Association of v-Mos with soluble vimentin *in vitro* and in transformed cells. Oncogene 8, 2207–2212.

Baribault, H., Price, J., Miyai, K., & Oshima, R.G. (1993). Mid-gestational lethality in mice lacking keratin 8. Genes and Development 7, 1191–1202.

Bartnik, E. & Weber, K. (1989). Widespread occurrence of intermediate filaments in invertebrates; common principles and aspects of diversion. Eur. J. Cell Biol. 50, 17–33.

Bastos, R., Engel, P., Pujades, C., Falchetto, R., Aligué, R., & Bachs, O. (1992). Increase of cytokeratin D during liver degeneration: Association with the nuclear matrix. Hepatology 16, 1434–1446.

Belin, M.-T. & Boulanger, P. (1987). Processing of vimentin occurs during the early stages of adenovirus infection. J. Virol. 61, 2559–2566.

Ben-Ze'ev, A. (1984). Inhibition of vimentin synthesis and disruption of intermediate filaments in simian virus 40-infected monkey kidney cells. Mol. Cell Biol. 4, 1880–1889.

Birkenberger, L. & Ip, W. (1990). Properties of the desmin tail domain: Studies using synthetic peptides and antipeptide antibodies. J. Cell Biol. 111, 2063–2075.

Blessing, M., Rüther, U., & Franke, W.W. (1993). Ectopic synthesis of epidermal cytokeratins in pancreatic islet cells of transgenic mice interferes with cytoskeletal order and insulin production. J. Cell Biol. 120, 743–755.

Bravo, R. & Celis, J.E. (1980). A search for differential polypeptide synthesis throughout the cell cycle of HeLa cells. J. Cell Biol. 84, 795–802.

Bravo, R., Small, J.V., Fey, S., Larsen, P.M., & Celis, J.E. (1982). Architecture and polypeptide composition of HeLa cytoskeletons. Modification and cytoarchitecture polypeptides during mitosis. J. Mol. Biol. 154, 121–143.

Brown, K.D. & Binder, L.I. (1992). Identification of the intermediate filament-associated protein gyronemin as filamin. Implications for a novel mechanism of cytoskeletal interaction. J. Cell Sci. 102, 19–30.

Bruce, J., Schwartz, M.L., Schneidman, P.S., & Schlaepfer, W.W. (1993). Methylation and expression of neurofilament genes in tissues and in cell lines of the mouse. Mol. Brain Res. 17, 269–278.

Capetanaki, Y., Kuisk, I., Rothblum, K., & Starnes, S. (1990). Mouse vimentin: Structural relationship to *fos, jun*, CREB and *tpr*. Oncogene 5, 645–655.

Capetanaki, Y., Smith, S., & Heath, J.P. (1989). Overexpression of the vimentin gene in transgenic mice inhibits normal lens cell differentiation. J. Cell Biol. 109, 1653–1664.

Carmo-Fonseca, M. & David-Ferreira, J.F. (1990). Interactions of intermediate filaments with cell structures. Electron Microsc. Rev. 3, 115–141.

Cartaud, A., Ludosky, M.A., Courvalin, J.C., & Cartaud, J. (1990). A protein antigenically related to nuclear lamin B mediates the association of intermediate filaments with desmosomes. J. Cell Biol. 111, 581–588.

Chan, D., Goate, A., & Puck, T.T. (1989). Involvement of vimentin in the reverse transformation reaction. Proc. Natl. Acad. Sci. USA 86, 2747–2751.

Chen, J.D. & Pirrotta, V. (1993). Multimerization of the *Drosophila zeste* protein is required for efficient DNA binding. EMBO J. 12, 2075–2083.

Chen, P.H., Ornelles, D.A., & Shenk, T. (1993). The adenovirus L3 23-kilodalton proteinase cleaves the amino-terminal head domain of cytokeratin 18 and disrupts the cytokeratin network of HeLa cells. J. Virol. 67, 3507–3514.

Ching, G.Y. & Liem, R.K.H. (1993). Assembly of type IV neuronal intermediate filaments in nonneuronal cells in the absence of preexisting cytoplasmic intermediate filaments. J. Cell Biol. 122, 1323–1335.

Chou, C.-F. & Omary, M.B. (1993). Mitotic arrest-associated enhancement of O-linked glycosylation and phosphorylation of human keratins 8 and 18. J. Biol. Chem. 268, 4465–4472.

Chou, R.-G. R., Stromer, M.R., Robson, R.M., & Huiatt, T.W. (1990). Determination of the critical concentration required for the desmin assembly. Biochem. J. 272, 139–145.

Choudhary, S.K. & Capetanaki, Y. (1990). Inhibition of desmin expression blocks myogenic differentiation. J. Cell Biol. 111, 44a.

Collard, J.-F. & Raymond, Y. (1992). Phorbol esters induce transient changes in the accessibility of the carboxy-terminal domain of nuclear lamin A. Exp. Cell Res. 201, 174–183.

Conway, J.F. & Parry, D.A.D. (1988). Intermediate filament structure: 3. analysis of sequence homologies. Int. J. Biol. Macromol. 10, 79–98.

Côte, F., Collard, J.-F., & Julien, J.-P. (1993). Progressive neuropathy in transgenic mice expressing the human neurofilament heavy gene: A mouse model of amyotrophic lateral schlerosis. Cell 73, 35–46.

Coulombe, P.A. (1993). The cellular and molecular biology of keratins: Beginning a new era. Curr. Opin. Cell Biology 5, 17–29.

Cress, A.E. & Kurath, K.M. (1988). Identification of attachment proteins for DNA in Chinese hamster ovary cells. J. Biol. Chem. 263, 19678–19683.

Dale, B.D., Resing, K.A., & Haydock, P.V. (1990). Filaggrins. In: Cellular and Molecular Biology of Intermediate Filaments (Goldman, R.D. & Steinert, P.M., Eds), pp. 393–412, Plenum Press, New York.

Dahlstrand, J., Zimmerman, L.B., McKay, R.D.G., & Lendahl, U. (1992). Characterization of the human nestin gene reveals a close evolutionary relationship to neurofilaments. J. Cell Sci. 103, 589–597.

Dawson, C.W., Rickinson, A.B., & Young, L.S. (1990). Epstein-Barr virus latent membrane protein inhibits human epithelial cell differentiation. Nature 344, 777–780.

De Lange, T. (1992). Human telomeres are attached to the nuclear matrix. EMBO J. 11, 717–724.

Djabali, K., Portier, M.M., Gros, F., Blobel, G., & Georgatos, S.D. (1991). Network antibodies identify nuclear lamin B as physiological attachment site for peripherin intermediate filaments. Cell 64, 109–121.

Dodemont, H., Riemer, D., & Weber, K. (1990). Structure of an invertebrate gene encoding cytoplasmic intermediate filament (IF) proteins: Implication for the origin and the diversification of IF proteins. EMBO J. 9, 4083–4094.

Döring, V. & Stick, R. (1990). Gene structure of nuclear lamin LIII of *Xenopus laevis*; a model for the evolution of IF proteins from a lamin-like ancestor. EMBO J. 9, 4073–4081.

Doherty, F.J., Wassell, J.A., & Mayer, R.J. (1987). A putative protein-sequestration site involving intermediate filaments for protein degradation by autophagy. Studies with microinjected purified glycolytic enzymes in 3T3-L1 cells. Biochem. J. 241, 793–800.

Dong, D.L.-Y., Xu, Z.-S., Chevrier, M.R., Cotter, R.J., Cleveland, D.W., & Hart, G.W. (1993). Glycosylation of mammalian neurofilaments. Localization of multiple O-linked N-acetylglucosamine moieties on neurofilament polypeptides L and M. J. Biol. Chem. 268, 16679–16687.

Doorbar, J., Ely, S., Sterling, J., McLean, C., & Crawford, L. (1991). Specific interaction between HPV-16 E1-E4 and cytokeratins results in collapse of the epithelial cell intermediate filament network. Nature 352, 824–827.

Dosemeci, A. & Pant, H.C. (1992). Association of cyclic-AMP-dependent protein kinase with neurofilaments. Biochem. J. 282, 477–481.

Downing, S.W., Spitzer, R.H., Koch, E.A., & Salo, W.L. (1984). The hagfish slime gland thread cell I. A unique cellular system for the study of intermediate filaments and intermediate filament-microtubule interactions. J. Cell Biol. 98, 653–699.

Dráberová, E. & Dráber, P. (1993). A microtubule-interacting protein involved in coalignment of vimentin intermediate filaments with microtubules. J. Cell Sci. 106, 1263–1273.

Dunia, I., Pieper, F., Manenti, S., van de Kemp, A., Devilliers, G., Benedetti, E.L., & Bloemendal, H. (1990). Plasma membrane-cytoskeleton damage in eye lenses of transgenic mice expressing desmin. Eur. J. Cell Biol. 53, 59–74.

Earl, R.T., Mangiapane, E.H., Billett, E.E., & Mayer, R.J. (1987). A putative protein-sequestration site involving intermediate filaments for protein degradation by autophagy. Studies with transplanted Sendai-viral envelope proteins in HTC cells. Biochem. J. 241, 809–815.

Eckelt, A., Herrmann, H., & Franke, W.W. (1992). Assembly of a tail-less mutant of the intermediate filament protein, vimentin, in vitro and in vivo. Eur. J. Cell Biol. 58, 319–330.

Eckert, B.S. & Yeagle, P.L. (1988). Acrylamide treatment of PtK1 cells causes dephosphorylation of keratin polypeptides. Cell Motil. Cytoskelel. 11, 24–30.

Eckert, B.S. & Yeagle, P.L. (1990). Modulation of keratin intermediate filament distribution in vivo by induced changes in cyclic AMP-dependent phosphorylation. Cell Motil. Cytoskelel. 17, 291–300.

Eichner, R., Sun, T.-T., & Aebi, U. (1986). The role of keratin subfamilies and keratin pairs in the formation of human epidermal intermediate filaments. J. Cell Biol. 102, 1767–1777.

Engel, A., Eichner, R., & Aebi, U. (1985). Polymorphism of reconstituted human epidermal keratin filaments: Determination of their mass-per-unit-length and width by scanning transmission electron microscopy. J. Ultrastruct. Res. 90, 323–335.

Erickson, P.A., Feinstein, S.C., Lewis, G.P., & Fisher, S.K. (1992). Glial fibrillary acidic protein and its mRNA: Ultrastructural detection and determination of changes after CNS injury. J. Struct. Biol. 108, 148–161.

Eriksson, A. & Thornell, L.-E. (1979). Intermediate (skeletin) filaments in heart purkinje fibers. J. Cell Biol. 80, 231–247.

Eriksson, J.E., Opal, P., & Goldman, R.D. (1992a). Intermediate filament dynamics. Curr. Opin. Cell Biol. 4, 99–104.

Eriksson, J.E., Brautigan, D.L., Vallee, R., Olmsted, J., Fujiki, H., & Goldman, R.D. (1992b). Cytoskeletal integrity in interphase cells requires protein phosphatase activity. Proc. Natl. Acad. Sci. USA 89, 11093–11907.

Fahraeus, R., Rymo, L., Rhim, J.S., & Klein, G. (1990). Morphological transformation of human keratinocytes expressing the LMP gene of Epstein-Barr virus. Nature 345, 447–449.

Fey, E.G., Wan, K.M., & Penman, S. (1984). Epithelial cytoskeletal framework and nuclear matrix-intermediate filament scaffold: Three-dimensional organization and protein composition. J. Cell Biol. 23, 1973–1984.

Fidlerová, H., Sovová, V., Krekule, I., Viklicky, V., & Levan, G. (1992). Immunofluorescence detection of the vimentin epitope in chromatin structures of cell nuclei and chromosomes. Herediatas 117, 265–273.

Foisner, R. & Wiche, G. (1991). Intermediate filament-associated proteins. Curr. Opin. Cell Biol. 3, 75–81.

Forry-Schaudies, S., Murray, J.M., Toyama, Y., & Holtzer, H. (1986). Effects of colcemid and taxol on microtubules and intermediate filaments in chick embryo fibroblasts. Cell Motil. Cytoskel. 6, 324–338.

Franke, W.W. (1971). Relationship of nuclear membranes with filaments and microtubules. Protoplasma 73, 263–292.

Franke, W.W., Hergt, M., & Grund, C. (1987). Rearrangement of the vimentin cytoskeleton during adipose conversion: formation of an intermediate filament cage around lipid globules. Cell 49, 131–141.

Franke, W.W., Schiller, D.L., & Grund, C. (1982a). Protofilamentous and annular structures as intermediates during reconstitution of cytokeratin filaments *in vitro*. Biol. Cell 46, 257–268.

Franke, W.W., Schmid, E., Grund, C., & Geiger, B. (1982b). Intermediate filament proteins in nonfilamentous structures: Transient disintegration and inclusion of subunit proteins in granular aggregates. Cell 30, 103–113.

Fraser, R.D.B., MacRae, T.P., & Parry, D.A.D. (1990). The three-dimensional structure of IF. In: Cellular and Molecular Biology of Intermediate Filaments (Goldman, R.D. & Steinert, P.M., Eds), pp. 205–231, Plenum Press, New York.

French, S.W., Kawahara, H., Katsuma, Y., Ohta, M., & Swierenga, S.H.H. (1989). Interaction of intermediate filaments with nuclear lamina and cell periphery. Electron Microsc. Rev. 2, 17–51.

Fuchs, E. & Coulombe, P. (1992). Of mice and men: Genetic skin diseases of keratin. Cell 69, 899–902.

Fulton, A.B. & Isaacs, W.B. (1991). Titin, a huge, elastic sarcomeric protein with a probable role in morphogenesis. Bio Essays 13, 157–161.

Fulton, A.B. & L'Ecuyer, T. (1993). Commentary. Cotranslational assembly of some cytoskeletal proteins: Implications and prospects. J. Cell Sci. 105, 867–871.

Galcheva-Gargova, Z. & Dessev, G.N. (1987). Crosslinking of DNA to nuclear lamina proteins by UV irradiation *in vivo*. J. Cell Biochem. 34, 163–168.

Garcia-Barreno, B., Jorcano, J.L., Aukenbauer, T., Lopez-Galindez, C., & Melero, J.A. (1988). Participation of cytoskeletal intermediate filaments in the infectious cycle of human respiratory syncytial virus (RSV). Virus Res. 9, 307–323.

Geisler, N., Hatzfeld, M., & Weber, K. (1989). Phosphorylation *in vitro* of vimentin by protein kinases A and C is restricted to the head domain. Identification of the phosphoserine sites and their influence on filament formation. Eur. J. Biochem. 183, 441–447.

Geisler, N., Heimburg, T., Schünemann, J., & Weber, K. (1993). Peptides from the conserved ends of the rod domain of desmin disassemble intermediate filaments and reveal unexpected structural features: A circular dichroism, fourier transform infrared, and electron microscopic study. J. Struct. Biol. 110, 205–214.

Geisler, N., Kaufmann, E., & Weber, K. (1982). Proteinchemical characterization of three structurally distinct domains along the protofilament unit of desmin 10 nm filaments. Cell 30, 277–286.

Geisler, N., Schünemann, J., & Weber, K. (1992). Chemical cross-linking indicates a staggered and antiparallel protofilament of desmin intermediate filaments and characterizes one higher-level complex between protofilaments. Eur. J. Biochem. 206, 841–852.

Georgatos, S.D. (1993). Dynamics of intermediate filaments. Recent progress and unanswered questions. FEBS Letters 318, 101–107.

Georgatos, S.D. & Blobel, G. (1987a). Two different attachment sites for vimentin along the plasma membrane and nuclear envelope in avian erythrocytes: A basis for a vectorial assembly of intermediate filaments. J. Cell Biol. 105, 105–115.

Georgatos, S.D. & Blobel, G. (1987b). Lamin B constitutes an intermediate filament attachment site at the nuclear envelope. J. Cell Biol. 105, 117–125.

Georgatos, S.D., Maroulakou, I., & Blobel, G. (1989). Lamin A, lamin B and lamin B receptor analogues in yeast. J. Cell Biol. 108, 2069–2082.

Georgatos, S.D., Weber, K., Geisler, N., & Blobel, G. (1987). Binding of two desmin derivatives to the plasma membrane and the nuclear envelope of avian erythrocytes: Evidence for a conserved site-specificity in intermediate filament-membrane interactions. Proc. Natl. Acad. Sci. USA 84, 6780–6784.

Giese, G. & Traub, P. (1986). Induction of vimentin synthesis in mouse myeloma cells MPC-11 by 12-O-tetradecanoylphorbol-13-acetate. Eur. J. Cell Biol. 40, 266–274.

Gilbert, M.R., Harding, B.L., Hoffman, P.N., Griffin, J.W., Price, D.L., & Troncoso, J.C. (1992). Aluminum-induced neurofilamentous changes in cultured rat dorsal ganglia explants. J. Neurosci. 12, 1763–1771.

Gill, S.R., Wong, P.C., Monteiro, M.J., & Cleveland, D.W. (1990). Assembly properties of dominant and recessive mutations in the small mouse neurofilament (NF-L) subunit. J. Cell Biol. 111, 2005–2019.

Gillard, B.K., Heath, J.P., Thurman, L.T., & Marcus, D.M. (1991). Association of glycosphingolipids with intermediate filaments of human umbilical vein endothelial cells. Exp. Cell Res. 192, 433–444.

Gillard, B.K., Thurman, L.T., & Marcus, D.M. (1992). Association of glycosphingolipids with intermediate filaments of mesenchymal, epithelial, glial and muscle cells. Cell Motil. Cytoskeleton 21, 255–271.

Gilson, E., Laroche, T., & Glasser, S.M. (1993). Telomeres and the functional architecture of the nucleus. Trends Cell Biol. 3, 128–134.

Glasgow, E., Druger, R.K., Levine, E.M., Fuchs, C., & Schechter, N. (1992). Plasticin, a novel type III neurofilament protein from goldfish retina: Increased expression during optic nerve regeneration. Neuron 9, 373–381.

Goldman, R.D., Chou, Y.-H., Dessev, C., Dessev, G., Eriksson, J., Goldman, A., Khuon, S., Kohnken, R., Lowy, M., Miller, R., Murphy, K., Opal, P., Skalli, O., & Straube, K. (1991). Dynamic aspects of cytoskeletal and karyoskeletal intermediate filament systems during the cell cycle. Cold Spring Harbor Symp. Quant. Biol. LVI, 629–642.

Goldman, R.D., Goldman, A.E., Green, K.C., Jones, J.C.R., Jones, S.M., & Yang, H.-Y. (1986). Intermediate filament networks: Organization and possible functions of a diverse group of cytoskeletal elements. J. Cell Sci. Suppl. 5, 69–97.

Goldman, R.D., Zackaroff, R.V., & Steinert, P.M. (1990). Intermediate filaments. An overview. In: Cellular and Molecular Biology of Intermediate Filaments (Goldman, R.D. & Steinert, P.M., Eds), pp. 3–17, Plenum Press, New York.

Gounari, F., Merdes, A., Quinlan, R., Hess, J., FitzGerald, P.G., Ouzounis, C.A., & Georgatos, S.D. (1993). Bovine filensin possesses primary and secondary structure similarity to intermediate filament proteins. J. Cell Biol. 121, 847–853.

Green, K.J. & Goldman, R.D. (1986). Evidence for an interaction between the cell surface and the intermediate filaments in cultured fibroblasts. Cell Motil. Cytoskel. 6, 389–405.

Gyoeva, F.K. & Gelfand, V.I. (1991). Coalignment of vimentin intermediate filaments with microtubules depends on kinesin. Nature 353, 445–448.

Haltiwanger, R.S., Kelly, W.G., Roquemore, E.P., Blomberg, M.A., Dong, L.-Y.D., Kreppel, L., Chou, T.-Y., & Hart, G.W. (1992). Glycosylation of nuclear and cytoplasmic proteins is ubiquitous and dynamic. Biochem. Soc. Transact. 20, 264–269.

Hatzfeld, M. & Franke, W.W. (1985). Pair formation and promiscuity of cytokeratins: Formation *in vitro* of heterotypic complexes and intermediate-sized filaments by homologous and heterologous recombinations of purified polypeptides. J. Cell Biol. 101, 1826–1841.

Hatzfeld, M. & Weber, K. (1991). Modulation of keratin intermediate filament assembly by single amino acid exchanges in the consensus sequence at the C-terminal end of the rod domain. J. Cell Sci. 99, 351–362.

Hatzfeld, M. & Weber, K. (1992). A synthetic peptide representing the consensus sequence motif at the carboxy-terminal end of the rod domain inhibits intermediate filament assembly and disassembles preformed filaments. J. Cell Biol. 116, 157–166.

Hay, M. & De Boni, U. (1991). Chromatin motion in neuronal interphase nuclei: Changes induced by disruption of intermediate filaments. Cell Motil. Cytoskel. 18, 63–75.

He, D., Nickerson, J.A., & Penman, S. (1990). Core filaments of the nuclear matrix. J. Cell Biol. 110, 569–580.

Hennekes, H., Peter, M., Weber, K., & Nigg, E.A. (1993). Phosphorylation on protein kinase C sites inhibits nuclear import of lamin B_2. J. Cell Biol. 120, 1293–1304.

Herrmann, H., Eikelt, A., Brettel, M., Grund, C., & Franke, W.W. (1993). Temperature-sensitive intermediate filament assembly. Alternative structures of *Xenopus laevis* vimentin *in vitro* and *in vivo*. J. Mol. Biol. 234, 99–113.

Herrmann H., Hofmann, I., & Franke, W.W. (1992). Identification of a nonapeptide motif in the vimentin head domain involved in intermediate filament assembly. J. Mol. Biol. 223, 637–650.

Hesketh, J.E. & Pryme, I.F. (1991). Interaction between mRNA, ribosomes and the cytoskeleton. Biochem. J. 277, 1–10.

Hess, J.F., Casselman, J.T., & FitzGerald, P.G. (1993). cDNA analysis of the 49 kDa lens fiber cell cytoskeleton protein: A new, lens-specific member of the intermediate filament family? Curr. Eye Res. 12, 77–88.

Hisanaga, S.-I., Gonda, Y., Inagaki, M., Ikai, A., & Hirokawa, N. (1990). Effects of phosphorylation of the neurofilament protein on filamentous structures. Cell Regul. 1, 237–248.

Höner, B., Shoeman, R.L., & Traub, P. (1991). Human immunodeficiency virus type 1 protease microinjected into cultured human skin fibroblasts cleaves vimentin and affects cytoskeletal and nuclear architecture. J. Cell Sci. 100, 799–807.

Hoffman, P.N. & Cleveland, D.W. (1988). Neurofilament and tubulin expression recapitulates the developmental program during axonal regeneration: Induction of a specific β-tubulin isotype. Proc. Natl. Acad. Sci. USA 85, 4530–4533.

Hofmann, I. & Herrmann H. (1992). Interference in vimentin assembly *in vitro* by synthetic peptides derived from the vimentin head domain. J. Cell Sci. 101, 687–700.

Hofmann, I., Herrmann, H., & Franke, W.W (1991). Assembly and structure of calcium-induced thick vimentin filaments. Eur. J. Cell Biology 56, 328–341.

Hollenbeck, P.J. (1989). The transport and assembly of the axonal cytoskeleton. J. Cell Biol. 108, 223–227.

Horkovics-Kovats, S. & Traub, P. (1990). Specific interaction of the intermediate filament protein vimentin and its isolated N-terminus with negatively charged phospholipids as determined by vesicle aggregation, fusion, and leakage measurements. Biochemistry 29, 8652–8659.

Huang, H.-Y., Graves, D.J., Robson, R.M., & Huiatt, T.W. (1993). ADP-ribosylation of the intermediate filament protein desmin by ADP-ribosyl-transferase. FASEB J. 7, A1078.

Inagaki, M., Gonda, Y., Nishizawa, K., Kitamura, S., Sato, C., Ando, S., Tanabe, K., Kikuchi, K., Tsuiki, S., & Nishi, Y. (1990). Phosphorylation sites linked to glial filament disassembly *in vitro* locate in a non-α-helical head domain. J. Biol. Chem. 265, 4722–4729.

Inagaki, M., Takahara, H., Nishi, Y., Sugawara, K., & Sato, C. (1989). Ca^{2+}-dependent deimination-induced disassembly of intermediate filaments involves specific modification of the amino-terminal head domain. J. Biol. Chem. 264, 18119–18127.

Ip, W., Hartzer, M.K., Pang, Y.-Y.S., & Robson, R.M. (1985). Assembly of vimentin *in vitro* and its implications concerning the structure of intermediate filaments. J. Mol. Biol. 183, 365–375.

Isaacs, W.B., Cook, R.K., Van Atta, J.C., Redmond, C.M., & Fulton, A.B. (1989). Assembly of vimentin in cultured cells varies with cell type. J. Biol. Chem. 264, 17953–17960.

Jackson, D.A. & Cook, P.R. (1988). Visualization of a filamentous nucleoskeleton with a 23 nm axial repeat. EMBO J. 7, 3667–3677.

Jackson, B.W., Grund, C., Schmid, E., Bürki, K., Franke, W.W., & Illmensee, K. (1980). Formation of cytoskeletal elements during mouse embryogenesis. I. Intermediate filaments of the cytokeratin type and desmosomes in preimplantation embryos. Differentiation 17, 161–179.

Jones, J.C.R., Goldman, A.E., Yang, H.Y., & Goldman, R.D. (1985). The organizational fate of intermediate filament networks in two epithelial cell types during mitosis. J. Cell Biol. 100, 93–102.

Kachi, K., Cadrin, M., & French, S.W. (1993). Synthesis of Mallory body, intermediate filament, and microfilament proteins in liver cell primary cultures. An electron microscopic autoradiography assay. Lab. Invest. 68, 71–81.

Kartasova, T., Roop, D.R., Holbrook, K.A., & Yuspa, S.H. (1993). Mouse-differentiation-specific keratins 1 and 10 require a preexisting keratin scaffold to form a filament network. J. Cell Biol. 120, 1251–1261.

Kaufmann, E., Weber, K., & Geisler, N. (1985). Intermediate filament forming ability of desmin derivatives lacking either the amino-terminal 67 or the carboxy-terminal 27 residues. J. Mol. Biol. 185, 733–742.

Klymkowsky, M.W. (1981). Intermediate filaments in 3T3 cells collapse after the intracellular injection of a monoclonal anti-intermediate filament antibody. Nature 291, 249–251.

Klymkowsky, M.W., Miller, R.H., & Lane, E.B. (1983). Morphology, behaviour, and interaction of cultured epithelial cells after antibody-induced disruption of keratin filament organization. J. Cell Biol. 17, 494–509.

Klymkowsky, M.W., Bachant, J.B., & Domingo, A. (1989). Functions of intermediate filaments. Cell Motil. Cytoskeleton 14, 309–331.

Koch, E.A., Spitzer, R.H., Pithawalla, R.B., & Downing, S.W. (1991). Keratin-like components of gland thread cells modulate the properties of mucus from hagfish (Eptatretus stouti). Cell Tissue Res. 264, 79–86.

Kulesh, D.A. & Oshima, R.G. (1988). Cloning of the human keratin 18 gene and its expression in nonepithelial mouse cells. Mol. Cell. Biol. 8, 1540–1550.

Lamb, P. & McKnight, S.L. (1991). Diversity and specificity in transcriptional regulation: the benefits of heterotypic dimerization. Trends Biochem. Sci. 16, 417–422.

Lane, E.B. & Alexander, C.M. (1990). Use of keratin antibodies in tumor diagnosis. Seminars in Cancer Biology 1, 165–179.

Lang, S. & Loidl, P. (1993). Identification of proteins immunologically related to vertebrate lamins in the nuclear matrix of the myxomycete Physarum polycephalum. Eur. J. Cell Biol. 61, 177–183.

Langley, Jr., R.C. & Cohen, C.M. (1986). Association of spectrin with desmin intermediate filaments. J. Cell Biochem. 30, 101–109.

Lazarides, E. (1980). Intermediate filaments as mechanical integrators of cellular space. Nature 283, 249–256.

Lee, Y.J., Hou, Z.-Z., Curetty, L., Armour, E.P., Al-Saadi, A., Bernstein, J., & Corry, P.M. (1992). Heat-resistant variants of the Chinese hamster ovary cell: Alteration of cellular structure and expression of vimentin. J. Cell. Physiol. 151, 138–146.

Lee, M.K., Xu, Z., Wong, P.C., & Cleveland, D.W. (1993). Neurofilaments are obligate heteropolmers in vivo. J. Cell Biol. 122, 1337–1350.

Leterrier, J.F., Langui, D., Probst, A., & Ulrich, A. (1992). A molecular mechanism for the initiation of neurofilament bundling by aluminum ions. J. Neurochem. 58, 2060–2070.

Lewis, S.A. & Cowan, N.J. (1986). Anomalous placement of introns in a member of intermediate filament multigene family: An evolutionary conundrum. Mol. Cell Biol. 6, 1529–1534.

Liebowitz, D., Kopan, R., Fuchs, E., Sample, J., & Kieff, E. (1987). An Epstein-Barr virus transforming protein associates with vimentin in lymphocytes. Molec. Cell. Biol. 7, 2299–2308.

Liem, R.K.H. (1993). Molecular biology of neuronal intermediate filaments. Curr. Opin. Cell Biol. 5, 12–16.

Lin, F. & Worman, H.J. (1993). Structural organization of the human gene encoding nuclear lamin A and nuclear lamin C. J. Biol. Chem. 268, 16321–16326.

Lindhofer, H., Nitschko, H., Wachinger, G., & von der Helm, K. (1993). Human-immunodeficiency-virus (HIV) proteinase in HIV-infected cell culture: Dual intracellular effects in the absence or presence of inhibitor. In: Proteolysis And Protein Turnover (Bond, J.S. & Barrett, A.J., Eds.), pp. 219–224. Portland Press, London.

Lockard, V.G. & Bloom, S. (1993). Trans-cellular desmin-lamin B intermediate filament network in cardiac myocytes. J. Mol. Cell Cardiol. 25, 303–309.

Lowe, J., Blanchard, A., Morrell, K., Lennox, G., Reynolds, L., Billet, M., Landon, M., & Mayer, R.J. (1988). Ubiquitin is a common factor in intermediate filament inclusion bodies of diverse type in man, including those of Parkinson's disease, Pick's disease, and Alzheimer's disease as well as Rosenthal fibres in cerebellar astrocytomas, cytoplasmic bodies in muscle, and Mallory bodies in alcoholic liver disease. J. Pathol. 155, 9–15.

Lowe, J., Morrell, K., Lennox, G., Landon, M., & Mayer, R.J. (1989). Rosenthal fibres are based on the ubiquitination of glial filaments. Neuropathol. Appl. Neurobiol. 15, 45–53.

Luftig, R.B. (1982). Does the cytoskeleton play a significant role in animal virus replication? J. Theor. Biol. 99, 173–191.

Mack, J.W., Steven, A.C., & Steinert, P.M. (1993). The mechanism of interaction of filaggrin with intermediate filaments. The ionic zipper hypothesis. J. Mol. Biol. 232, 50–66.

Maduke, M. & Roise, D. (1993). Import of a mitochondrial presequence into protein-free phospholipid vesicles. Science 260, 364–367.

Maison, C., Horstmann, H., & Georgatos, S.D. (1993). Regulated docking of nuclear membrane vesicles to vimentin filaments during mitosis. J. Cell Biol. 123, 1491–1505.

Manabe, M., Sanchez, M., Sun, T.-T., & Dale, B.A. (1991). Interaction of filaggrin with keratin filaments during advanced stages of normal human epidermal differentiation and in *Ichthyosis vulgaris*. Differentiation 48, 43–50.

Manuelidis, L. & Borden, J. (1988). Reproducible compartmentalization of individual chromosome domains in human CNS cells revealed by in situ hybridization and three-dimensional reconstruction. Chromosoma 96, 397–410.

Marugg, R.A. (1992). Transient storage of a nuclear matrix protein along intermediate-type filaments during mitosis: a novel function of cytoplasmic intermediate filaments. J. Struct. Biol. 108, 129–139.

Masaki, S. & Watanabe, T. (1992). cDNA sequence analysis of CP94: Rat lens fiber cell beaded-filament structural protein shows homology to cytokeratins. Biochem. Biophys. Res. Commun. 186, 190–198.

McCormick, M.B., Coulombe, P.A., & Fuchs, E. (1991). Sorting out IF networks: Consequences of domain swapping on IF recognition and assembly. J. Cell Biol. 113, 1111–1124.

McCormick, M.B., Kouklis, P., Syder, A., & Fuchs, E. (1993). The roles of the rod end and the tail in vimentin IF assembly and IF network formation. J. Cell Biol. 122, 395–407.

Menzel, D. (1993). Chasing coiled coils: Intermediate filaments in plants. Botanica Acta 106, 294–300.

Merdes, A., Brunkener, M., Horstmann, H., & Georgatos, S.D. (1991). Filensin: A new vimentin-binding, polymerization-competent and membrane-associated protein of the lens fiber cell. J. Cell Biol. 115, 397–410.

Merdes, A., Gounari, F., & Georgatos, S.D. (1993). The 47 kDa lens-specific protein phakinin is a member of the intermediate filament family and an assembly partner of filensin. J. Cell Biol. 123, 1507–1516.

Miller, R.K., Vikstrom, K., & Goldman, R.D. (1991). Keratin incorporation into intermediate filaments is a rapid process. J. Cell Biol. 113, 843–855.

Miyasaka, H., Okabe, S., Ishiguro, K., Uchida, T., & Hirokawa, N. (1993). Interaction of the tail domain of high molecular weight subunits of neurofilaments with the COOH-terminal region of tubulin and its regulation by τ protein kinase II. J. Biol. Chem. 268, 22695–22702.

Moir, R.D., Donaldson, A.D., & Stewart, M. (1991). Expression in *E. coli* of human lamins A and C: Influence of head and tail domains on assembly properties and paracrystal formation. J. Cell Sci. 99, 363–372.

Moir, R.D. & Goldman, R.D. (1993). Lamin dynamics. Curr. Opin. Cell Biol. 5, 408–411.

Monteiro, M.J., Hoffmann, P.N., Gearhart, J.D., & Cleveland, D.W. (1990). Expression of NF-L in both neuronal and non-neuronal cells of transgenic mice: Increased neurofilament density in axons without affecting caliber. J. Cell Biol. 111, 1543–1557.

Murti, K.G., Goorha, R., & Klymkowsky, M.W. (1988). A functional role for intermediate filaments in the formation of frog virus 3 assembly sites. Virology 162, 264–269.

Murti, K.G., Kaur, K., & Goorha, R.M. (1992). Protein kinase C associates with intermediate filaments and stress fibers. Exp. Cell Res. 202, 36–44.

Nanavati, C. & Fernandez, J.M. (1993). The secretory granule matrix: A fast-acting smart polymer. Science 259, 963–965.

Nakamura, Y., Takeda, M., Aimoto, S., Hojo, H., Takao, T., Shimonishi, Y., Hariguchi, S., & Nishimura, T. (1992). Assembly regulatory domain of glial fibrillary acidic protein. A single phosphorylation diminishes its assembly-accelerating property. J. Biol. Chem. 267, 23269–23274.

Nelson, W.J., Vorgias, C.E., & Traub, P. (1982). A rapid method for the large scale purification of the intermediate filament protein vimentin by single-stranded DNA-cellulose affinity chromatography. Biochem. Biophys. Res. Commun. 106, 1141–1147.

Nigg, E.A. (1992a). Assembly-disassembly of the nuclear lamina. Curr. Opin. Cell Biol. 4, 105–109.

Nigg, E.A. (1992b). Assembly and cell cycle dynamics of the nuclear lamina. Semin. Cell Biol. 3, 245–253.

Nixon, R.A., Brown, B.A., & Marotta, C.A. (1983). Limited proteolytic modification of a neurofilament protein involves a proteinase activated by endogenous levels of calcium. Brain Res. 275, 384–388.

Ohara, O., Gahara, Y., Miyake, T., Teraoka, H., & Kitamura, T. (1993). Neurofilament deficiency in quail caused by nonsense mutation in neurofilament-L gene. J. Cell Biol. 121, 387–395.

Olink-Coux, M., Huesca, M., & Scherrer, K. (1992). Specific types of prosomes are associated to subnetworks of the intermediate filaments in PtK$_1$ cells. Eur. J. Cell Biol. 59, 148–159.

Osborn, M. & Weber, K. (1983). Biology of disease-tumor diagnosis by intermediate filament typing: A novel tool for surgical pathology. Lab. Invest. 48, 372–394.

Ouellet, T., Levac, P., & Royal, A. (1988). Complete sequence of the mouse type-II keratin Endo A: Its amino-terminal region resembles mitochondrial signal peptides. Gene 70, 75–84.

Parry, D.A.D., Conway, J.F., Goldman, A.E., Goldman, R.D., & Steinert, P.M. (1987). Nuclear lamin protein: Comment. Structures for paracrystalline, filamentous and lattice forms. Int. J. Biol. Macromol. 137–145.

Parry, D.A.D. & Fraser, R.D.B. (1985). Intermediate filament structure: 1. Analysis of IF protein sequence data. Int. J. Biol. Macromol. 7, 203–213.

Perides, G., Scherbarth, A., & Traub, P. (1986). Influence of phospholipids on the formation and stability of vimentin-type intermediate filaments. Eur. J. Cell Biol. 42, 268–280.

Perides, G., Harter, C., & Traub, P. (1987a). Electrostatic and hydrophobic interactions of the intermediate filament protein vimentin and its amino terminus with lipid bilayers. J. Biol. Chem. 262, 13742–13749.

Perides, G., Kühn, S., Scherbarth, A., & Traub, P. (1987b). Probing the structural stability of vimentin and desmin-type intermediate filaments with Ca^{+2}-activated proteinase, thrombin and lysine-specific endoproteinase Lys-C. Eur. J. Cell Biol. 43, 450–458.

Pieper, F.R., Schaart, G., Krimpenfort, P.J., Henderik, J.B., Moshage, H.J., van de Kemp, A., Ramaekers, F.C., Berns, A., & Bloemendal, H. (1989). Transgenic expression of the muscle-specific intermediate filament protein desmin in nonmuscle cells. J. Cell Biol. 108, 1009–1024.

Puck, T.T. & Krystosek, A. (1992). Role of the cytoskeleton in genome regulation and cancer. Int. Rev. Cytol. 132, 75–108.

Pruss, R.M., Mirsky, R., Raff, M.C., Thorpe, R., Dowding, A.J., & Anderton, B.H. (1981). All classes of intermediate filaments share a common antigenic determinant defined by a monoclonal antibody. Cell 27, 419–428.

Quinlan, R.A. (1993). Assembly of intermediate filament proteins is modulated by the chaperone activity of alpha-crystallins. Invest. Ophthalmol. Visual Sci. 34, 989.

Quinlan, R.A., Moir, R.D., & Stewart, M. (1989). Expression in *Escherichia coli* of fragments of glial fibrillary acidic protein: Characterization, assembly properties and paracrystal formation. J. Cell Sci. 93, 71–83.

Ramaekers, F.C.S., Dunia, I., Dodemont, H.J., Benedetti, E.L., & Bloemendal, H. (1982). Lenticular intermediate-sized filaments: Biosynthesis and interaction with plasma membrane. Proc. Natl. Acad. Sci. USA 79, 3208–3212.

Raats, J.M.H., Gerards, W.L.H., Schreuder, M.I., Grund, C., Henderik, J.B.J., Hendriks, I.L.A.M., Ramaekers, F.C.S., & Bloemendal, H. (1992). Biochemical and structural aspects of transiently and stably expressed mutant desmin in vimentin-free and vimentin-containing cells. Eur. J. Cell Biol. 58, 108–127.

Remington, S.G. (1993). Chicken filensin: A lens fiber cell protein that exhibits sequence similarity to intermediate filament proteins. J. Cell Sci. 105, 1057–1068.

Riemer, D., Dodemont, H., & Weber, K. (1991). Cloning of the non-neuronal intermediate filament protein of the gastropod *Aplysia californica*; identification of an amino acid residue essential for the IFA epitope. Eur. J. Cell Biol. 56, 351–357.

Riopel, C.L., Butt, I., & Omary, M.B. (1993). Method of cell handling affects leakiness of cell surface labelling and detection of intracellular keratins. Cell Motil. Cytoskel. 26, 77–87.

Roise, D., Horvath, S.J., Tomich, J.M., Richards, J.H., & Schatz, G. (1986). A chemically synthesized pre-sequence of an imported mitochondrial protein can form an amphiphilic helix and perturb natural and artificial phospholipid bilayers. EMBO J. 5, 1327–1334.

Robson, R.M. (1989). Intermediate filaments. Curr. Opin. Cell Biol. 1, 36–43.

Rorke, E.A., Crish, J., & Eckert, R.L. (1992). Central rod domain insertion and carboxy-terminal fusion mutants of human cytokeratin K19 are incorporated into endogenous keratin filaments. J. Invest. Dermatol. 98, 17–23.

Rosevear, E.R., Reynolds, M., & Goldman, R.D. (1990). Dynamic properties of intermediate filaments: Disassembly and reassembly during mitosis in baby hamster kidney cells. Cell Motil. Cytoskel. 17, 150–166.

Sarria, A.J. & Evans, R.M. (1989). Intermediate filaments in biology and disease. RBC Cell Biol Rev. 22, 1–118.

Sarria, A.J., Nordeen, S.K., & Evans, R.M. (1990). Regulated expression of vimentin cDNA in cells in the presence and absence of a preexisting vimentin filament network. J. Cell Biol. 111, 553–565.

Sarria, A.J., Sankhavaram, R.P., & Evans, R.M. (1992). A functional role for vimentin intermediate filaments in the metabolism of lipoprotein-derived cholesterol in human SW-13 cells. J. Biol. Chem. 267, 19455–19463.

Schaafsma, H.E., Van der Velden, L.-A., Manni, J.J., Peters, H., Link, M., Ruiter, D.J., & Ramaekers, F.C.S. (1993). Increased expression of cytokeratins 8, 18 and vimentin in the invasion front of mucosal squamous cell carcinoma. J. Pathol. 170, 77–86.

Schlaepfer, W.W., Lee, C., Lee, V.M.-Y., & Zimmerman, U.-J.P. (1985). An immunoblot study of neurofilament degradation *in situ* and during calcium-activated proteolysis. J. Neurochem. 44, 502–509.

Schlaepfer, W.W. & Micko, S. (1978). Chemical and structural changes in neurofilaments in transected rat sciatic nerve. J. Cell Biol. 78, 369–378.

Schliwa, M. & Euteneuer, U. (1979). Structural transformation of epidermal tonofilaments upon cold treatment. Exp. Cell Res. 122, 93–101.

Schroeder, W.T., Thacher, S.M., Stewart-Galetka, S., Annarella, M., Chema, D., Siciliano, M.J., Davies, P.J.A., Tang, H.-Y., Sowa, B.A., & Duvic, M. (1992). Type I keratinocyte transglutaminase: Expression in human skin and psoriasis. J. Invest. Dermatol. 99, 27–34.

Schüssler, M.H., Skoudy, A., Ramaekers, F., & Real, F.X. (1992). Intermediate filaments as differentiation markers of normal pancreas and pancreas cancer. Am. J. Pathol. 140, 559–568.

Schultheiss, T., Lin, Z., Ishikawa, H., Zamir, I., Stoeckert, C.J., & Holtzer, H. (1991). Desmin/vimentin intermediate filaments are dispensable for many aspects of myogenesis. J. Cell Biol. 114, 953–966.

Sejersen, T. & Lendahl, U. (1993). Transient expression of the intermediate filament nestin during skeletal muscle development. J. Cell Sci. 106, 1291–1300.

Shea, T.B., Beermann, M.L., & Fischer, I. (1993). Transient requirement for vimentin in neuritogenesis: Intracellular delivery of anti-vimentin antibodies and antisense oligonucleotides inhibit neurite initiation but not elongation of existing neurites in neuroblastoma. J. Neurosci. Res. 36, 66–76.

Shestakova, E.A., Motuz, L.P., & Gavrilova, L.P. (1993). Co-localization of components of the protein-synthesizing machinery with the cytoskeleton in G0-arrested cells. Cell Biol. Int. 17, 417–424.

Shoeman, R.L., Höner, B., Mothes, E., & Traub, P. (1992). Potential role of the viral protease in human immunodeficiency virus type 1 associated pathogenesis. Medical Hypotheses 37, 137–150.

Shoeman, R.L., Höner, B., Stoller, T.J., Kesselmeier, C., Miedel, M.C., Traub, P., & Graves, M.C. (1990). Human immunodeficiency virus type 1 protease cleaves the intermediate filament proteins vimentin, desmin and glial fibrillary acidic protein. Proc. Natl. Acad. Sci. USA 87, 6336–6340.

Shoeman, R.L., Wadle, S., Scherbarth, A., & Traub, P. (1988). The binding in vitro of the intermediate filament protein vimentin to synthetic oligonucleotides containing telomere sequences. J. Biol. Chem. 263, 18744–18749.

Shoeman, R.L. & Traub, P. (1990a). Calpains and the cytoskeleton. In: Intracellular Calcium-Dependent Proteolysis. (Mellgren R.L. & Murachi T., Eds.), pp. 191–209. CRC Press, Inc.

Shoeman, R.L. & Traub, P. (1990b). The in vitro DNA-binding properties of purified nuclear lamin proteins and vimentin. J. Biol. Chem. 265, 9055–9061.

Shoeman, R.L. & Traub, P. (1993). Assembly of intermediate filaments. Bio Essays 15, 605–611.

Shyy, T.-T., Asch, B.B., & Asch, H.L. (1989). Concurrent collapse of keratin filaments, aggregation of organelles, and inhibition of protein synthesis during the heat shock response in mammary epithelial cells. J. Cell Biol. 108, 997–1008.

Singh, B. & Arlinghaus, R.B. (1989). Vimentin phosphorylation by p37[mos] protein kinase in vitro and generation of a 50 kda cleavage product in V-mos-transformed cells. Virol. 173, 144–156.

Skalli, O., Chou, Y.-H., & Goldman, R.D. (1992). Intermediate filaments: Not so tough after all. Trends Cell Biol. 2, 308–312.

Skalli, O. & Goldman, R.D. (1991). Recent insights into the assembly, dynamics and function of intermediate filament networks. Cell Motil. Cytoskel. 19, 67–79.

Small, J.V. & Squire, J.M. (1972). Structural basis of contraction in vertebrate smooth muscle. J. Mol. Biol. 67, 117–149.

Spitzer, R.H., Downing, S.W., Koch, E.A., Salo, W.L., & Saidel, L.J. (1984). Hagfish slime gland thread cells. II. Isolation and characterization of intermediate filament components associated with the thread. J. Cell Biol. 98, 670–677.

Spitzer, R.H., Koch, E.A., & Downing, S.W. (1988). Maturation of hagfish gland thread cells: Composition and characterization of intermediate filament polypeptides. Cell Motil. Cytoskeleton 11, 31–45.

Stappenbeck, T.S. & Green, K.J. (1992). The desmoplakin carboxyl terminus coaligns with and specifically disrupts intermediate filament networks when expressed in cultured cells. J. Cell Biol. 116, 1197–1209.

Stappenbeck, T.S., Bornslaeger, E.A., Corcoran, C.M., Luu, H.H., Virata, M.L.A., & Green, K.J. (1993). Functional analysis of desmoplakin domains: Specification of the interaction with keratin versus vimentin intermediate filament networks. J. Cell Biol. 123, 691–705.

Steinert, P.M. (1991). Analysis of the mechanism of assembly of mouse keratin 1/keratin 10 intermediate filaments in vitro suggest that intermediate filaments are built from multiple oligomeric units rather than a unique tetrameric building block. J. Struct. Biol. 107, 175–188.

Steinert, P.M. (1993). Structure, function and dynamics of keratin intermediate filaments. J. Invest. Dermatol. 100, 729–734.

Steinert, P.M. & Bale, S.J. (1993). Genetic skin diseases caused by mutations in keratin intermediate filaments. Trends in Genetics 9, 280–284.

Steinert, P.M., Marekov, L.N., Fraser, R.D.B., & Parry, D.A.D. (1993a). Keratin intermediate filament structure. Crosslinking studies yield quantitative information on molecular dimensions and mechanism of assembly. J. Mol. Biol. 230, 436–452.

Steinert, P.M., Marekov, L.N., & Parry, D.A.D. (1993b). Conservation of the structure of keratin intermediate filaments: Molecular mechanism by which different keratin molecules integrate into preexisting keratin intermediate filaments during differentiation. Biochem. 32, 10046–10056.

Steinert, P.M. & Parry, D.A.D. (1993). The conserved H1 domain of the type II keratin 1 chain plays an essential role in the alignment of nearest neighbor molecules in mouse and human keratin 1 / keratin 10 intermediate filaments at the two- to four-molecule level of structure. J. Biol. Chem. 268, 2878–2887.

Steinert, P.M. & Roop, D.R. (1988). The molecular and cellular biology of intermediate filaments. Annu. Rev. Biochem. 57, 593–625.

Steinert, P.M. & Roop, D.R. (1990). The structure, complexity, and evolution of intermediate filament genes. In: Cellular and Molecular Biology of Intermediate Filaments (Goldman, R.D. & Steinert, P.M., Eds), pp. 353–364, Plenum Press, New York.

Stewart, M. (1993). Intermediate filament structure and assembly. Curr. Opin. Cell Biol 5, 3–11.

Steven, A.C., Hainfeld, J.F., Trus, B.L., Wall, J.S., & Steinert, P.M. (1983). Epidermal keratins have masses-per-unit-length that scale according to average subunit mass: Structural basis for homologous packing of subunits in intermediate filaments. J. Cell Biol. 97, 1939–1944.

Steven, A.C., Hainfeld, J.F., Trus, B.L., Wall, J.S., & Steinert, P.M. (1984). Radia distribution of density within macromolecular complexes determined from dark-field electron micrographs. Proc. Natl. Acad. Sci. USA 81, 6363–6367.

Stick, R. (1992). The gene structure of Xenopus nuclear lamin A: A model for the evolution of A-type from B-type lamins by exon shuffling. Chromosoma 101, 566–574.

Stromer, M.H., Ritter, M.A., Pang, Y.-Y.S., & Robson, R.M. (1987). Effects of cations and temperature on kinetics of desmin assembly. Biochem. J. 246, 75–81.

Suzuki, K. & Ohno, S. (1990). Calcium-activated neutral protease-structure-function relationship and functional implications. Cell Struct. Funct. 15, 1–6.

Szaro, B.G., Pant, H.C., Way, J., & Battey, J. (1991). Squid low molecular weight neurofilament proteins are a novel class of neurofilament protein. J. Biol. Chem. 266, 15035–15041.

Tölle, H.-G., Weber, K., & Osborn, M. (1985). Microinjection of monoclonal antibodies specific for one intermediate filament protein in cells containing multiple keratins allow insight into the composition of particular 10 nm filaments. Eur. J. Cell Biol. 38, 234–244.

Traub, P. (1985). Intermediate Filaments. A Review. Springer Verlag, Heidelberg.

Traub, P., Kühn, S., & Grüb, S. (1993). Separation and characterization of homo- and hetero-oligomers of the intermediate filament proteins desmin and vimentin. J. Mol. Biol. 230, 837–856.

Traub, P., Mothes, E., Shoeman, R.L., Kühn, S., & Scherbarth, A. (1992a). Characterization of the nucleic acid-binding activities of the isolated amino-terminal head domain of the intermediate filament protein vimentin reveals its close relationship to the DNA-binding regions of some prokaryotic single-stranded DNA-binding proteins. J. Mol. Biol. 228, 41–57.

Traub, P., Mothes, E., Shoeman, R.L., Schröder, R., & Scherbarth, A. (1992b). Binding of nucleic acids to intermediate filaments of the vimentin type and their effects on filament formation and stability. J. Biomol. Struct. Dynamics 10, 505–531.

Traub, P., Perides, G., Kühn, S., & Scherbarth, A. (1986). Interaction *in vitro* of non-epithelial intermediate filament proteins with histones. Z. Naturforsch. 42c, 47–63.

Traub, P., Perides, G., Kühn, S., & Scherbarth, A. (1987a). Efficient interaction of nonpolar lipids with intermediate filaments of the vimentin type. Eur. J. Cell Biol. 43, 55–64.

Traub, P., Plagens, U., Kühn, S., & Perides, G. (1987b). Function of intermediate filaments. A novel hypothesis. Fortschritte der Zoologie (Progress in Zoology) 34, 275–287.

Traub, P., Scherbarth, A., Wiegers, W., & Shoeman, R.L. (1992c). Salt-stable interaction of the amino-terminal head region of vimentin with the alpha-helical rod domain of cytoplasmic intermediate filament proteins and its relevance to protofilament structure and filament formation and stability. J. Cell Sci. 101, 363–381.

Traub, P. & Shoeman, R. (1994). Intermediate filament proteins: Cytoskeletal elements with a gene regulatory function? Int. Rev. Cytology 154, 1–103.

Trevor, K., Linney, E., & Oshima, R.G. (1987). Suppression of endo B cytokeratin by its antisense RNA inhibits the normal coexpression of endo A cytokeratin. Proc. Natl. Acad. Sci. USA 84, 1040–1044.

Troyanovsky, S.M., Eshkind, L.G., Troyanovsky, R.B., Leube, R.E., & Franke, W.W. (1993). Contributions of cytoplasmic domains of desmosomal cadherins to desmosome assembly and intermediate filament anchorage. Cell 72, 561–574.

van de Klundert, F.A.J.M., Raats, J.M.H., & Bloemendal, H. (1993). Intermediate filaments: Regulation of gene expression and assembly. Eur. J. Biochem. 214, 351–366.

van der Ven, P.F., Schaart, G., Croes, H.J.E., Hap, P.H.K., Ginsel, L.A., & Ramaekers, F.C.S. (1993). Titin aggregates associated with intermediate filaments align along stress fiber-like structures during human skeletal muscle cell differentiation. J. Cell Sci. 106, 749–759.

Vikstrom, K.L., Lim, S.-S., Goldman, R.D., & Borisy, G.G. (1992). Steady state dynamics of intermediate filament networks. J. Cell Biol. 118, 121–129.

Vogt, P. (1990). Potential genetic functions of tandem repeated DNA sequence blocks in the human genome are based on a highly conserved "chromatin folding code". Hum. Genet. 84, 301–336.

Wang, X. & Traub, P. (1991). Resinless section immunogold electron microscopy of karyo-cytoskeletal frameworks of eukaryotic cells cultured in vitro. Absence of a salt-stable nuclear matrix from mouse plasmacytoma MPC-11 cells. J. Cell Sci. 98, 107–122.

Way, J., Hellmich, M.R., Jaffe, H., Szaro, G., Pant, H.C., Gainer, H., & Battey, J. (1992). A high-molecular-weight squid neurofilament protein contains a lamin-like rod domain and a tail domain with Lys-Ser-Pro repeats. Proc. Natl. Acad. Sci. USA 89, 6963–6967.

Weber, K., Riemer, D., & Dodemont, H. (1991). Aspects of the evolution of the lamin/intermediate filament protein family: A current analysis of invertebrate intermediate filament proteins. Biochem. Soc. Trans. 19, 1021–1023.

Wedrychowski, A., Schmidt, W.N., Ward, W.S., & Hnilica, L.S. (1986a). Cross-linking of cytokeratins to DNA in vivo by chromium salt and cis-diamminedichloroplatinum (II). Biochemistry 25, 1–9.

Wedrychowski, A., Schmidt, W.N., & Hnilica, L.S. (1986b). The in vitro cross-linking of proteins and DNA by heavy metals. J. Biol. Chem. 261, 3370–3376.

Weinstein, D.E., Shelanski, M.L., & Liem, R.K.H. (1991). Suppression by antisense mRNA demonstrates a requirement for the glial fibrillary acidic protein in the formation of stable astrocytic processes in response to neurons. J. Cell Biol. 112, 1205–1213.

Welch, W.J. & Suhan, J.P. (1985). Morphological study of the mammalian response: Characterization of changes in cytoplasmic organelles, cytoskeleton, and nucleoli, and appearance of intranuclear actin fibroblasts in rat fibroblasts after heat-shock treatment. J. Cell Biol. 101, 1198–1211.

White, E. & Cipriani, R. (1989). Specific disruption of intermediate filaments and the nuclear lamina by the 19-kDa product of the adenovirus E1B oncogene. Proc. Natl. Acad. Sci. USA 86, 9886–9890.

White, E. & Cipriani, R. (1990). Role of adenovirus E1B proteins in transformation: Altered organization of intermediate filaments in transformed cells that express the 19-kilodalton protein. Molec. Cell. Biol. 10, 120–130.

Whitesides, G.M., Mathias, J.P., & Seto, C.T. (1991). Molecular self-assembly and nanochemistry: A chemical strategy for the synthesis of nanostructures. Science 254, 1312–1319.

Wiche, G. (1989). Plectin: General overview and appraisal of its potential role as a subunit protein of the cytomatrix. CRC Crit. Rev. Biochem. 24, 41–67.

Wiche, G., Gromov, D., Donovan, A., Castañón, M.J., & Fuchs, E. (1993). Expression of plectin mutant cDNA in cultured cells indicates a role of COOH-terminal domain in intermediate filament association. J. Cell Biol. 121, 607–619.

Wiche, G., Krepler, R., Antlieb, U., Pytela, R., & Denk, H. (1983). Occurrence and immunolocalization of plectin in tissues. J. Cell Biol. 97, 887–901.

Wong, P.C. & Cleveland, D.W. (1990). Characterization of dominant and recessive assembly-defective mutations in mouse neurofilament NF-M. J. Cell Biol. 111, 1987–2003.

Xu, Z., Cork, L., Griffin, J.W., & Cleveland, D.W. (1993). Increased expression of NF-L in both neuronal and nonneuronal cells of transgenic mice: increased neurofilament density in axons without affecting caliber. J. Cell Biol. 111, 1543–1547.

Yang, Z.W. & Babitch, J.A. (1988). Factors modulating filament formation by bovine glial fibrillary acidic protein, the intermediate filament component of astroglial cells. Biochemistry 27, 7038–7045.

Yang, Z.W., Kong, C.F., & Babitch, J.A. (1988). Characterization and localization of divalent cation binding sites in bovine glial fibrillary acidic protein. Biochemistry 27, 7045–7050.

Yang, H.-Y., Lieska, N., & Goldman, R.D. (1990). Intermediate filament-associated proteins. In: Cellular and Molecular Biology of Intermediate Filaments (Goldman, R.D. & Steinert, P.M., Eds), pp. 371–391, Plenum Press, New York.

Yano, T., Tokui, T., Nishi, Y., Nishizawa, K., Shibata, M., Kikuchi, K., Tsuiki, S., Yamauchi, T., & Inagaki, M. (1991). Phosphorylation of keratin intermediate filaments by protein kinase C, by calmodulin-dependent protein kinase and by cAMP-dependent protein kinase. Eur. J. Biochem. 197, 281–290.

Yen, S.-H., Dickson, D.W., Peterson, C., & Goldman, J.E. (1986). Cytoskeletal abnormalities in neuropathology. Prog. Neuropathol. 6, 63–90.

Zackroff, R.V. & Goldman, R.D. (1980). *In vitro* reassembly of squid brain intermediate filaments (neurofilaments): Purification by assembly-disassembly. Science 208, 1152–1155.

Zamansky, G.B. & Chou, I.-N. (1990). Disruption of keratin intermediate filaments by ultraviolet radiation in cultured human keratinocytes. Photochem. Photobiol. 52, 903–906.

Zatloukal, K., Fesus, L., Denk, H., Tarsca, E., Spurej, G., & Böck, G. (1992). High amount of ε-(γ-glutamyl) lysine cross-links in Mallory bodies. Lab. Invest. 66, 774–777.

Zhao, J.X., Ohnishi, A., Itakura, C., Mizutani, M., Yamamoto, T., Hayashi, H., & Murai, Y. (1993). Smaller number of large myelinated fibers and focal myelin thickening in mutant quail deficient in neurofilaments. Acta Neuropathol. 86, 242–248.

NUCLEAR LAMINS AND THE NUCLEOSKELETON

Reimer Stick

The Cytoskeleton, Volume 1
Structure and Assembly, pages 257–296.
Copyright © 1995 by JAI Press Inc.
All rights of reproduction in any form reserved.
ISBN: 1-55938-687-8

I. STRUCTURE AND EVOLUTION OF
NUCLEAR LAMIN PROTEINS

A wealth of primary sequence information on vertebrate lamins has accumulated over the past years. Sequences for most of the different vertebrate lamin subtypes were initially obtained by cDNA expression cloning using lamin-specific poly-clonal or monoclonal antibodies (McKeon et al., 1986; Fisher et al., 1986; Stick, 1988; Peter et al., 1989; Vorburger et al., 1989a; Höger et al., 1990a; Pollard et al., 1990). Low stringency screening has allowed only the isolation of subtype homo-logues within the class of vertebrates (Wolin et al., 1987; Höger et al., 1988; Riedel and Werner, 1989; Höger et al., 1990a; Furukawa and Hotta, 1993; Nakajima and Sado, 1993). A notable exception is the cloning of *Xenopus* lamin LI (a B1-type lamin) by screening with a human lamin C probe (an A-type lamin; Krohne et al., 1987).

    In contrast to the wealth of information on vertebrate lamins, relatively little is known about lamins from other eukaryotes. Currently only three invertebrate lamin sequences are available. Interestingly, two of these sequences, the *Drosophila* lamin Dmo (Gruenbaum et al., 1988) and a sequence encoding a laminlike intermediate filament protein (Bossie and Sanders, 1993) have been isolated by expression screening while the third one, encoding the *Caenorhabditis elegans* lamin CeLam-1 (Riemer et al., 1993), was initially found in a random selection of expressed sequence tags (McCombie et al., 1992).

    Sequence comparison as well as structural analysis (see below) has classified lamins as intermediate filament (IF) proteins. Lamins show the structural features diagnostic for members of the intermediate filament protein family, that is, a tripartite domain structure with a central rod domain subdivided into four α-helical coils characterized by a heptad repeat of hydrophobic amino acids (see Figure 1; for review see Steinert and Roop, 1988). This sequence principle is responsible for the coiled-coil forming ability of IF proteins. Sequence similarity within the lamin subfamily is high, reaching values of 60% to 80% amino acid identity among members of the vertebrate lamins. Noticeable sequence similarity between cyto-plasmic IFs and nuclear lamins, on the other hand, is seen only at the ends of the rod domains, while along most of the rod domain conservation of the heptad sequence principle rather than actual sequences is found. The central rod domain of lamins is about 360 amino acids long, and alignment of the lamin rod sequences can be done without introducing any gap. A notable exception is the recently

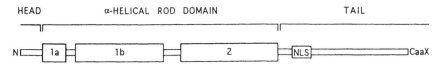

Figure 1. Schematic view of the primary structure of nuclear lamins. The α-helical rod domain is flanked by nonhelical head and tail domains. The rod domain is further subdivided into coils 1a, 1b, and 2. Subdivision of Coil 2 into coil 2a and 2b is less pronounced and has not been indicated in the figure. NLS: nuclear localization signal; CaaX: C-terminal target sequence for posttranslational prenylation.

described lamin CeLam-1 of *C. elegans* that has a deletion of 14 amino acids (two heptades) in the coil 2 domain.

The tail domain of nuclear lamins shows two hallmark sequences by which lamins are distinguished from cytoplasmic IF proteins and that are related to lamin functionality. These are a nuclear localization signal and a C-terminal CaaX motif (Figure 1). These two topogenic sequences in conjunction are necessary for targeting newly synthesized lamins to the inner nuclear envelope membrane (see below). The nuclear localization signal is located early in the tail approximately 20 to 30 amino acid residues past the carboxyl end of the rod domain. It has both structural and functional similarities with that of the simian virus 40 large T antigen (Loewinger and McKeon, 1988). The CaaX tetrapeptide is the substrate for a series of posttranslational modifications including isoprenylation, proteolytic removal of the last three amino acids, followed by carboxyl methylation of the resulting C-terminal cysteine residue (for review see Clarke, 1992).

A previous classification of lamins had been based on biochemical and immunological data, as for example, isoelectric point (neutral vs. acidic), cross-reactivity with certain monoclonal antibodies, and the subcellular distribution (membrane associated vs. cytosolic) during mitosis (Lehner et al., 1986; Krohne and Benavente, 1986). Analysis of more than a dozen vertebrate lamin sequences has placed this classification on a more solid ground. Sequence comparison clearly shows that there are at least three different subtypes of B lamins but only a single type of A lamin. A-type lamins, while resembling B lamins in the head and rod domain, possess a larger C-terminal domain with an unique region of about 70 to 90 amino acids in length. Sequence information on A-type lamins is available for man, mouse, chicken, and *Xenopus* (McKeon et al., 1986; Fisher et al., 1986; Wolin et al., 1987; Peter et al., 1989; Nakajima and Sado, 1993).

The protein structures of the different B-type lamins are very similar. The classification is based mainly on overall sequence similarity and the presence of short subtype-specific sequence motifs within the rod domain as well as small sequence insertions restricted to specific sites in the C-terminal domain. This is in line with the notion that the non-α-helical N- and C-terminal domains of IF proteins in general show greater variability in sequence and length. The overall sequence similarity within the rod domains of lamins of the same subtype is substantially higher than the similarity between different subtypes of the same species. Identity values within a subtype are in the range of 70% to 80% and similarity values are as high as 90%. Homologues of lamin subtypes B1 and B2 have been isolated for man, B1 (Pollard et al., 1990), B2 (partial sequence only; Biamonti et al., 1992), mouse, B1 (Höger et al., 1988), B2 (Höger et al., 1990a), chicken, B1 (Peter et al., 1989), B2 (Vorburger et al., 1989a), *Xenopus* LI (Krohne et al., 1987), *Xenopus* LII (Höger et al., 1990a). To express the observed homologies the nomenclature of the *Xenopus* B-type lamins should be unified as previously suggested (Stick, 1992). In the following all lamins belonging to subtype B1, including *Xenopus* LI, will be named B1, and correspondingly all lamins belonging to subtype B2, including *Xenopus* LII, will be designated B2.

In *Xenopus* another lamin has been isolated (lamin LIII). Its general structure is of the B type. Its sequence, however, matches neither the B1- nor the B2-subtype classification, and it does not show pronounced sequence similarity to either of the subtypes described so far. It therefore represents a third subtype of B lamins (Stick, 1988). In accordance with the classification mentioned above it will be named lamin B3 (Stick, 1992). In *Xenopus laevis* two closely related B3 genes exist, B3α and B3β. They show regions of about 90% sequence identity in the coding parts as well as in the 5' and 3' untranslated regions (Raab and Stick, unpublished). Two copies of each B3 gene are present per haploid genome in *Xenopus laevis* reflecting a duplication of the whole genome that has taken place some 30 million years ago in the genus *Xenopus* (for a more detailed discussion see Döring and Stick, 1990). Whether both of these gene pairs are expressed at the protein level remains at present an unresolved issue because none of the lamin specific monoclonal antibodies distinguishes between lamin B3α and B3β. The diversity of lamin B3 polypeptides might even be greater since cDNAs have been isolated that encode alternatively spliced versions of B3α (see below). For *Xenopus* lamin B3 no avian nor mammalian counterpart has been found, and it is tempting to speculate that B3 serves specialized functions in the giant oocyte nucleus and the rapidly dividing nuclei in early amphibian development.

A.   Gene Structure of Vertebrate Lamins

The gene structure of lamin genes encoding A-type lamins (Stick, 1992; Lin and Worman, 1993), as well as representatives of the three different B-type lamins (Döring and Stick, 1990; Zewe et al, 1991; Stick, 1994), has been analyzed. Ten introns found in vertebrate lamin genes are strictly conserved with respect to the homologue sequence position and codon phase. Six introns interrupt the central rod domain while four introns are found in the C-terminal domain (Figure 2). There are numerous examples where introns demarcate either folding modules (Go, 1981) or functional domains of proteins (for review see Holland and Blake, 1990). In lamin genes no clear correlation between protein structure and gene organization is seen.

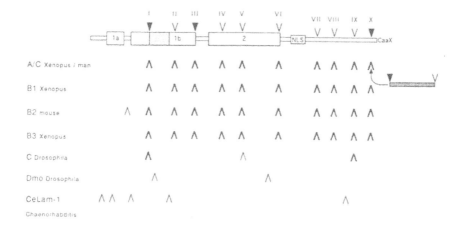

Figure 2. Comparison of intron positions in lamin genes. The conserved intron positions in vertebrate lamin genes are indicated above the schematic representation of the lamin primary structure and are numbered with roman numbers. Intron positions of individual lamin genes that match these positions are given in bold. Introns for which alternative splicing has been reported are marked by filled symbols. The region of the 42 amino acids in the central rod domain that is absent in vertebrate cytoplasmic IFs is shown as a stippled box. Intron I precisely coincides with the 5' end of this deletion. Generation of lamin A by insertion of an extra exon with its flanking sequences into intron X of a lamin B gene is indicated by a hatched box (for details see text). The alternative splice product of lamin A (i.e., vertebrate lamin C) has only been found in mammals. Intron patterns in invertebrate lamin genes greatly diverge from that found in vertebrates. Of the three known cases only the *Drosophila* lamin C gene has two of the three introns in positions that are found in all currently known vertebrate lamin genes. Data are taken from Döring and Stick (1990), *Xenopus* B3; Osman et al., 1990, *Drosophila* Dmo; Zewe et al. (1991), mouse B2; Stick (1992), *Xenopus* A; Lin and Worman (1993), human A; Riemer et al. (1993), *Chaenorhabditis*; Riemer et al. (1994), *Drosophila* C; and Stick (1994), *Xenopus* B1. NLS: nuclear localization signal; CaaX: C-terminal target sequence for posttranslational prenylation.

The central rod domain consists of two α-helical subdomains of approximately equal length (coil 1 and coil 2). Each of the subdomains is interrupted by three introns. While introns are found at the C termini of both of these subdomains, the relative positions of the two other introns are different in coil 1 and coil 2, respectively. However, the gene structure of lamin genes allows insight into how the diversity within the vertebrate lamin protein family might have been generated and how cytoplasmic IF proteins might have evolved from a laminlike ancestor.

B. The Generation of a Germ-Cell-Specific Lamin by Differential Splicing

Furukawa and Hotta (1993) have recently reported the cloning of a new mouse lamin that is selectively expressed in spermatocytes. Sequence analysis revealed that the mRNA encoding this lamin probably is generated by differential splicing and alternative polyadenylation from the lamin B2 gene. It has been named lamin B3 and should, therefore, not be confused with the *Xenopus* lamin B3, which is encoded by a different gene (Döring and Stick, 1990). In the spermatocyte-specific lamin exon 1-4 of lamin B2, encoding the 3-kDa N-terminal head and coil 1 of the rod domain are replaced by a novel 13-kDa amino acid sequence. In nuclear lamins the highly conserved sequences at the N terminus of the rod domain and conserved phosphorylation sites in the head domain are involved in the assembly and disassembly, respectively, of the lamina (Heald and McKeon, 1990; Peter et al., 1990). In the unique N-terminal region of the spermatocyte lamin no such sequences are present, and no homology in this domain is identified with previously described proteins. The presence of nine proline residues appearing semiperiodically in a segment corresponding to amino acids 19 to 85 seems to rule out the possibility that this domain forms an α-helical structure. The coil 2 domain and the C-terminal tail domain encoding the nuclear localization signal and the CaaX motif of this lamin are identical to those of lamin B2. This lamin is present in nuclear matrix fractions of spermatocytes and was found to be concentrated in the nuclear envelope of transfected cells. Moreover, transient expression of the spermatocyte-specific lamin in somatic cells resulted in interesting changes in nuclear morphology (Furukawa and Hotta, 1993). It can be assumed that this protein retained the ability to assemble into a supramolecular structure despite the deletion of half of the central rod domain. A direct comparison of the *in vitro* assembly properties of this protein with that of other lamin proteins and cytoplasmic IF proteins might extend our knowledge about the mechanisms of IF filament formation.

C. Evolution of Lamin A from a B-Type Lamin

As mentioned above A-type lamins possess a larger C-terminal domain with a unique part of about 70 to 90 amino acids in length (Fisher et al., 1986; Wolin et al., 1987; Peter et al., 1989; Nakajima and Sado, 1993). This lamin A-specific domain is encoded in a single exon and is flanked by introns on both sides (Stick, 1992; Lin and Worman, 1993). The positions of these flanking introns precisely define the borders of this domain and allow proper assignment of homologous regions in sequence alignments of A- and B-type lamins.

It is supposed that B-type lamins are involved in basic metabolic functions while lamin A might play a more specialized part dispensable for basic cellular activity. This has been inferred from the observation that B-type lamins are constitutively expressed while in contrast expression of A-lamins is highly regulated during development. The acquisition of lamin A during vertebrate embryogenesis is tissue dependent and appears to be associated with particular developmental stages. Certain cell types do not express A-type lamins at all, even in their fully differentiated state (for a more detailed discussion see below). These observations are in favor of the assumption that B-type lamins represent the ancestral type of lamins and that A-type lamins are derived therefrom. Furthermore, the comparison of the gene structure of vertebrate A- and B-type lamins strongly suggests that lamin A evolved by integration of a complete, preexisting exon into the last intron of a lamin B gene by exon shuffling (see Figure 2; Stick, 1992). The origin of this protein domain is not known and no significant similarity to known proteins has been observed. A lamins are proteolytically processed at their C terminus concomitant with or shortly after integration into the lamina structure (Weber et al., 1989; Beck et al., 1990). Interestingly, the site of proteolytic cleavage of A lamins lies within the lamin A-specific exon. It has been speculated that a cryptic protease cleavage site may exist in this domain and that maturation of lamin A is a consequence of the exon integration rather than a functional adaptation.

In addition to lamin A, mammals express a closely related but smaller lamin, namely lamin C. cDNA cloning and sequencing have shown that mammalian lamins A and C are identical for the first 566 amino acids (McKeon et al., 1986; Fisher et al., 1986). While lamin A has about 100 carboxyl-terminal amino acids including a CaaX motif, lamin C has only six unique carboxyl-terminal amino acids. It lacks a CaaX motif and cannot be modified by farnesylation. Analysis of the human lamin A/C gene shows how lamins A and C are generated by alternative splicing (Lin and Worman, 1993). The last nucleotides of lamin C mRNA encoding the six lamin C-specific amino acids and its unique 3' untranslated region derive from DNA contiguous with the codon for amino acid 566, which is the 3' most codon common to lamin A and C. To generate prelamin A mRNA, a splice joins the 3' end of codon 566 with the 5' side of exon 11, the lamin A-specific exon. There seems to be no differential regulation of splicing of the two forms as both mRNAs and their respective proteins appear to be expressed equally or not at all.

No evidence has been found for the existence of lamin C in vertebrates other than mammals. While the six carboxyl-terminal amino acids of lamin C are perfectly conserved between man and mouse (McKeon et al., 1986; Fisher et al., 1986; Riedel and Werner, 1989), hypothetical in-frame translation of intron 10 of the *Xenopus* gene does not show any resemblance to the mammalian sequence (Stick, 1992). It can therefore be assumed that lamin C is restricted to mammals rather than lack of detection in other vertebrate species. This poses the question of the functional significance of this lamin (see also below).

In all currently known vertebrate lamin genes the CaaX motif is encoded by a mini-exon only 9 to 13 amino acids long. At least one case exists where this genomic organization has been used to generate sequence variation of the CaaX motif (Döring and Stick, 1990). *Xenopus* oocyte lamin B3 is expressed in two splice variants, B3a and B3b, that differ in their last 12 amino acids. Both variants encode a CaaX motif; however, B3b shows additional sequence motifs that are reminiscent of splice variants in the *ras* protooncogene family. It contains an extra cysteine residue, a possible site of palmitoylation, and a polybasic stretch of six amino acids. In *ras* proteins such a sequence mediates strong membrane association in conjunction with the CaaX dependent prenylation (Hancock et al., 1990). The two splice variants of lamin B3 are expressed at the RNA level in oocytes; however, it is not known whether both forms are translated in oocytes or eggs.

D. The Gene Structure of Vertebrate B2-Type Lamins

B2-type lamin genes possess an extra intron in the first linker region of the rod domain while all other intron positions are identical to those of type B1, B3, and A-type lamins (see Figure 2; Zewe et al., 1991; Döring and Stick, 1990; Stick, 1992; Stick, 1994). The position of this extra intron is conserved between mouse and *Xenopus* B2 and therefore is probably B2 specific. One cannot, however, yet decide whether this intron is an ancient feature of lamins or was introduced during evolution of the lamin B2 subtype.

In conclusion, in the course of lamin evolution extensive use has been made of the given gene organization, both at the level of RNA splicing and at the gene level to generate diversity of lamin proteins as well as new lamin genes. Further analysis might uncover an even greater variability of this protein family generated by these mechanisms.

E. Invertebrate Lamin Genes

Sequence information of invertebrate lamins is available only from two species, the arthropod *Drosophila melanogaster* and the nematode *Caenorhabditis elegans* (Gruenbaum et al., 1988; Bossie and Sanders, 1993; Riemer et al., 1993). In *Drosophila* cDNA cloning documents two different lamins. *Drosophila* lamin Dmo represents a B-type lamin. It shows all the sequence characteristics of lamins, that is, a nuclear localization signal, a CaaX motif, and a cdc2 kinase phosphorylation site (Gruenbaum et al., 1988). A second laminlike protein from *Drosophila* has recently been isolated via expression cloning. It shows about 50% sequence identity with lamin Dmo. The predicted protein contains a cdc2 kinase phosphorylation site in front of the rod domain and a putative nuclear localization signal but, interestingly, lacks a C-terminal CaaX motif (Bossie and Sanders, 1993). Since this is reminiscent of mammalian lamin C, this protein has been tentatively designated as a *Drosophila* C-type lamin. However, given the fact that mammalian lamin C is generated by alternative splicing from a lamin A gene, this designation should probably not imply more than the lack of the CaaX motif. The protein seems to be expressed at low levels only, and future studies will reveal whether it is localized in the nuclear envelope lamina as expected for a class V IF protein and, if so, whether it can do so on its own or is dependent on the coexpression of the CaaX-containing lamin Dmo.

The nematode lamin CeLam-1 shows some other exceptional sequence features (Riemer et al., 1993). Its general structure resembles that of the various vertebrate B-type lamins and the *Drosophila* lamin Dmo. It contains a nuclear localization signal and a CaaX motif but it lacks the SPTR sequence in front of the rod domain that constitutes the major mitotic cdc2 kinase phosphorylation site (Peter et al., 1990). *In vitro* studies have shown that mitotic disassembly of lamins is triggered by phosphorylation of the highly conserved serine residue in the SPTR sequence motif. The absence of this cdc2 kinase site in CeLam-1 raises the question whether this *C. elegans* lamin is controlled by a different protein kinase or whether it forms complexes with an additional lamin that shows cdc2 dependent mitotic regulation.

In contrast to the highly conserved exon–intron organization of vertebrate lamin genes, both the *Drosophila* Dmo and the *C. elegans* CeLam-1 genes show unique intron patterns that are neither related to each other nor to those of vertebrate lamin genes (Figure 2). The same holds true when the intron positions of invertebrate lamin genes are compared with those of cytoplasmic IF genes.

F. Evolution of IF Proteins from a Laminlike Ancestor

In light of the latter findings the high conservation of the exon–intron organization of vertebrate lamin genes and genes of cytoplasmic IF proteins is striking (Döring and Stick, 1990; Steinert and Roop, 1988). The intron pattern of the vertebrate lamin rod domain is nearly identical to that of the vertebrate type III IF genes. The conservation of the gene structure is even more impressive for invertebrate IF genes and vertebrate lamin genes (Dodemont et al., 1990). In the gene of a cytoplasmic IF of the mollusc *Helix aspersa* and the lamin B3 gene of *Xenopus*, which were the first to be compared, 8 out of 10 introns occur at the same position. Of the six introns interrupting the coding sequence of the rod, intron 2 in the lamin gene is placed 30 nucleotides upstream of the position of the second intron of the invertebrate IF. This position is a feature common to all known vertebrate lamin genes. The coding region of the tail domains is interrupted by four introns in each gene. Three of these are found at homologue positions. Intron 7 in the mollusc IF gene has no counterpart in the lamin gene; it occurs in a region that in lamin genes encodes the nuclear localization signal. Conversely, intron 10, which delineates the last exon of the lamin gene, is absent in the IF gene. This last exon encodes the CaaX motif that is involved in the posttranslational modifications necessary for membrane binding of lamins. Cytoplasmic IF proteins lack both of these sequence motifs. The comparison of the corresponding genes, the lamin gene and the mollusc IF gene, suggests how evolution from a nuclear to a cytoplasmic IF protein might have occurred. After gene duplication generation of new splice donor and acceptor sites flanking the nuclear localization signal would lead to the elimination of this

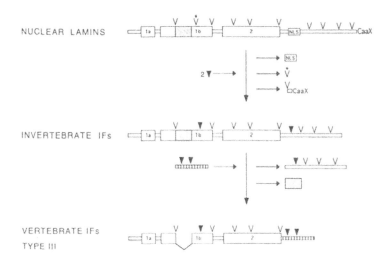

Figure 3. Model of IF gene evolution from a nuclear lamin ancestor gene. For details see text. Common intron positions are marked by V. The generation of the invertebrate IF gene is accompanied by loss of the second lamin intron, the nuclear localization signal (NLS), C-terminal CaaX encoding exon together with the last lamin intron (CaaX: C-terminal target sequence for posttranslational prenylation), and the gain of two new introns (filled triangles). During the evolution of vertebrate IFs 42 amino acids of coil Ib (stippled box) are deleted. The complete tail domain is replaced by new exons (hatched box). The different types of vertebrate genes might have evolved by shuffling of different tail domains. Shown here is the gene structure of a vertebrate type III IF gene. For further discussion of the IF gene evolution see Dodemont et al. (1990).

signal sequence as a result of RNA splicing. Indeed, for proper sequence alignment between lamins and IF proteins a gap has to be introduced in IF sequences at the positions of the lamin nuclear localization signal. The loss of the CaaX motif together with the last intron of lamin genes can simply be explained by the generation of a new stop codon upstream of the last lamin exon. By combining these two mutations, a cytoplasmic IF would have been generated from a nuclear IF progenitor (see Figure 3; Döring and Stick, 1990; Dodemont et al., 1990).

This evolutionary scenario anticipates that lamins are the ancestral members of the IF protein family. This assumption is supported by the observation that the nuclear lamina is a ubiquitous component of the eukaryotic nucleus. It might have emerged early in eukaryotic evolution during the transition from the prokaryotic state. Moreover, lamins of the B type are constitutively expressed, supporting the assumption that they serve housekeeping functions. This contrasts with the tissue-specific expression of cytoplasmic IFs as well as the notion that certain cell types lack IF proteins (for review see Franke, 1987).

II. STRUCTURE OF THE NUCLEAR LAMINA AND *IN VITRO* ASSEMBLY OF LAMINS

Electron microscopic evidence for the existence of a nuclear lamina was reported nearly forty years ago (Pappas, 1956; Beams et al., 1957). These reports described a complex supporting "layer of fine filaments" on the inner aspect of the nuclear envelope of protozoa. The filamentous nature of the lamina has been shown by thin sectioning of a variety of vertebrate cell types, and the structure has therefore been called the "fibrous lamina" (Fawcett, 1966). Detailed ultrastructural information of the native nuclear lamina is available only for a few cell types (Aebi et al., 1986; Stewart and Whytock, 1988; Hill and Whytock, 1993). On-face view of the lamina of amphibian oocyte nuclei shows a quasi-regular meshwork of lamin filaments. In the best preserved cases this meshwork consists of two approximately orthogonal sets of 11-nm diameter filaments with an average cross-over spacing of 52 nm (Aebi et al., 1986; Stewart and Whytock, 1988). Under certain conditions of preparation a pronounced axial beading is seen every 25 nm along the filaments. Similar structures have been observed in salivary gland nuclei of *Drosophila* (Hill and Whytock, 1993). The amphibian oocyte lamina consists of a single layer of orthogonally oriented IFs. The thickness of the lamina varies in different cell types, and some cells contain an unusually thick lamina (up to 200 nm in diameter) (Fawcett, 1966). In these cases the lamina may contain many layers of filaments and ultrastructural analysis might therefore be impeded. Moreover, in most nuclei the chromatin is tightly associated with the nuclear envelope and the lamina is hidden from view, complicating the interpretation of comparable studies in these cells. The major difference between the lamina of amphibian oocyte and insect salivary gland nuclei and that of diploid somatic cells might indeed be the difference in lamin–chromatin interactions. The amphibian oocyte lamina therefore may not represent other cell types.

From the ultrastructural analysis as well as from numerous antilamin fluorescence light microscopic and immunogold electron microscopic images, the lamina appears as a relatively continuous layer fenestrated only at sites of nuclear pore complexes. (Fawcett, 1966; Stick, 1987; Höger et al., 1990b). This view has been challenged recently by three-dimensional light microscopy and electron microscopy methodologies in conjunction with lamin-specific antibody staining (Paddy et al., 1990). The spatial organization of lamins in interphase cells appears to be highly discontinuous. Lamins form a fibrillar network that leaves large voids of little or no lamin staining in the nuclear periphery. Spacing between adjacent fibers of about 0.5 μm and fiber widths of 0.25 μm have been reported. According to these images no more than half of the nuclear surface area is covered by lamin fibers. Such a picture was found for different cell types, and it was largely unaffected by a variety of preparation methods. Moreover, close contact between the lamina and the peripheral chromatin was found to be limited to less than 20% of the inner nuclear surface (Paddy et al., 1990). These findings fit the "open cagelike" network lamina structure described by Penman and coworkers (Capco et al., 1982), rather than the dense orthogonal IF network described for amphibian oocytes (Aebi et al., 1986; Stewart and Whytock, 1988; Akey, 1989) and *Drosophila* salivary gland cells (Hill and Whytock, 1993; see also above). The obvious discrepancies in the appearance of the nuclear lamina as revealed by different methodologies cannot be resolved at present but may initiate reexamination of what was thought to be a settled issue.

A. *In Vitro* Assembly of Lamins

In vitro assembly of lamins has been studied both with proteins purified from eukaryotic tissues (Aebi et al., 1986; Goldman et al., 1986) and with wild-type and mutant lamin proteins expressed in *E. coli* (Moir et al., 1990; 1991; Heitlinger et al., 1991; 1992; Peter et al., 1991; Gieffers and Krohne, 1991). At a first level of structural organization both A-type and B-type lamins form "myosin-like" dimers consisting of a ~52-nm long rod flanked on one end by two globular domains. It has been interpreted that these molecules represent two lamin polypeptides interacting via their central α-helical rod domain to form a parallel and unstaggered 52-nm long two-stranded coiled coil. The length of the coiled-coil of lamin dimers is longer than that of vertebrate IF proteins. It correlates well with the longer central rod domain of lamins that exceed that of vertebrate IF proteins by an extra six heptades.

The carboxy-terminal domains correspond to the globular end domains, as clearly evidenced by the analysis of lamin mutants lacking the complete C-terminal domain (Moir et al., 1991; Gieffers and Krohne, 1991). During reconstitution, at a second level of organization, lamin dimers form polar head-to-tail polymers (Heitlinger et al., 1991; 1992). This longitudinal mode of association is in striking contrast to the lateral mode of association observed for cytoplasmic IFs, which form tetramers and octameres before undergoing longitudinal association (for review see Aebi et al., 1988). Finally, at a third level of structural organization, lamins form paracrystalline arrays exhibiting distinct transverse banding patterns with axial repeats of 24 to 25, 5 nm, rather than assembling into stable 10-nm filaments. The tendency to form paracrystals is quite extensive and IF-like structures are only formed transiently. These filaments appear as loose webs rather than as typical 10-nm lamin filaments that have been visualized in the lamin structure of amphibian oocytes (Heitlinger et al., 1991; Aebi et al., 1986).

Further analysis has yet to reveal whether additional factors are needed for the assembly of lamins into IFs. It might well be that the nuclear membrane or receptor molecules within the nuclear membrane play a role in lamin filament assembly *in vivo*. As shown by the analysis of N and C terminally truncated lamins, these domains make different contributions to the assembly properties of lamin polypeptides. Deletion of the complete C-terminal domain of either lamin A or lamin B2 does not interfere with the longitudinal head-to-tail association and the formation

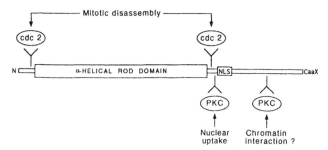

Figure 4. Modulation of nuclear lamins by phosphorylation. Phosphorylation sites for $p34^{cdc}$ 2 kinase (cdc 2) and protein kinase C (PKC) and their possible role in lamin function are indicated. Data are taken from Heald and McKeon (1990), Peter et al. (1990), Eggert et al. (1993), and Haas and Jost (1993).

of paracrystals (Moir et al., 1991; Gieffers and Krohne, 1991; Heitlinger et al., 1992). Instead, tailless lamin B2 molecules revealed an enhanced tendency for longitudinal assembly (Heitlinger et al., 1992). Truncation of the small head domain of lamin B2, on the other hand, abolishes the formation of any distinct oligomeres, indicating that the head domain is critically involved in lamin assembly. While headless lamin B2 is able to form paracrystals under appropriate conditions, it does not reveal a significant amount of filamentous structures at intermediate time points during paracrystal formation. The influence of the head domain in lamin assembly has been demonstrated elegantly by analysis of the structural consequences of lamin phosphorylation by cdc2 kinase. Head-to-tail polymers reconstituted *in vitro* from bacterially expressed lamin B2 have been shown to disassemble into myosin-like dimers upon treatment with purified cdc2 kinase (Peter et al., 1991). Subsequent dephosphorylation of lamin dimers allows the reformation of lamin head-to-tail oligomers. The phosphorylation sites are located within SPTP motifs N- and C-terminal to the central rod domain (Figure 4). These sites are phosphorylated during mitosis (Peter et al., 1990). While mutations in the C-terminal phosphoacceptor site have no profound effect on the *in vitro* disassembly properties, a single point mutation in the N-terminal SPTR motif renders head-to-tail polymers resistant to disassembly, stressing the importance of this phosphorylation site for controlling the state of assembly.

III. *IN VIVO* ASSEMBLY OF LAMINS AT THE INNER NUCLEAR ENVELOPE MEMBRANE

In vitro assembly of lamins is independent of posttranslational processing of lamin polypeptides, as evidenced by the fact that proteins produced in *E. coli* show the same properties in assembly reaction as lamins isolated from tissues (Aebi et al.,

1986; Goldman et al., 1986). Within living cells the targeting of newly synthesized lamins to and their assembly at the inner nuclear membrane depend on the presence of two signal sequences, the nuclear localization signal and the C-terminal CaaX motif (Loewinger and McKeon, 1988; Holtz et al., 1989; Krohne et al., 1989; Kitten and Nigg, 1991; Firmbach-Kraft and Stick; 1993; Hennekes and Nigg, 1993). Such a motif is found at the C termini of several cytoplasmic as well as secreted proteins. Examples are the yeast-mating type pheromones, the *ras* oncogene proteins, and other GTP-binding proteins. All known members of the lamin protein family, with the exception of human lamin C and a recently described lamin of *Drosophila* (Bossie and Sanders, 1993), contain a CaaX motif (for review see Schafer and Rine, 1992). The CaaX tetrapeptide (C = cysteine, a = aliphatic, X = any amino acid) is the substrate for a series of posttranslational modifications including isoprenylation of the cysteine via a thioether bond, proteolytic removal of the last three amino acids (aaX), followed by carboxyl methylation of the resulting C-terminal cysteine residue (Schafer and Rine, 1992).

There are two types of prenyl groups, C15 farnesyl and, more abundantly C20 geranylgeranyl, that can be linked to CaaX proteins. The nature of the attached substituent depends on specific sequence information in the carboxyl terminus of the protein. Lamins have been shown to be farnesylated (Casey et al., 1989; Wolda and Glomset, 1988; Lutz et al., 1992). The importance of these modifications for correct targeting of lamins to the nuclear envelope and for stable integration into the lamina has been shown by several experimental approaches. Treatment of cultured cells with Lovastatin, which blocks protein prenylation, results in the accumulation of prelamin A in the nucleoplasm. If the block in prenylation is circumvented by the addition of mevalonate, as a precursor of the prenyl group, prelamin A is incorporated into the lamina structure within a short period of time (Lutz et al., 1992). Similar results have been obtained with lamin B3 in *Xenopus* oocytes (Firmbach-Kraft and Stick, 1993). Mutant lamins in which the CaaX cysteine had been replaced by an amino acid that cannot function as an isoprenyl acceptor are prenylated neither *in vitro* nor *in vivo* (Vorburger et al., 1989b; Kitten and Nigg, 1991). When transfected in cultured cells these mutant lamins remain in the nucleoplasm and sometimes give rise to the formation of nucleoplasmic particles. In a third approach, *in vitro* synthesized wild-type and mutant lamin proteins were directly injected into the cytoplasm of *Xenopus* oocytes (Krohne et al., 1989). Microdissection of oocyte nuclei into nuclear envelope and nucleoplasm allowed the effect of the various mutations to be analyzed. In agreement with the above-mentioned findings, lamins that were blocked in prenylation remained in a soluble form in the nucleoplasm. The CaaX-dependent modifications result in an increased hydrophobicity of the carboxyl terminus and may explain the affinity of prenylated proteins for membranes. The experiments mentioned are in line with the assumption that lamins would concentrate at the inner nuclear membrane due to lipophilic modifications and that filament formation would then be initiated.

The role of isoprenylation for membrane association has been directly demonstrated by analyzing the subcellular distribution of mitotic wild-type and mutant lamins. A-type and B-type lamins differ in their fate during cell division. Upon mitotic nuclear envelope breakdown, A-type lamins become freely soluble while B-type lamins, although depolymerized during mitosis, remain associated with remnants of nuclear envelope membranes. This has been explained by the fact that B-type lamins remain permanently isoprenylated while A-type lamins lose the C-terminal modifications concomitantly or shortly after incorporation into the lamina by an additional proteolytic processing event (Weber et al., 1989; Beck et al., 1988; 1990). Interestingly, isoprenylation is a prerequisite for the proteolytic processing of prelamin A into the mature form C (Beck et al., 1990; Lutz et al., 1992). B lamins mutated in the CaaX motif are no longer associated with mitotic membranes (Kitten and Nigg, 1991). Moreover, cleavage mutants of lamin A, in which a single amino acid adjacent to the putative cleavage site has been changed, become readily incorporated into the nuclear envelope lamina after transfection. This permanently farnesylated lamin A remains associated with mitotic membranes similar to endogenous wild-type B lamins (Hennekes and Nigg, 1993). The results reported so far indicate that the differential membrane association of mature A- and B-type lamins is determined primarily by the presence or absence of a hydrophobically modified C-terminal domain.

Analysis of the *Xenopus* oocyte lamin B3, however, shows that prenylation of lamins, although necessary for targeting and *in vivo* assembly of lamins, is not sufficient for stable membrane association. Lamin B3 is the major lamin of amphibian oocytes. As mentioned earlier, it belongs to the B-type lamins and contains a C-terminal CaaX motif. In contrast to B-type lamins in somatic cells, lamin B3 is soluble rather than membrane-associated in egg cytosol after meiotic envelope breakdown (Benavente et al., 1985; Stick, 1987; Newport et al., 1990) and only a minor fraction of lamin B3 might be associated with membranes in eggs (Lourim and Krohne, 1993; Firmbach-Kraft and Stick, 1995). While this is reminiscent of the fate of A-type lamins, biochemical analysis shows that lamin B3, similar to somatic B-type lamins, is permanently isoprenylated in oocytes (interphase) and eggs (meiotic metaphase) (Firmbach-Kraft and Stick, 1993). In contrast, in transfected mouse cells a significant portion of B3 remains associated with membranes during mitosis. This latter finding indicates that *Xenopus* lamin B3, like the endogenous B-type lamins, has the ability to remain associated with nuclear envelope membranes. From these data it has been concluded that, similar to the situation with *ras* oncogenes, additional factors must be involved in membrane anchoring of lamins. Several possible candidates capable of mediating the specific membrane binding of lamins have been described (Worman et al., 1988; Senior and Gerace, 1988; Padan et al., 1990; Bailer et al., 1991; Simos and Georgatos, 1992; Foisner and Gerace, 1993). They are integral membrane proteins located in the inner nuclear membrane. Most of the evidence for their interaction with lamins is indirect. Lamins and putative receptors colocalize in interphase cells (Worman et al., 1988;

Senior and Gerace, 1988; Padan et al., 1990; Bailer et al., 1991; Chaudhary and Courvalin, 1993). For one of the membrane proteins, avian p58, tight association of lamin A and B was demonstrated in a complex isolated immunochemically from a whole cell lysate (Simos and Georgatos, 1992). p58 is phosphorylated *in vivo*, and removal of the phosphate moieties affects lamin binding (Appelbaum et al., 1990; Courvalin et al., 1992). More recently the *in vitro* interaction of a series of integral membrane proteins (lamin-associated proteins LAPs 1A, 1B, 1C, and LAP2) with lamins and chromosomes has been described; one of these proteins (LAB1C) is identical to a previously characterized nuclear envelope protein (p55; Senior and Gerace, 1988; Foisner and Gerace, 1993). LAPs 1A and 1B bind specifically to both lamins A and B, while LAP2 associates only with lamin B1. LAP2 also binds to mitotic chromosomes. These lamin-associated proteins are phosphorylated during mitosis, and phosphorylation of LAP2 inhibits its binding to both lamin B1 and chromosomes (Foisner and Gerace, 1993).

Since all of these membrane-spanning polypeptides bind to lamins, it is conceivable that these proteins could function as "receptors" for attaching lamins to the inner nuclear membrane. They could be important for directing the assembly of lamins and for maintaining the structural integrity of the nuclear envelope. An interesting observation in this respect stems from experiments which analyzed the distribution of p55 (LAP1C) between nuclei of interspecies heterokaryons (Powell and Burke, 1990). The equilibration of p55 in nuclei of heterokaryons was shown to depend on the presence of lamin A in the recipient nucleus. In this case the lamin-binding ability of the membrane protein may reflect a mechanism for achieving localization at the inner nuclear membrane, rather than p55 functioning as a lamin receptor.

Conflicting results have been published concerning determinants that sort p58, the putative avian lamin receptor, from its site of integration (rough endoplasmic reticulum and outer nuclear membrane) to the inner nuclear membrane. While Smith and Blobel (1993) demonstrated that the first membrane-spanning region of p58 is necessary and sufficient for sorting chimeric proteins to the inner nuclear envelope, Soullam and Worman (1993) have assigned this property to the N-terminal region of the polypeptide excluding the first membrane domain.

Based on the analysis of farnesylation and the CaaX mutant lamins, the farnesylated C terminus of lamins should be a major site of interaction between lamins and the membrane receptor(s). No direct evidence for such interaction has been reported for a candidate lamin receptor.

A. Control of Lamina Assembly by Cell-Cycle-Dependent Phosphorylation

The assembly and disassembly of nuclear lamins during the cell cycle represent conserved features of higher eukaryotes. Disassembly of mitotic lamins is accompanied by hyperphosphorylation and reassembly around daughter nuclei by dephosphorylation (Gerace and Blobel, 1980; Burke and Gerace, 1986; Ottaviano and Gerace, 1985). The mitosis-specific phosphorylation sites in lamins have been mapped to two regions flanking the α-helical rod domain on either side (see Figure 4; Ward and Kirschner, 1990; Peter et al., 1990; Peter et al., 1991). Mutations in these sites that prevent phosphorylation block the disassembly of the nuclear lamina during mitosis. Detailed analysis reveals that mutation of the N-terminal phosphorylation site alone is sufficient to prevent lamina disassembly, while double mutants affecting both the N-terminal and the C-terminal site have more severe effects (Heald and McKeon, 1990). Several lines of evidence strongly support the notion that the $p34^{cdc2}$-cyclin B complex, the cdc2 kinase, a major regulator of eukaryotic cell cycle, is responsible for mitotic lamin phosphorylation (Peter et al., 1990). The M-phase-specific phosphorylation sites match the consensus sequence of cdc2 kinase. Purified cdc2 kinase phosphorylates lamins on M-phase-specific sites. Incubation of isolated nuclei with cdc2 kinase results in solubilization of lamins and *in vitro* assembled head-to-tail polymers of purified lamins are disaggregated upon cdc2 kinase treatment but dimer formation is not inhibited. Furthermore, when expressed in the fission yeast *Schizosaccharomyces pombe*, chicken lamin B2 is phosphorylated by a mitotically activated yeast kinase. This kinase was shown to be temperature sensitive in a strain carrying a temperature-sensitive mutation in the cdc2 gene (Enoch et al., 1991). Meanwhile, mitogen-activated kinases have been shown to phosphorylate lamins on a site overlapping that of mitotic cdc2 kinase (Peter et al., 1992). Moreover, these kinases cause depolymerization of *in vitro* assembled head-to-tail lamin polymers. This raises the possibility that some of the purported substrates of $p34^{cdc2}$ may also be substrates of mitogen-activated kinases. In conclusion, these experiments strongly support the notion that cdc2 kinase activity regulates lamin disassembly. The function of cdc2 kinase on lamin disassembly seems to be antagonized by protein kinase A (Molloy and Little, 1992).

However, it has been reported that not all aspects of lamina disassembly might be achieved by cdc2 kinase. These arguments are based on experiments with phosphatase inhibitors *in vivo*. Although inhibition of phosphatases results in the appearance of M-phase-specific phosphorylation and concomitantly increases the solubility of lamins in nonionic detergent, this phosphorylation is not sufficient to disassemble lamin B2 from the nuclear lamina as judged by immunofluorescence analysis of intact cells (Lüscher et al., 1991). On the contrary, in a well-documented study by Goldman and colleagues a single protein kinase activity, the clam MPF, identical with $p34^{cdc2}$-H1 kinase, has been shown to phosphorylate clam lamin, to disassemble clam oocyte nuclei, and to cause chromosome condensation (Dessev et al., 1991). Notably, and in contrast to other cell free systems, these results were obtained with a homologous system, that is, nuclei as well as extracts prepared from the same type of cells were used, both to reproduce nuclear lamina disassembly and as a source for the purification of lamin kinase activity.

Lamins are phosphorylated in response to a number of other stimuli (Hornbeck et al., 1988; Martell et al., 1992; Krachmarov and Traub, 1993) and several phosphorylation sites in addition to the M-phase-specific cdc2 kinase sites have been mapped (see Figure 4; Hennekes et al., 1993; Hocevar et al., 1993; Eggert et al., 1993). Most of these are protein kinase C phosphorylation sites. They are located in the tail domain in the vicinity of the nuclear localization signal (Hennekes et al., 1993). It has been shown that phosphorylation of lamin B2 with protein kinase C significantly inhibits uptake into the nucleus in an *in vitro* nuclear transport system. This effect is specific since phosphorylation at two cdc2 kinase sites at the C terminus of the central rod domain, approximately 30 amino acids upstream of the nuclear localization signal, did not interfere with nuclear transport of lamin B2. Moreover, these sites are phosphorylated *in vivo* upon stimulation of protein kinase C by phorbol ester. Although phorbol ester stimulation *in vivo* does not have such a profound effect as the *in vitro* phosphorylation with protein kinase C, it leads to a transient reduction in the nuclear uptake of lamins. These findings might have implications for the regulation of the structure of the nuclear lamina and its dynamics during interphase.

The interconversion of *Drosophila* nuclear lamin isoforms during oogenesis and upon entry into mitosis is accompanied by a specific rearrangement of phosphate groups indicative of M-phase-specific phosphorylation. This interconversion into a soluble, mitosis-specific lamin form Dmo_{mit} does not, however, result in a dramatic net change in the level of lamin phosphorylation (Smith and Fisher, 1989). While there seems to be no marked differences between the mitotic lamins of vertebrates and *Drosophila*, interphase *Drosophila* lamins are phosphorylated to about 10 times the level of the vertebrate counterparts (Smith et al., 1987). It has been speculated that this may maintain nuclear plasticity in the rapidly developing *Drosophila* embryo.

IV. DIFFERENTIAL EXPRESSION OF LAMINS IN DEVELOPMENT AND TISSUE DIFFERENTIATION

Nuclear lamins are differentially expressed during development and tissue differentiation in vertebrates. Numerous studies have been carried out over the past decade, and attempts have been made to relate lamin expression with basic cellular functions as well as developmental stages and processes of cell differentiation. It is beyond the scope of this review to give a complete summary of the wealth of information that has accumulated in this field. For more detailed reading the reader is referred to the literature cited.

A few general conclusions can be drawn: B-type lamins seem to be constitutively expressed. In every nucleated vertebrate cell at least one subtype of B lamin is present (Benavente et al., 1985; Stick and Hausen, 1985; Stewart and Burke, 1987; Lehner et al., 1987; Paulin-Levasseur et al., 1988; Röber et al., 1989; 1990; Cance et al., 1992; Kubiak et al., 1991; Guilly et al., 1987; Houliston et al., 1988; Vester et al., 1993; Moss et al., 1987). Early development in *Xenopus* represents an example of differential expression of B-type lamins. Lamin B3 is the major lamin in female germ cells and in early cleavage embryos, lamin B1 becomes detectable during the midblastula transition, and lamin B2 at the gastrula stages (Stick and Hausen, 1985; Benavente et al., 1985; see also Lourim and Krohne, 1993). Appearance of nuclear lamins B1 and B2 corresponds to the profound change of the cell cycle (from embryonic to somatic-like during the midblastula transition) and to processes of differentiation (gastrulation), respectively. Expression of B1 and B2 is independent of embryonic transcription at this stage of development, it is rather under translational control (Stick and Hausen, 1985).

In birds and mammals two B-type lamins are expressed in somatic cells, lamin B1 and B2. They are coexpressed in most cell types but the ratio of the two B forms may vary with the cell type (Lehner et al., 1987; Röber et al., 1989). Furthermore, while B2 represents the major component in avian cells, it is a quantitatively minor one in certain mammalian nuclei.

Expression of A-type lamins, on the contrary, is highly regulated during development. The acquisition of lamin A during vertebrate embryogenesis is tissue dependent and appears to be associated with particular developmental stages. Lamin A is absent in early amphibian and chicken embryos (Wolin et al., 1987; Lehner et al., 1987). The presence of A lamins in preimplantation embryos of the mouse has been a matter of debate (Schatten et al., 1985; Stewart and Burke, 1987; Houliston et al., 1988; Kubiak et al., 1991). It appears that lamin A might be present throughout preimplantation development but the organization of the protein may somehow be altered in the early stages rendering it inaccessible to some antibodies. In general, expression of lamin A is absent or low in undifferentiated cells and increases during the course of differentiation. In several cell lines that can be induced to differentiate *in vitro*, this differentiation is accompanied by expression of A-type lamins (Stewart and Burke, 1987; Lebel et al., 1987; Lourim and Lin, 1989; Röber et al., 1990; Mattia et al., 1992; Lanoix et al., 1992; Beug and Stick, unpublished). These cell types have been used to analyze at which level lamin A expression is regulated. For mouse embryonic carcinoma cells conflicting results have been reported. Mattia et al. (1992) presented data from embryonic carcinoma cells which suggests that expression of lamin A is regulated at the transcriptional level. Lanoix et al. (1992), on the other hand, showed that the level of lamin A transcription measured by run-on transcription assays remains unaltered upon induction of differentiation and that lamin A and lamin C mRNAs are present at detectable levels prior to induction. This indicates that regulation is posttranscriptional. Similar results have been obtained by analyzing the *in vitro* differentiation of avian monocytes, which express lamin A upon terminal differentiation (Beug and Stick, unpublished). More recently the expression pattern of lamins has been analyzed in neoplastic tissues. Similar to normal tissue, B lamins are constitutively expressed in tumor tissue (for exceptions see Broers et al., 1993) while lamin A shows differential expression depending on the type of cancer (Cance et al., 1992; Hytiroglou et al., 1993; Broers et al., 1993). Whether or not the expression pattern of A-type lamins will be useful as a tumor marker or prognostic indicator has to await further analysis.

Does the expression pattern allow any conclusions to be drawn on the functional implications of lamin A? The concomitant expression of lamin A with cell differentiation led to the hypothesis that A-lamins might play a role in establishing or stabilizing differentiation-specific differences in nuclear organization, and this might relate to the developmental potential of a cell (Nigg, 1989; Burke, 1990). Such a correlation, however, is not universal and should be taken with some reservation. Certain cell types of the hemopoetic system of birds and mammals do not express A-type lamins at all, even in their fully differentiated state (Röber et al., 1989; Stick and Beug, unpublished). More detailed inspection reveals interspecies differences that are hard to explain in functional terms. While, for example, lamin A is prominent in mature, circulating erythrocytes in chicken, it is absent in the same cell type in amphibians (Lehner et al., 1987; Wolin et al., 1987).

Embryonic carcinoma cells have been used to express ectopically lamin A and C (Collard and Raymond, 1990; Peter and Nigg, 1991; Horton et al., 1992). These analyses revealed interesting differences between lamin A and C in their abilities to incorporate into the nuclear lamina. While lamin A is assembled and processed in a normal fashion, lamin C, which lacks the C-terminal CaaX motif, requires progression through the cell cycle in order that assembly into the lamina may occur (Horton et al., 1992). Moreover, when stably transfected, it could be shown that ectopically expressed lamin A does not induce differentiation, nor does the expression of lamin A interfere with retinoic acid differentiation (Peter and Nigg, 1991). It may well be that expression of lamin A leads to more subtle changes in nuclear organization or that other factors interacting with lamin A are involved in these processes and that these factors have to be expressed in concert with lamin A.

A. Lamin Expression in Germ Cells

The existence of a nuclear lamina in germ cells has been a matter of debate for some time. Ultrastructural analysis does not reveal a distinct nuclear lamina structure either in male and female pachytene cells or in postmeiotic male germ cells (Stick and Schwarz, 1982; 1983). The characterization of a germ-cell-specific lamin by cDNA cloning, however, as well as the demonstration of B-type lamins in spermatogenic cells by immunological methods shows that lamins are expressed in spermatogenesis (Furukawa and Hotta, 1993; Maul et al., 1986; Moss et al., 1987; Biggiogera et al., 1991; Moss et al., 1993; Smith and Benavente, 1992; Vester et al., 1993). As described in detail above, the germ-cell-specific lamin, that has a unique N terminus, represents a splice variant of the mouse somatic lamin B2 and has a lower molecular weight than somatic lamins. A protein of similar molecular weight, which is also expressed in pachytene spermatocytes, has been detected in the rat. Surprisingly the antibody reactive with this germ cell protein cross-reacts with lamin A/C in somatic cells (Smith and Benavente, 1992).

It has been consistently found that A-type lamins are not detected in spermatogenic cells. Several B-lamin-specific antibodies fail to react with spermatogenic cells. Remarkably, those that do react show weaker labeling than with somatic cells (Vester et al., 1993). Furthermore, it has been shown that the detection of lamins in certain stages of spermiogenesis depends on chemical treatment prior to staining, indicating a masking of antigenic sites (Moss et al., 1993). The staining pattern of meiotic and postmeiotic cells has often been reported to be distinctly different from that of somatic cells (Moss et al., 1993; Sudhakar et al, 1992). One may speculate therefore that the organization of lamin filaments in germ cells is different from that of somatic cells. This, however, demands a detailed ultrastructural analysis.

Spermatids and sperm of *Xenopus* express a spermatogenesis-specific lamin protein, lamin LIV (Benavente and Krohne, 1985). A homologous lamin has not been detected in other vertebrates so far, and structural comparison of this protein with other lamins will have to await sequence information.

V. LAMIN–CHROMATIN INTERACTION

Morphological observations suggest a close association between chromatin and the nuclear lamina. To study possible interactions of lamins with chromatin, several *in vitro* systems have been established (Burke, 1990, Glass and Gerace, 1990; Höger et al., 1991; Yuan et al., 1991). In one type of experiment the assembly of lamins A and C at mitotic chromosome surfaces has been studied *in vitro* (Burke, 1990; Glass and Gerace, 1990). It could be shown that under appropriate conditions mitotic lamins A and C, as well as purified interphase lamins A and C, bind to the surface of mitotic chromosomes and that in both cases a lamin-containing supra-molecular structure is formed that is resistant to extraction with high salt and digestion with DNases (Glass and Gerace, 1990). Specific and saturable binding of lamin A to interphase polynucleosomes has been demonstrated by Yuan et al. (1991). In the assay developed by these authors, binding was specific for lamin A while lamin B did not bind. The association of lamins A and C with chromatin seems to be mediated by proteins as evidenced by the sensitivity against pretreatment of the chromatin with proteases (Glass and Gerace, 1990; Yuan et al., 1991). Using a different experimental approach, Kleinschmidt and coworkers (Höger et al., 1991) have assigned chromatin-binding capacity to specific regions of lamins. In this test system the interaction between minichromosomes, assembled in *Xenopus* oocyte nuclear extracts, and lamins derived from coupled *in vitro* transcription and translation of corresponding cDNAs was studied. Binding of *Xenopus* lamins A and B2, but not human lamin C and *Xenopus* lamin B1, was detected. Using site-directed mutagenesis as well as competition experiments with synthetic peptides the capacity of lamin A and B2 to bind chromatin could be assigned to specific regions in the carboxyl-terminal domain rich in serine, threonine, and glycine residues (Höger et al., 1991). At present it is difficult to explain the differences in the specificity of lamin–chromatin binding in these assays, that is, the differences between A-type and B-type lamins in general and lamin A and C in particular. It remains difficult to relate *in vitro* binding studies to the *in vivo* situation. One should also bear in mind that in those cells that exclusively express B-type lamins (see above) chromatin is in close contact with the lamina, as judged by morphological criteria, and therefore that interaction between B-lamins and chromatin should exist *in vivo*. Lamin–chromatin interaction might be indirect, mediated by lamin associated proteins (see below) and, therefore, *in vitro* binding studies with isolated lamins might not reflect the *in vivo* situation.

The *in vitro* binding properties of purified lamins to naked DNA have also been analyzed (Schoeman and Traub, 1990). In these studies the capacity of individual lamins (lamin A, C, and B) as well as vimentin to bind to single-stranded DNA homopolymers was analyzed. Surprisingly, higher affinity binding of nucleic acids was determined for the cytoplasmic protein vimentin than for nuclear lamins, posing the question of biological significance of these findings. The possible interaction of matrix attachment regions (MAR) with B-type lamins will be discussed below.

VI. THE ROLE OF LAMINS IN NUCLEAR FORMATION AT THE END OF CELL DIVISION

When daughter cell nuclei re-form at the end of mitosis and meiosis, nuclear membranes, pore complexes, and the nuclear lamina reassemble at the surface of telophase chromosomes (Conner et al., 1980; Zeligs and Wollman, 1979; Maul, 1977; Gerace et al., 1978; for review see Lohka, 1988; Dabauvalle and Scheer, 1991). These assembly processes have been studied in cell-free preparations of the cytoplasm from amphibian eggs, pioneered by Lohka and Masui (1983; Forbes et al., 1983; Newport, 1987; Dabauvalle et al., 1991), *Drosophila* embryos (Ulitzur and Gruenbaum, 1989) and mitotic cultured cells (Burke and Gerace, 1986; Nakagawa et al., 1989). From these studies, two models of nuclear envelope reassembly have been proposed that differ with respect to the role that lamins play in the reassembly process. According to one model, early binding of lamins to the surface of chromosomes is a prerequisite for the subsequent binding of membrane vesicles to chromatin (Burke and Gerace, 1986; Glass and Gerace, 1990; Burke, 1990; Höger et al., 1991; Ulitzur et al., 1992). This model is supported by studies with somatic cell extracts, with extracts from *Drosophila* embryos and from *Xenopus* eggs from which lamins had been functionally depleted by adding lamin-specific antibodies (Burke and Gerace, 1986; Dabauvalle et al., 1991; Ulitzur et al., 1992).

In an alternative model a lamin-independent targeting of membrane vesicles to chromosomes has been proposed as an initial event of nuclear envelope reconstruction. This model is based on studies with *Xenopus* egg extracts from which lamin B3 had been immunodepleted (Newport et al., 1990; Meier et al., 1991) and on microinjection experiments into living cells (Benavente and Krohne, 1986). In the latter study lamin-specific antibodies, injected into mitotic cultured cells, inhibited chromatin decondensation but did not impair nuclear envelope formation. In the experiments using lamin-depleted *Xenopus* egg extracts, nuclei did form that are surrounded by a nuclear envelope consisting of membranes and functional pores that, however, lack a lamina structure. These lamina-free nuclei are extremely fragile and fail to grow beyond a limited extent. Moreover, while these nuclei decondense their chromatin and accumulate proteins required for DNA replication, these proteins apparently do not organize into replicon clusters and the nuclei fail to initiate DNA replication (Newport et al., 1990; Meier et al., 1991). Recently the presence of minor amounts of lamin B2 or a closely related lamin has been reported in *Xenopus* oocyte nuclei and eggs. A fraction of this lamin (~20%), as well as a small fraction of lamin B3 (~5%), seems to be associated with membrane vesicles in egg extracts (Lourim and Krohne, 1993). The presence of B2 in egg extracts and the fact that a small fraction of the egg lamins is associated with membranes has been taken to explain the discrepancies between the results of Dabauvalle et al. (1991) and Newport et al. (1990; see also Lourim and Krohne, 1993). The two experimental approaches differ with respect to the lamin-specific antibodies used and the way lamins were depleted in the extract (functional inactivation in complete extract versus immunodepletion of the soluble fraction). However, in the experiments reported by Meier et al. (1991) functional lamin depletion had been achieved by incubating complete *Xenopus* egg extracts with antibodies reactive with both lamin B3 and B2. Moreover, efficient lamin depletion of unfractionated extracts has also been carried out by repeated incubation with antilamin antibodies bound to magnetic beads and subsequent elimination of the beads with a magnet, avoiding loss of membrane precursors by centrifugation (Hollemann and Stick, unpublished). In both cases assembly of double membranes and pores occurred in a normal fashion. The latter nuclear assembly model predicts that nuclear membranes interact with the chromosome surface via other receptors (Wilson and Newport, 1988) and candidate proteins have been characterized that differ in size from lamins (Wiese and Wilson, 1993).

Analysis of the chronology of nuclear envelope assembly obtained *in vivo* or *in vitro* shows that formation of nuclear lamina is a relatively late event. At the completion of mitosis, the inner nuclear membrane-derived vesicles that contain the lamin B receptor associate with chromatin first, whereas pore membranes and the lamina assemble later (Bailer et al., 1991; Chaudhary and Courvalin, 1993). In the latter study the time course of reassembly of both lamin B and lamins A/C was found to be late, not before early telophase. The late association of lamin A with chromosomes is in variance to the observation by Höger et al. (1991). These authors report binding of lamin A to metaphase chromosomes, indicating that association of lamin A with chromosomes precedes nuclear membrane reformation. In *Xenopus* extracts nuclear reassembly can occur under conditions where nuclear transport is efficiently blocked by wheat germ agglutinin (WGA). Nuclei formed in these extracts lack lamins. When transport is restored by addition of a competing sugar, nuclei begin to accumulate lamins and to form a lamina (Newport et al., 1990). These observations support a model in which lamin assembly occurs only after a nuclear envelope has formed around chromatin and the conditions of nuclear transport are operational.

Despite the discrepancies mentioned above, the experiments have consistently shown that nuclear pore formation is independent of the presence of lamins. Annulate lamellae, the flattened membrane cisternae containing densely spaced pore complexes, do not contain lamins (Chen and Merisko, 1988). Annulate lamellae are most commonly observed in the cytoplasm of germ cells and rapidly dividing somatic cells (Kessel, 1989). Upon maturation of *Xenopus* oocytes annulate lamellae disassemble concomitantly with the oocyte nucleus. In the absence of chromatin, reformation of annulate lamellae can be observed in extracts of activated eggs. This reassembly is independent of the presence of functional lamins in the extract (Dabauvalle et al., 1991).

A. Lamin-Rich Structures Within the Nucleoplasm

Nuclear lamins were originally thought to be localized exclusively in the lamina at the periphery of the nucleus, closely underlying the inner nuclear membrane. Furthermore it was thought that the lamina was a static structure during interphase. Recent immunofluorescence studies show that there are exceptions to this general observation and that lamins are more dynamic than originally assumed. As described in more detail above, certain cell lines express lamin A (and C) when induced to differentiate while other cells express lamins A and C before induction. Two such cases have been analyzed in detail by indirect immunofluorescence (Collard and Raymond, 1992; Collard et al., 1992). When human epithelial cells in culture are treated with phorbol esters, the carboxyl-terminal domain of lamin A becomes transiently inaccessible to antibodies. This effect is specific, since no change in the accessibility of epitopes present in the common domain of lamins A and C is observed. It is not due to selective proteolysis or charge modifications of the C terminus. It is speculated that a transient conformational change of the carboxy-terminal domain of lamin A or an interaction of this domain with another nuclear component is responsible for this effect (Collard and Raymond, 1992). Human promyelocytic leukemia HL-60 cells show another interesting redistribution upon induction of differentiation. In undifferentiated cells lamin A is present as a nuclear cap and is redistributed such that a full peripheral nuclear localization is observed early after induction of differentiation by phorbol esters. Again, this effect is specific for lamin A, while no changes were observed in the uniform location of lamin C (Collard et al., 1992). These results suggest that the assembly and reorganization of each of the nuclear lamins can take place independently. Analysis of the distribution of lamin A in quiescent and proliferating cells has led to the surprising observation that during a restricted phase of the cell cycle (G1) lamin A is found in a series of spots and fibers within the nucleus and only to a lesser degree at the nuclear periphery (Bridger et al., 1993). The lamin A containing structures were shown to be closely associated with areas of condensed chromatin but not with nuclear membranes. These internal lamin foci are transient; they disappear as cells progress towards S phase. A recent study followed the fate of microinjected biotinylated lamin A into the cytoplasm of cell culture cells (Goldman et al., 1992). The injected protein was rapidly translocated into the nucleus where it formed discrete foci within the nucleoplasm. Subsequently it became incorporated into the peripheral lamina. It has been suggested that the internal lamin structures of lamin A in G1 phase may represent pools of newly synthesized lamins that are required for nuclear growth. If so, one should expect to detect similar foci containing lamin B; this has not been described so far. They may also represent sites at which lamins undergo various posttranslational modifications. In fact, cells treated with the isoprenylation inhibitor Lovastatin accumulate lamin A in nucleoplasmic foci (Lutz et al., 1992).

The distribution of the lamin foci is similar to the nuclear staining patterns seen with a variety of other antibodies or DNA probes including those specific for RNP particles and replication foci. It is therefore tempting to speculate that these foci represent common nucleoplasmic domains that might be connected to the nucleoskeleton and that lamins might be involved in organization of these domains (see also below).

VII. THE NUCLEOSKELETON

In previous sections we have discussed the molecular structure and dynamics of nuclear lamins. Lamins are involved in maintaining the structural integrity of the nucleus, and the *in vitro* nuclear assembly experiments mentioned above show that lamins may play a critical role in DNA replication (Newport et al., 1990; Meier et al., 1991). The lamina represents, however, only one element of the nucleoskeleton, and lamins are located exclusively at the periphery of the nucleus (for exceptions see above). It is widely believed that the nuclear architecture may be determined by an internal proteinaceous skeleton. Numerous publications over the last 20 years have dealt with the isolation and characterization of such nuclear residue preparations. It is beyond the scope of this article to review all these findings. Here I will briefly discuss concepts and recent advances in the field. For a more detailed discussion the reader is referred to recent reviews (Berezney, 1991; de Jong et al., 1990; Gasser et al., 1989; Verheijen et al., 1988).

Biochemical fractionation procedures, ultrastructural studies, and more recently optical sectioning at the light microscopic level in conjunction with specific fluorescence labeling techniques indicate that the nucleus is a highly organized organelle. Specific functions are localized to discrete nuclear domains, rather than being diffusely distributed (for review see Spector, 1993; Swedlow et al., 1993a). The best known example is the nucleolus. The nucleolus is a highly specialized structure readily discernible at the light microscopic level. It is the site of pre-rRNA transcription, processing and assembly of precursors of ribosomal subunits (for review see Scheer and Benavente, 1990).

The three-dimensional organization of the chromosomes within the interphase nucleus is more difficult to discern. Substantial progress has been made in the analysis of the basic units of chromatin structure, the nucleosomes and the 30-nm chromatin fiber (for review see Kornberg and Lor, 1992; Bradbury, 1992; Felsenfeld, 1992; Morse, 1992). The use of electron microscopic tomography has allowed the visualization of 30-nm fibers *in situ,* circumventing the problems of structural rearrangements during fiber isolation for conventional electron microscopical analysis (Woodcock et al., 1991; for review see Swedlow et al., 1993a). This approach allows chromatin structure to be examined in unprecedented detail and shows that the 30-nm chromatin fiber is more variable than previously anticipated. It twists, loops, and folds along its path rather than appearing as a solenoid. These variations might be influenced by even subtle modifications of histones or non-nucleosomal proteins, affecting nucleosome linker packing and hence modulating the local structure of the 30-nm fiber (Swedlow et al., 1993b; Bradbury, 1992). Detailed structural analysis of the higher order organization of chromatin fibers into condensed heterochromatin and chromosomes, however, remains to be determined. Currently, there are a number of models for higher order chromatin organization, but these are based on ultrastructural data of partially extracted nuclei or chromosomes and are therefore extremely sensitive to artifacts occurring during preparation.

A. The DNA Loop Model

According to the DNA loop model the chromatin fiber forms topologically sequestered loop domains (Cook and Brazell, 1978; Lebkowsky and Laemmli, 1982). Loop domains are generated by the binding of defined sequences to the nucleoskeleton. The size of these loops is in the range of 30 to 150 kb DNA. Scaffold (matrix) attachment regions (SARs or MARs) have been identified in a number of eukaryotic loci (for review see Gasser and Laemmli, 1987). They are 200 to 1000-bp long stretches of DNA, normally rich in AT bases and often containing consensus binding sites for topoisomerase II. Attempts have been made to characterize proteins that bind to these regions. At present three SAR-binding proteins have been identified: an attachment region binding protein (ARBP; von Kries et al., 1991), the enzyme topoisomerase II (Adachi et al., 1991), and lamin B1 (Ludérus et al., 1992). ARBP and topoisomerase II bind SARs in a cooperative manner and are part of the inner nucleoskeleton, while lamin B1 is a constituent of the peripheral lamina (for further discussion of nuclear skeletal proteins see below).

Much effort has been made to demonstrate the *in vivo* existence of a karyoskeleton that organizes basic nuclear activities. These attempts have been ongoing for nearly two decades, and much conflicting data have been published. The search for nuclear skeletal structures has been guided conceptually by analogies to the analysis of the cytoskeleton. Protocols used for preparation of nuclear skeletal structures closely resemble those for the isolation of cytoplasmic intermediate filaments. And, in fact, under certain conditions the resulting "skeleton" is composed predominantly, if not exclusively, of nuclear IF proteins, the nuclear lamins (Kaufmann and Shaper, 1984; Lebkowsky and Laemmli, 1982). Nucleoskeletons with internal structures have been isolated in different ways and have been given different names (for reviews see Kaufmann et al., 1986; Gasser et al., 1989; Nigg, 1989; Fisher, 1990; van Driel et al., 1991; Stuurman et al., 1992b). The term *nuclear matrix*, introduced by Berezney and Coffey (1974), is used to denote preparations of nuclei treated with nucleases, high ionic strength buffers, and nonionic detergent. Various parameters in the isolation procedure have been altered including the use of RNase and reducing agents and the effect on the composition of the resulting matrix has been analyzed (Kaufmann et al., 1981; Stuurman et al., 1992a).

Penman and coworkers (Nickerson et al., 1989) have used ammonium sulfate instead of sodium chloride for the high salt extraction under RNase-free conditions (for review see Nickerson et al., 1989). The resulting "nuclear matrix-intermediate filament complexes" also contain cytoskeletal structures. They have been extensively analyzed by high-voltage electron microscopy drawing special attention to interconnections between the nucleo- and the cytoskeleton (Capco et al., 1982). These authors also found a dramatic change in the internal matrix structure when RNase treatment was included in the preparation (Fey et al., 1986).

Jackson and Cook (1985; 1986) have used electroelution of DNA at physiological ionic strength to isolate "nucleoskeletons." In this technique agarose embedded cells are permeabilized with nonionic detergent, DNA is digested with restriction nucleases and subsequently removed by electrophoresis. The resulting nucleoskeletons retain replication and transcription activity. The latter observation has been taken as evidence that the enzymatic machinery for these processes is bound to the nucleoskeleton rather than being soluble (for review see Jackson, 1991). With this preparation technique a network of filaments has been demonstrated that has characteristics of intermediate filaments (i.e., ~10-nm wide) with an axial repeat of 23 nm (Jackson and Cook, 1988). The molecular nature of these nuclear filaments however, has not been identified.

In attempts to isolate DNA that is specifically associated with the nucleoskeleton Laemmli and coworkers (Mirkovitch et al., 1984) have used lithium diiodosalicylate to extract chromosomal proteins under conditions that stabilize a nuclear framework. DNA was then digested with restriction nucleases. The resulting residual structures were termed *nuclear scaffolds* (for review see Gasser and Laemmli, 1987; Gasser et al., 1989; Garrard, 1990). Analysis of the scaffold-associated DNA (SAR or MAR) led to the characterization of DNA sequences, which are thought to anchor DNA loops to an intranuclear framework (Mirkovitch et al., 1984; see also above).

The above listing of only four of the different preparative methods provides evidence that the term *nuclear matrix* is only an operational one. Comparison of the protein composition of these preparations shows in most cases a very complex pattern. Lamins, although present in all preparations, often represent only a minor fraction (for review see Stuurman et al., 1992b). In recent years some of these proteins have been characterized either by immunological methods, by enzymatic activity, or by their property to bind scaffold-attachment regions (SAR's). The ARBP (von Kries et al., 1992), topoisomerase II (Adachi et al., 1991) and lamin B1 (Ludérus et al., 1993) have already been mentioned above as SAR-binding proteins. A 62-kDa protein has been located in matrix-associated nuclear bodies by specific monoclonal antibodies (Stuurman et al., 1992c). Its biological function at present is unknown. One criterion for a nuclear scaffolding protein would be that it forms continuous structures within the nucleus or within chromosomes and, in analogy to cytoplasmic filament systems, might also do so *in vitro*. With the exception of lamin B1, which is restricted to the peripheral lamina, a role in organizing an intranuclear framework has not been demonstrated for any of the above proteins. Topoisomerase II might exemplify the difficulties to assess a candidate protein as a nuclear skeletal component. Topoisomerase II has been described as a major component of the chromosome scaffold by Lewis and Laemmli (1982). Later, Earnshaw and Heck localized topoisomerase II in mitotic chromosomes (Earnshaw et al., 1985; Earnshaw and Heck, 1985). The location closely resembled the chromosome scaffolds previously visualized by electron microscopy (Paulson and Laemmli, 1977). More recently it has been shown that topoisomerase II is necessary for chromosome assembly and chromatin condensation *in vitro* (Adachi et al., 1991; Hirano and Mitchison, 1993; Wood and Earnshaw, 1990). However, topoisomerase II is not confined to an axial chromosome core, but is distributed throughout the chromosome in the *in vitro* extracts. Moreover, extraction of most of the enzyme does not alter the gross morphology of the chromosomes (Hirano and Mitchison, 1993). Topoisomerase II therefore probably does not play a scaffolding role in the organization of chromosomes. A similar conclusion has been drawn from *in vivo* localization of fluorescently labeled microinjected topoisomerase II in early mitotic cycles of *Drosophila* embryos (Swedlow et al., 1993).

In conclusion, at present there is no definitive proof for the existence of an intranuclear skeleton *in vivo*. Possible strategies for the identification of intranuclear framework components have recently been discussed by van Driel and coworkers (Stuurman et al., 1992b; deJong et al., 1990). Finally, one should even envisage the possibility that no internal nucleoskeleton exists at all and that we are dealing with supramolecular components within the nucleus, the replication complexes, and RNP processing complexes, which are too large to be extracted from the nucleus and which tend to lead to precipitation upon extraction of chromatin. Newly developed high-resolution light microscopic techniques, which enable the observation of individual types of fluorescently labeled molecules *in vivo*, might overcome some of the dilemmas inherent in nuclear extraction procedures.

B. RNA Synthesis and Processing and the Nucleoskeleton

Conceptually it has been distinguished between an interphase chromatin scaffold, that may organize the three-dimensional distribution of interphase chromatin (see above) and an RNP processing matrix, that may contribute to processing and transport of RNP particles (Nigg, 1988). Several lines of evidence have led to the assumption that hnRNA synthesis and processing might occur in association with a nucleoskeleton. All studies that deal with the analysis of components that remain associated with a nucleoskeleton after isolation suffer from the same shortcomings discussed above for the chromatin scaffold. Therefore, I will concentrate only on recent studies on *in situ* localization of components of the mRNA processing machinery and its relation to a nucleoskeleton (for review see Spector, 1993). Subnuclear localization of nascent RNA by fluorescence microscopy reveals that nascent hnRNA transcripts are found in several hundred defined area, but scattered throughout the nucleus rather than being randomly distributed (Wansink et al., 1993). snRNPs involved in RNA splicing are similarly concentrated in "nuclear speckles" in addition to being distributed diffusely as evidenced by immunofluorescence detection of the snRNP proteins and *in situ* hybridization with the respective snRNAs; several splicing components that concentrate in speckled nuclear regions contain arginine/serine rich domains, which function as targeting sequences for these subnuclear compartments (Li and Bingham, 1991). At the light microscopic level, the speckled pattern is organized in a latticework that interacts at various places with the surface of the nuclear envelope and the nucleolus. The latticework corresponds to nuclear regions rich in interchromatin granules and perichromatin fibrils. A ribonucleoprotein network has been identified previously by cytochemical staining (Puvion and Bernard, 1975). These structures can be correlated by comparing immunofluorescence and cytochemical data.

More recently *in situ* hybridization methods have been optimized for fluorescence detection of specific nuclear RNAs. These analyses reveal striking localization within the nucleus. Transcripts are often restricted to a small region of the nucleus, frequently in a curvilinear "track." These tracks extend from an internal region into the periphery, but in most cases do not directly contact the nuclear envelope (Lawrence et al., 1989). When exon and intron specific probes are used in the *in situ* hybridization, it is evident that splicing occurs directly within these RNA tracks because intron-containing and spliced transcripts can be detected in spatial separation along the tracks. Excised introns appear, in contrast, to diffuse freely (Xing et al., 1993). Spliceosome assembly factors colocalize with these RNA tracks (Carter et al., 1993).

All these observations have been interpreted as evidence that RNA metabolism is architecturally organized in the nucleus, thereby supporting the previous view of an RNP processing matrix that was based on the analysis of isolated nuclear matrices. However, the concept of a solid phase model of nuclear structures in which RNA molecules are actively transported from their site of synthesis to the nuclear pores is not unopposed. In an elegant experimental approach, making use of resources uniquely available in *Drosophila*, Bingham and coworkers have studied the movement of a specific pre-mRNA between its gene and the nuclear surface in polyploid nuclei of salivary glands (Zechar et al., 1993). The results show that the movement of pre-mRNA in the nucleus is isotropic, at rates consistent with diffusion, and is restricted to a small nuclear subcompartment defined by exclusion from the chromosomal axis and the nucleolus. Bulk polyadenylated nuclear pre-mRNA precisely localizes in this extrachromosomal compartment allowing the assumption that most pre-mRNA uses the same route of intranuclear movement.

Clearly, many more gene transcripts have to be analyzed and more cell types in different states of differentiation have to be assessed before general conclusions can be drawn. However, with the use of new, sensitive, high-resolution *in situ* hybridization, antibody staining, and microinjection techniques, in conjunction with improved light microscopy, we are at the beginning of what will be exciting developments in molecular cytology.

ACKNOWLEDGMENT

I would like to thank Dr. Klaus Weber and Dr. Irm Huttenlauch (Max-Planck-Institute for biophysical Chemistry, Göttingen, FRG) for critically reading the manuscript and Dr. David Capco (Arizona State University, Tempe, USA) for encouragement. I thank Gillian Paterson for typing the manuscript. The author's work is supported by the Deutsche Forschungs Gemeinschaft (Bonn, FRG) (DFG grants Sti 98/ 3-2; Sti 98/4-4).

REFERENCES

Adachi, Y., Luke, M., & Laemmli, U.K. (1991). Chromosome assembly *in vitro*: Topoisomerase II is required for condensation. Cell 64, 137–148.

Aebi, U., Cohn, J., Buhle, L., & Gerace, L (1986). The nuclear lamina is a meshwork of intermediate filaments. Nature 323, 560–564.

Aebi, U., Häner, M., Troncoso, J., Eichner, R., & Engel, A. (1988). Unifying principles in intermediate filament (IF) structure and assembly. Protoplasma 145, 73–81.

Akey, C.W. (1989). Interactions and structure of the nuclear pore complex revealed by cryo-electron microscopy. J. Cell Biol. 109, 955–970.

Appelbaum, J., Blobel, G., & Georgatos, S.D. (1990). *In vivo* phosphorylation of the lamin B receptor. J. Biol. Chem. 265, 4181–4184.

Bailer, S.M., Eppenberger, H.M., Griffiths G., & Nigg, E.A. (1991). Characterization of a 54-kD protein of the inner nuclear membrane: Evidence for cell cycle-dependent interaction with the nuclear lamina. J. Cell Biol. 114, 389–400.

Beams, H.W., Tamisian, T.N., Devine R., & Anderson, E. (1957). Ultrastructure of the nuclear membrane of a Gregarine parasitic in grasshoppers. Exp. Cell Res. 13, 200–204.

Beck, L.A., Hosick, T.J., & Sinensky, M. (1988). Incorporation of a product of melvalonic acid metabolism into proteins of Chinese hamster ovary cell nuclei. J. Cell Biol. 107, 1307–1316.

Beck, L.A., Hosick, T.J., & Sinensky, M. (1990). Isoprenylation is required for the processing of the lamin A precursor. J. Cell Biol. 110, 1489–1499.

Benavente, R. & Krohne, G. (1985). Change of karyoskeleton during spermatogenesis of *Xenopus*: Expression of lamin L$_{IV}$, a nuclear lamina protein specific for the male germ line. Proc. Natl. Acad. Sci. USA 82, 6176–6180.

Benavente, R. & Krohne, G. (1986). Involvement of nuclear lamins in postmitotic reorganization of chromatin as demonstrated by microinjection of lamin antibodies. J. Cell Biol. 103, 1847–1854.

Benavente, R., Krohne, G., & Franke, W.W. (1985). Cell type-specific expression of nuclear lamin proteins during development of *Xenopus laevis*. Cell 41, 177–190.

Berezney, R. (1991). The nuclear matrix: A heuristic model for investigating genomic organization and function in the cell nucleus. J. Cell. Biochem. 47, 109–123.

Berezney, R. & Coffey, D.S. (1974). Identification of a nuclear protein matrix. Biochem. Biophys. Res. Comm. 60, 1410–1417.

Biamonti, G., Giacca, M., Perini, G., Contreas, G., Zentilin, L., Weighardt, F., Guerra, M., Della Valle, G., Saccone, S., Riva, S., & Falaschi, A. (1992). The gene for a novel human lamin maps at a highly transcribed locus of chromosome 19 which replicates at the onset of S-phase. Mol. Cell. Biol. 12, 3499–3506.

Biggiogera, M., Kaufmann, S.H., Shaper, J.H., Gas, N., Amalric, F., & Fakan, S. (1991). Distribution of nucleolar proteins B23 and nucleolin during mouse spermatogenesis. Chromosoma 100, 162–172.

Bossie, C.A. & Sanders, M.M. (1993). A cDNA from *Drosophila melanogaster* encodes a lamin C-like intermediate filament protein. J. Cell Sci. 104, 1263–1272.

Bradbury, E.M. (1992). Histone modifications and the chromosome cell cycle. BioEssays 14, 9–16.

Bridger, J.M., Kill, I.R., O'Farrell, M., & Hutchison, C.J. (1993). Internal lamin structures within the G1 nuclei of human dermal fibroblasts. J. Cell Sci. 104, 297–306.

Broers, J.L., Raymond, Y., Rot, M.K., Kuijpers, H., Wagenaar, S.S., & Ramaekers, F.C.S. (1993). Nuclear A-type lamins are differentially expressed in human lung cancer subtypes. Am. J. Pathol. 143, 211–220.

Burke, B. (1990). On the cell-free association of lamins A and C with metaphase chromosomes. Exp. Cell Res. 186, 169–176.

Burke, B. & Gerace, L. (1986). A cell free system to study reassembly of the nuclear envelope at the end of mitosis. Cell 44, 639–652.

Cance, W.G., Chaudhary, N., Worman, H.J., Blobel, G., & Cordon-Cardo, C. (1992). Expression of the nuclear lamins in normal and neoplastic human tissues. J. Exp. Clin. Cancer Res. 11, 233–246.

Capco, D.G., Wan, K.M., & Penman, S. (1982). The nuclear matrix: Three-dimensional architecture and protein composition. Cell 29, 847–858.

Carter, K.C., Bowman, D., Carrington, W., Fogarty, K., McNeil, J.A., Fay, F.S., & Lawrence, J.B. (1993). A three-dimensional view of precursor messenger RNA metabolism within the mammalian nucleus. Science 259, 1330–1335.

Casey, P.J., Solski, P.A., Der, C.J.H., & Buss, J.E. (1989). p21^{ras} is modified by a farnesyl isoprenoid. Proc. Natl. Acad. Sci. USA 86, 8323–8327.

Chaudhary, N. & Courvalin, J.-C. (1993). Stepwise reassembly of the nuclear envelope at the end of mitosis. J. Cell Biol. 122, 295–306.

Chen, T.-Y. & Merisko, M.M. (1988). Annulate lamellae: Comparison of antigenic epitopes of annulate lamellae membranes with the nuclear envelope. J. Cell Biol. 107, 1299–1306.

Clarke, S. (1992). Protein isoprenylation and methylation at carboxyl-terminal cysteine residues. Ann. Rev. Biochem. 61, 355–386.

Collard, J.-F. & Raymond, Y. (1990). Transfection of human lamins A and C into mouse embryonal carcinoma cells possessing only lamin B. Exp. Cell Res. 186, 182–187.

Collard, J.-F. & Raymond, Y. (1992). Phorbol esters induce transient changes in the accessibility of the carboxy-terminal domain of nuclear Lamin A. Exp. Cell Res. 201, 174–183.

Collard, J.-F., Senecal, J.-L., & Raymond, Y. (1992). Redistribution of nuclear lamin A is an early event associated with differentiation of human promyelocytic leukemia HL-60 cells. J. Cell Sci. 101, 657–670.

Conner, G.E., Noonan, N.E., & Noonan, K.D. (1980). Nuclear envelope of Chinese hamster ovary cells. Re-formation of the nuclear envelope following mitosis. Biochemistry 19, 277–289.

Cook, P.R. & Brazell, I.A. (1978). Spectro-fluorometric measurement of the binding of ethidium to superhelical DNA from cell nuclei. Eur. J. Biochm. 84, 465–477.

Courvalin, J.-C., Segil, N., Blobel, G., & Worman, H.J. (1992). The lamin B receptor of the inner nuclear membrane undergoes mitosis-specific phosphorylation and is a substrate for $p34^{cdc2}$-type protein kinase. J. Biol. Chem. 267, 19035–19038.

Dabauvalle, M.-C., Loos, K., Merkert, H., & Scheer, U. (1991). Spontaneous assembly of pore complex-containing membranes ("Annulate Lamellae") in *Xenopus* egg extract in the absence of chromatin. J. Cell Biol. 112, 1073–1082.

Dabauvalle, M.-C. & Scheer, U. (1991). Assembly of nuclear pore complexes in *Xenopus* egg extract. Biol. Cell. 72, 25–29.

deJong, L., van Driel, R., Stuurman, N., Meijne, A.M.L., & van Renswoude, J. (1990). Principles of nuclear organization. Cell Biol. Int. Rep. 14, 1051–1074.

Dessev, G., Iovcheva-Dessev, C., Bischoff, J.R., Beach, D., & Goldman, R. (1991). A complex containing $p34^{cdc2}$ and cyclin B phosphorylates the nuclear lamin and disassembles nuclei of clam oocytes *in vitro*. J. Cell Biol. 112, 523–533.

Dodemont, H., Riemer, D., & Weber, K. (1990). Structure of an invertebrate gene encoding cytoplasmic intermediate filament (IF) proteins: Implications for the origin and the diversification of IF proteins. EMBO J. 9, 4083–4094.

Döring, V. & Stick, R. (1990). Gene structure of nuclear lamin LIII of *Xenopus laevis*; A model for the evolution of IF proteins from a lamin-like ancestor. EMBO J. 9, 4073–4081.

Earnshaw, W.C., Halligan, B., Cooke, C.A., Heck, M.M.S., & Liu L.F. (1985). Topoisomerase II is a structural component of mitotic chromosome scaffolds. J. Cell Biol. 100, 1706–1715.

Earnshaw, W.C. & Heck, M.M.S. (1985). Localization of topoisomerase II in mitotic chromosomes. J. Cell Biol. 100, 1716–1725.

Eggert, M., Radomski, N., Linder, D., Tripier, D., Traub, P., & Jost, E. (1993). Identification of novel phosphorylation sites in murine A-type lamins. Eur. J. Biochem. 213, 659–671.

Enoch, T., Peter, M., Nurse, P., & Nigg, E.A. (1991). $p34^{cdc2}$ acts as a lamin kinase in fission yeast. J. Cell Biol. 112, 797–807.

Fawcett, D.W. (1966). On the occurrence of a fibrous lamina on the inner aspect of the nuclear envelope in certain cells of vertebrates. Am. J. Anat. 119, 129–146.

Felsenfeld G. (1992). Chromatin as an essential part of the transcriptional mechanism. Nature 355, 219–224.

Fey, E.G., Krochmalnic, G., & Penman, S. (1986). The non-chromatin structure of the nucleus: The ribonucleo-protein (RNP)-containing and RNP-depleted matrix analyzed by sequential fractionation and resinless section microscopy. J.Cell Biol. 102, 1654–1665.

Firmbach-Kraft, I., & Stick, R. (1993). The role of CaaX dependent modifications in membrane association of *Xenopus* nuclear lamin B3 during meiosis and the fate of B3 in transfected mitotic cells. J. Cell Biol. 123, 501–512.

Fisher, D.Z., Chaudhary, N., & Blobel, G. (1986). cDNA sequencing of nuclear lamins A and C reveals primary and secondary structural homology to intermediate filament proteins. Proc. Natl. Acad. Sci. USA. 83, 6450–6454.

Fisher, P.A. (1990). Effects of thermal stress on the karyoskeleton: Insights into the possible role of karyoskeletal elements in DNA replication and transcription. In: The Eukaryotic Nucleus: Molecular Biochemistry and Macromolecular Assemblies (Strauss, P.R. and Wilson, S.H., Eds.). Vol. 2, pp. 737–762. Telford Press, Caldwell.

Foisner, R. & Gerace, L. (1993). Integral membrane proteins of the nuclear envelope interact with lamins and chromosomes, and binding is modulated by mitotic phosphorylation. Cell 73, 1267–1279.

Forbes, D., Kirschner, M., & Newport, J. (1983). Spontaneous formation of nucleus-like structures around bacteriophage DNA microinjected into *Xenopus* eggs. Cell 34, 13–23.

Franke, W.W. (1987). Nuclear lamins and cytoplasmic intermediate filament proteins: A growing multigene family. Cell 48, 3–4.

Furukawa, K. & Hotta, Y. (1993). cDNA cloning of a germ cell specific lamin B_3 from mouse spermatocytes and analysis of its function by ectopic expression in somatic cells. EMBO J. 12, 97–106.

Garrard, W.T. (1990). Chromosomal loop organization in eukaryotic genomes. Nucl. Acids Mol. Biol. 4, 163–175.

Gasser, S.M. & Laemmli, U.K. (1987). A glimpse at chromosomal order. Trends Genet. 3, 16–22.

Gasser, S.M., Amati B.B., Cardenas M.E., & Hofmann J.F.X. (1989). Studies on scaffold attachment sites and their relation to genome function. Int. Rev. Cytol. 119, 57–96.

Gerace, L. & Blobel, G. (1980). The nuclear envelope lamina is reversible depolymerized during mitosis. Cell 19, 277–287.

Gerace, L., Blum, A., & Blobel, G. (1978). Immunocytochemical localization of the major polypeptides of the nuclear pore complex-lamina fraction. Interphase and mitotic distribution. J. Cell Biol. 79, 546–566.

Gieffers, C. & Krohne, G. (1991). *In vitro* reconstitution of recombinant lamin A and a lamin A mutant lacking the carboxy-terminal tail. Eur. J. Cell Biol. 55, 191–199.

Glass, J.R. & Gerace, L. (1990). Lamins A and C bind and assemble at the mitotic chromosomes. J. Cell Biol. 111, 1047–1057.

Go, M. (1981). Correlation of DNA exonic regions with protein structural units in haemoglobin. Nature 291, 90–92.

Goldman, A.E., Maul, G., Steinert, P.M., Yang, H.-Y., & Goldman, R.D. (1986). Keratin-like proteins that coisolate with intermediate filaments of BHK-21 cells are nuclear lamins. Proc. Natl. Acad. Sci. USA 83, 3839–3843.

Goldman, A.E., Moir, R.D., Lowy-Montag, M., Stewart, M., & Goldman R.D. (1992). Pathway of incorporation of microinjected lamin A into the nuclear envelope. J. Cell Biol. 119, 725–735.

Gruenbaum, Y., Landesman, Y., Drees, B., Bare, J.W., Saumweber, H., Paddy, M.R., Sedat, J.W., Smith, D.E., Benton, B.M., & Fisher, P.A. (1988). *Drosophila* nuclear lamin precursor Dmo is translated from either of two developmentally regulated mRNA species apparently encoded by a single gene. J. Cell Biol. 106, 585–596.

Guilly, M.N., Bensussan, A., Bourge, J.F., Bornens, M., & Courvalin, J.C. (1987). A human T lymphoblastic cell line lacks lamins A and C. EMBO J. 6, 3795–3799.

Hancock, J.F., Paterson, H., & Marshall, C.J. (1990). A polybasic domain or palmitoylation is required in addition to the CaaX motif to localize $p21^{ras}$ to the plasma membrane. Cell 63, 133–139.

Heald, R. & McKeon, F. (1990). Mutations of phosphorylation sites in lamin A that prevent nuclear lamina disassembly in mitosis. Cell 61, 579–589.

Heitlinger, E., Peter, M., Häner, M., Lustig, A., Aebi, U., & Nigg, E.A. (1991). Expression of chicken lamin B_2 in *Escherichia coli*: Characterization of its structure, assembly and molecular interactions. J. Cell Biol. 113, 485–495.

Heitlinger, E., Peter, M., Lustig, A., Villiger, W., Nigg, E.A., & Aebi, U. (1992). The role of the head and tail domain in lamin structure and assembly: Analysis of bacterially expressed chicken lamin A and truncated B_2 lamins. J. Struct. Biol. 108, 74–91.

Hennekes, H. & Nigg, E.A. (1994). The role of isoprenylation in membrane attachment of nuclear lamins. A single point mutation prevents proteolytic cleavage of the lamin A precursor and confers membrane binding properties. J. Cell Sci. 107, 1019–1029.

Hennekes, H., Peter, M., Weber, K., & Nigg, E.A. (1993). Phosphorylation on protein kinase C sites inhibits nuclear import of lamin B_2. J. Cell Biol. 120, 1293–1304.

Hill, R.J. & Whytock, S. (1993). Cytological structure of the native polytene salivary gland nucleus of *Drosophila melanogaster*: A microsurgical analysis. Chromosoma 102, 446–456.

Hirano, T. & Mitchison, T.J. (1993). Topoisomerase II does not play a scaffolding role in the organization of mitotic chromosomes assembled in *Xenopus* egg extracts. J. Cell Biol. 120, 601–612.

Hocevar, B.A., Burns, D.J., & Fields, A.P. (1993). Identification of protein kinase C (PKC) phosphorylation sites on human lamin B. J. Biol. Chem. 266, 7545–7552.

Höger, T.H., Krohne, G., & Franke, W.W. (1988). Amino acid sequence and molecular characterization of murine lamin B as deduced from cDNA clones. Eur. J. Cell Biol. 47, 283–290.

Höger, T.H., Krohne, G., & Kleinschmdt, J.A. (1991). Interaction of *Xenopus* lamins A and LII with chromatin *in vitro* mediated by a sequence element in the carboxyterminal domain. Exp. Cell Res. 197, 280–289.

Höger, T. H., Zatloukal, K., Waizenegger, I., & Krohne, G. (1990a). Characterization of a second highly conserved B-type lamin present in cells previously thought to contain only a single B-type lamin. Chromosoma. 99, 379–390.

Höger, T.H., Grund, C., Franke, W.W., & Krohne, G. (1990b). Immunolocalization of lamins in the thick nuclear lamina of human synovial cells. Eur. J. Cell Biol. 54, 150–156.

Holland, S.K. & Blake, C.C. (1990). Proteins, exons and molecular evolution. In: Intervening Sequences In Evolution And Development (Stone, E.M., & Schwartz, R.J., Eds.) pp. 10–42. Oxford University Press, Oxford.

Holtz, D., Tanaka, R.A., Hartwig, J., & McKeon, F. (1989). The CaaX motif of lamin A functions in conjunction with the nuclear localization signal to target assembly to the nuclear envelope. Cell 59, 969–977.

Hornbeck, P., Huang, K.-P., & Paul, W.E. (1988). Lamin B is rapidly phosphorylated in lymphocytes after activation of protein kinase C. Proc. Natl. Acad. Sci. USA 85, 2279–2283.

Horton, H., McMorrow, I., & Burke, B. (1992). Independent expression and assembly properties of heterologous lamins A and C in murine embryonal carcinomas. Eur. J. Cell Biol. 57, 172–183.

Houliston, E., Guilly, M.-N., Courvalin, J.-C., & Maro, B. (1988). Expression of nuclear lamins during mouse preimplantation development. Development 102, 271–278.

Hytiroglou, P., Choi, S.-W., Theise, N.D., Chaudhary, N., Worman, H.J., & Thung, S.N. (1993). The expression of nuclear lamins in human liver. Hum. Pathol. 24, 169–172.

Jackson, D.A. (1991). Structure-function relationships in eukaryotic nuclei. BioEssays 13, 1–10.

Jackson, D.A. & Cook, P.R. (1985). Transcription occurs at a nucleoskeleton. EMBO J. 4, 919–925.

Jackson, D.A. & Cook, P.R. (1986). Replication occurs at a nucleoskeleton. EMBO J. 5, 1403–1410.

Jackson, D.A. & Cook, P.R. (1988). Visualization of a filamentous nucleoskeleton with a 23 nm axial repeat. EMBO J. 7, 3667–3677.

Kaufmann, S.H., Coffey D.S., & Shaper, J.H. (1981). Considerations in the isolation of rat liver nuclear matrix, nuclear envelope and pore complex lamina. Exp. Cell Res. 132, 105–123.

Kaufmann, S.H. & Shaper, J.H. (1984). A subset of non-histone nuclear proteins reversibly stabilized by the sulfhydryl cross-linking reagent tetrathionate. Exp. Cell Res. 155, 477–495.

Kaufmann, S.H., Okret, S., Wikström, A.C., Gustafson, J.A., & Shaper, J.H. (1986). Binding of the glucocorticoid receptor to the rat liver nuclear matrix. J. Biol. Chem. 261, 11962–11967.

Kessel, R.G. (1989). The annulate lamellae—from obscurity to spotlight. Electron Microsc. Rev. 2, 257–348.

Kitten, G.T. & Nigg, E.A. (1991). The CaaX motif is required for isoprenylation, carboxyl methylation and nuclear membrane association of lamin B2. J. Cell Biol. 113, 13–23.

Kornberg, R.D., Lor, H.Y. (1992). Chromatin structure and transcription. Annu. Rev. Cell Biol. 8, 563–587.

Krachmarov, C.P. & Traub, P. (1993). Heat-induced morphological and biochemical changes in the nuclear lamina from ehrlich ascites tumor cells *in vivo*. J. Cell. Biochem. 52, 308–319.

Krohne, G. & Benavente, R. (1986). The nuclear lamins. Exp. Cell Res. 162, 1–10.

Krohne, G., Waizenegger, I., & Höger, T.H. (1989). The conserved carboxy-terminal cystine of nuclear lamins is essential for lamin association with the nuclear envelope. J. Cell Biol. 109, 2003–2013.

Krohne, G., Wolin, S.L., McKeon, F., Franke, W.W., & Kirschner, M.W. (1987). Nuclear lamin L1 of *Xenopus laevis*: cDNA cloning, amino acid sequence and binding specificity of a member of the lamin B subfamily. EMBO J. 6, 3801–3808.

Kubiak, J.Z., Prather, R.S., Maul, G.G., & Schatten, G. (1991). Cytoplasmic modification of the nuclear lamina during pronuclear-like transformation of mouse blastomere nuclei. Mech. Dev. 35, 103–111.

Lanoix, J., Skup, D., Collard, J.-F., & Raymond, Y. (1992). Regulation of the expression of lamins A and C is post-transcriptional in P19 embryonal carcinoma cells. Biochem. Biophys. Res. Commun. 189, 1639–1644.

Lawrence, J.B., Singer, R.H., & Marselle, L.M. (1989). Highly localized tracks of specific transcripts within interphase nuclei visualized by in situ hybridization. Cell 57, 493–502.

Lebel, S., Lampron, C., Royal, A., Raymond, Y. (1987). Lamins A and C appear during retinoic acid-induced differentiation of mouse embryonal carcinoma cells. J. Cell Biol. 105, 1099–1104.

Lebkowsky, J.S. & Laemmli, U.K. (1982). Evidence for two levels of DNA folding in histone-depleted HeLa interphase nuclei. J. Mol. Biol. 156, 309–324.

Lehner, C.F., Kurer, H.M., Eppenberger, H.M., & Nigg, E.A. (1986). The nuclear lamin protein family in higher vertebrates: Identification of quantitative minor lamin proteins by monoclonal antibodies. J. Biol. Chem. 261, 13292–13301.

Lehner, C.F., Stick, R., Eppenberger, H.M., & Nigg, E.A. (1987). Differential expression of nuclear lamin proteins during chicken development. J. Cell Biol. 105, 577–587.

Lewis, C.D. & Laemmli, U.K. (1982). Higher order metaphase chromosome structure: Evidence for metalloprotein interactions. Cell 29, 171–181.

Li, H. & Bingham, P.M. (1991). Arginine/Serine-rich domains of the $su(w^a)$ and *tra* RNA processing regulators target proteins to a subnuclear compartment implicated in splicing. Cell 67, 335–342.

Lin, F. & Worman, H.J. (1993). Structural organization of the human gene encoding nuclear lamin A and nuclear lamin C. J. Biol. Chem. 268, 16321–16326.

Loewinger, L. & McKeon, F. (1988). Mutations in the nuclear lamin proteins resulting in their aberrant assembly in the cytoplasm. EMBO J. 7, 2301–2308.

Lohka, M.J. (1988). The reconstitution of nuclear envelopes in cell-free extracts. Cell Biol. Int. Rep. 12, 833–848.

Lohka, M. & Masui, Y. (1983). Formation *in vitro* of sperm pronuclei and mitotic chromosomes induced by amphibian ooplasmic components. Science 220, 719–721.

Lourim, D. & Krohne, G. (1993). Membrane associated lamins in *Xenopus* egg extracts: Identification of two vesicle populations. J. Cell Biol. 123, 501–512.

Lourim, D. & Lin, J.J.-C. (1989). Expression of nuclear lamin A and muscle-specific proteins in differentiating muscle cells *in ovo* and *in vitro*. J. Cell Biol. 109, 495–504.

Ludérus E.M.E., de Graaf, A., Mattia, E., den Blaauwen, J.L., Grande, M.A., de Jong, L., & van Driel, R. (1992). Binding of matrix attachment regions to lamin B. Cell 70, 949–959.

Lüscher, B., Brizuela, L., Beach, D., & Eisenman, R.N. (1991). A role for the $p34^{cdc2}$ kinase and phosphatases in the regulation of phosphorylation and disassembly of lamin B2 during the cell cycle. EMBO J. 10, 865–875.

Lutz, R.J., Trujillo, M.A., Denham, K.S., Wenger, L., & Sinensky, M. (1992). Nucleoplasmic localization of prelamin A: Implications for prenylation-dependent lamin A assembly into the nuclear lamina. Proc. Natl. Acad. Sci. USA 89, 3000–3004.

McCombie, W.R., Adams, M.D., Kelley, J.M., Fitzgerald, M.G., Utterback, T.R., Khan, M., Dubnick, M., Kerlavage, A.R., Venter, J.C., & Fields, C. (1992). *Caenorhabditis elegans* expressed sequence tags identify gene families and potential disease gene homologues. Nature Genet. 1, 124–131.

McKeon, F.D., Kirschner, M.W., & Caput, D. (1986). Homologies in both primary and secondary structure between nuclear envelope and intermediate filament proteins. Nature 319, 463–468.

Martell, R.E., Strahler, J.R., & Simpson, R.U. (1992). Identification of lamin B and histones as 1,25-Dihydroxyvitamin D3-regulated nuclear phosphoproteins in HL-60 cells. J. Biol. Chem. 267, 7511–7519.

Mattia, E., Hoff, W.D., Blaauwen, J.d., Meijne, A.M.L., Stuurman, N., & Renswoude, J.V. (1992). Induction of nuclear lamins A/C during in vitro-induced differentiation of F9 and P19 embryonal carcinoma cells. Exp. Cell Res. 203, 449–455.

Maul, G.G. (1977). The nuclear and the cytoplasmic pore complex: Structure, dynamics, distribution and evolution. Int. Rev. Cytol. Suppl. 6, 75–186.

Maul, G.G., French, B.T., & Bechtol, K.B. (1986). Identification and redistribution of lamins during nuclear differentiation in mouse spermatogenesis. Dev. Biol. 115, 68–77.

Meier, J. Campbell, K.H.S., Ford, C.C., Stick, R., & Hutchison, C.J. (1991). The role of lamin LIII in nuclear assembly and DNA replication, in cell-free extracts of Xenopus eggs. J. Cell Sci. 98, 271–279.

Mirkovitch, J., Mirault, M.E., & Laemmli, U.K. (1984). Organization of the higher order chromatin loop: Specific DNA attachment sites on nuclear scaffold. Cell 39, 223–232.

Moir, R.D., Quinlan, R.A., & Stewart, M. (1990). Expression and characterization of human lamin C. FEBS Lett. 268, 301–305.

Moir, R.D., Donaldson, A.D., & Stewart, M. (1991). Expression in Escherichia coli of human lamins A and C: Influence of head and tail domains on assembly properties and paracrystal formation. J. Cell Sci. 99, 363–372.

Molloy, S. & Little, M. (1992). p34[cdc2] kinase-mediated release of lamins from nuclear ghosts is inhibited by cAMP-dependent protein kinase. Exp. Cell Res. 201, 494–499.

Morse, R. (1992). Transcribed chromatin. Trends Biochem. Sci. 17, 23–26.

Moss, S.B., Burnham, B.L., & Bellvé, A.R. (1993). The differential expression of lamin epitopes during mouse spermatogenesis. Mol. Rep. Dev. 34, 164–174.

Moss, S.B., Donovan, M.J., & Bellvé, A.R. (1987). The occurrence and distribution of lamin proteins during mammalian spermatogenesis and early embryonic development. Ann. N.Y. Acad. Sci. 513, 74–89.

Nakagawa, J., Kitten G.T., & Nigg, E.A. (1989). A somatic cell-derived system for studying both early and late mitotic events in vitro. J. Cell Sci. 94, 449–462.

Nakajima, N. & Sado, T. (1993). Nucleotide sequence of a mouse lamin A cDNA and its deduced amino acid sequence. Biochim. Biophys. Acta 1171, 311–314.

Newport, J. (1987). Nuclear reconstitution in vitro: Stages of assembly around protein-free DNA. Cell 48, 205–217.

Newport, J.W., Wilson, K.L., & Dunphy, W.G. (1990). A lamin-independent pathway for nuclear envelope assembly. J. Cell Biol. 111, 2247–2259.

Nickerson, J.A., Krochmalnic, G., Wan, K.M., & Penman, S. (1989). Chromatin architecture and nuclear RNA. Proc. Natl. Acad. Sci. USA 86, 177–181.

Nigg, E.A. (1988). Nuclear function and organization: The potential of immunochemical approaches. Int. Rev. Cytol. 110, 27–92.

Nigg, E.A. (1989). The nuclear envelope. Curr. Op. Cell Biol. 1, 435–440.

Ottaviano, Y. & Gerace, L. (1985). Phosphorylation of the nuclear lamins during interphase and mitosis. J. Biol. Chem. 260, 624–632.

Padan, R., Nainudel-Epszteyn, S., Goitein, R., Fainsod, A., & Gruenbaum, Y. (1990). Isolation and characterization of the Drosophila nuclear envelope olefin cDNA. J. Biol. Chem. 265, 7808–7813.

Paddy, M.R., Belmont, A.S., Saumweber, H., Agard, D.A., & Sedat, J.W. (1990). Interphase nuclear envelope lamins form a discontinuous network that interacts with only a fraction of the chromatin in the nuclear periphery. Cell 62, 89–106.

Pappas, G.D. (1956). The fine structure of the nuclear envelope of *Ameoba proteus*. J. Biophys. Biochem. Cytol. 2, 431–435.

Paulin-Levasseur, M., Scherbarth, A., Traub, U., & Traub, P. (1988). Lack of lamins A and C in mammalian hemopoietic cell lines devoid of intermediate filament proteins. Eur. J. Cell Biol. 47, 121–131.

Paulson, J.R. & Laemmli, U.K. (1977). The structure of histon-depleted metaphase chromosomes. Cell 12, 817–828.

Peter, M. & Nigg, E.A. (1991). Ectopic expression of an A-type lamin does not interfere with differentiation of lamin-A-negative embryonal carcinoma cells. J. Cell Sci. 100, 589–598.

Peter, M., Heitlinger, E., Häner, M., Aebi, U., & Nigg, E.A. (1991). Disassembly of *in vitro* formed lamin head-to-tail polymers by CDC2 kinase. EMBO J. 10, 1535–1544.

Peter, M., Sanghera, J.S., Pelech, S.L., & Nigg, E.A. (1992). Mitogen-activated protein kinases phosphorylate nuclear lamins and display sequence specificity overlapping that of mitotic protein kinase p34^{cdc2}. Eur. J. Biochem. 205, 287–294.

Peter, M., Nakagawa, J., Dorée, M., Labbé, J.C., & Nigg, E.A. (1990). *In vitro* disassembly of the nuclear lamina and M phase-specific phosphorylation of lamins by cdc2 kinase. Cell 61, 591–602.

Peter, M., Kitten, G.T., Lehner, C.F., Vorburger, K., Bailer, S.M., Maridor, G., & Nigg, E.A. (1989). Cloning and sequencing of cDNA clones encoding chicken lamins A and B1 and comparison of the primary structures of vertebrate A- and B-type lamins. J. Mol. Biol. 208, 393–404.

Pollard, K.M., Chan, E.K.L., Grant, B.J., Sullivan, K.F., Tan, E.M., & Glass, C.A. (1990). *In vitro* posttranslational modification of lamin B cloned from human T-cell line. Mol. Cell. Biol. 10, 2164–2175.

Powell, L. & Burke, B. (1990). Internuclear exchange of an inner nuclear membrane protein (p55) in heterokaryons; *In vivo* evidence for the interaction of p55 with the nuclear lamina. J. Cell Biol. 111, 2225–2234.

Puvion, E. & Bernhard, W. (1975). Ribonucleoprotein components in liver cell nuclei as visualized by cryoultramicrotomy. J. Cell Biol. 67, 200–214.

Riedel, W. & Werner, D. (1989). Nucleotide sequence of the full-length mouse lamin C cDNA and its deduced amino-acid sequence. Biochim. Biophys. Acta 1008, 119–122.

Riemer, D., Dodemont, H., & Weber, K. (1993). A nuclear lamin of the nematode *Caenorhabditis elegans* with unusual features; cDNA cloning and genomic organization. Eur. J. Cell Biol. 62, 214–223.

Röber, R.-A., Gieseler, R.K.H., Peters, H.J., Weber, K., & Osborn, M. (1990). Induction of nuclear lamins A/C in macrophages in *in vitro* cultures of rat bone marrow precursor cells and human blood monocytes and in macrophages elicited *in vivo* by thioglycolate stimulation. Exp. Cell Res. 190, 185–194.

Röber, R.-A., Weber, K., & Osborn, M. (1989). Differential timing of nuclear lamin A/C expression in the various organs of the mouse embryo and the young animal: A developmental study. Development 105, 365–378.

Schafer, W.R. & Rine, J. (1992). Protein prenylation: Genes, enzymes, targets, and functions. Ann. Rev. Genet. 30, 2089–2234.

Schatten, G., Maul, G.G., Schatten, H., Chaly, N., Simerly, C., Balczon, R., & Brown, D.L. (1985). Nuclear lamins and peripheral nuclear antigens during fertilization and embryogenesis in mice and sea urchins. Proc. Natl. Acad. Sci. USA 82, 4727–4731.

Scheer, U. & Benavente, R. (1990). Functional and dynamic aspects of the mammalian nucleolus. BioEssays 12, 14–21.

Senior, A. & Gerace, L. (1988). Integral membrane proteins specific to the inner nuclear membrane and associated with the nuclear lamina. J. Cell Biol. 107, 2029–2036.

Shoeman, R.L. & Traub, P. (1990). The *in vitro* DNA-binding properties of purified nuclear lamin proteins and vimentin. J. Biol. Chem. 265, 9055–9061.

Simos, G. & Georgatos, S.D. (1992). The inner nuclear membrane protein p58 associates *in vivo* with a p58-kinase and the nuclar lamins. EMBO J. 11, 4027–4036.

Smith, A. & Benavente, R. (1992). Identification of a short nuclear lamin protein selectively expressed during meiotic stages of rat spermatogenesis. Differentiation 52, 55–60.

Smith, D.E. & Fisher, P.A. (1989). Interconversion of *Drosophila* nuclear lamin isoforms during oogenesis, early embryogenesis, and upon entry of cultured cells into mitosis. J. Cell Biol. 108, 255–265.

Smith, D.E., Gruenbaum, Y., Berrios, M., & Fisher, P.A. (1987). Biosynthesis and interconversion of *Drosophila* nuclear lamin isoforms during normal growth and in response to heat shock. J. Cell Biol. 105, 771–790.

Smith, S. & Blobel, G. (1993). The first membrane spanning region of the lamin B receptor is sufficient for sorting to the inner nuclear membrane. J. Cell Biol 120, 631–637.

Soullam, B. & Worman, H. J. (1993). The amino-terminal domain of the lamin B receptor is a nuclear envelope targeting signal. J. Cell Biol. 120, 1093–1100.

Spector, D.L. (1993). Nuclear organization of pre-mRNA processing. Curr. Op. Cell Biol. 5, 442–448.

Steinert, P.M. & Roop, D.R. (1988). Molecular and cellular biology of intermediate filaments. Annu. Rev. Biochem. 57, 593–625.

Stewart, C. & Burke, B. (1987). Teratocarcinoma stem cells and early mouse embryos contain only a single major lamin polypeptide closely resembling lamin B. Cell 51, 383–392.

Stewart, M. & Whytock, S. (1988). The structure and interactions of components of nuclear envelopes from *Xenopus* germinal vesicles observed by heavy metal shadowing. J. Cell Sci. 90, 409–423.

Stick, R. (1987). Dynamics of the nuclear lamina during mitosis and meiosis. In: Molecular Regulation of Nuclear Events in Mitosis and Meiosis (Schlegel, R.A., Halleck, M.S., & Rao, P.N., Eds.), pp. 43–66. Academic Press, New York.

Stick, R. (1988). cDNA cloning of the developmentally regulated lamin LIII of *Xenopus laevis*. EMBO J. 7, 3189–3197.

Stick, R. (1992). The gene structure of *Xenopus* nuclear lamin A: A model for the evolution of A-type from B-type lamins by exon shuffling. Chromosoma 101, 566–574.

Stick, R. & Hausen, P. (1985). Changes in the nuclear lamina composition during early development of *Xenopus laevis*. Cell 41, 191–200.

Stick, R. & Schwarz, H. (1982). The disappearance of the nuclear lamina during spermatogenesis: An electron microscopic and immunofluorescence study. Cell Differ. 11, 235–243.

Stick, R. & Schwarz, H. (1983). Disappearance and reformation of the nuclear lamina structure during specific stages of meiosis in oocytes. Cell 33, 949–958.

Stuurman N., Floore, A., Colen, A., de Jong, L., & van Driel, R. (1992a). Stabilization of the nuclear matrix by disulfide bridges: Identification of matrix polypeptides that form disulfides. Exp. Cell Res. 200, 285–294.

Stuurman N, de Graaf, A., Floore, A., Josso, A., Humbel, B., de Jong, L., & van Driel, R. (1992c). A monoclonal antibody recognizing nuclear matrix-associated nuclear bodies. J. Cell Sci. 101, 773–784.

Stuurman N., de Jong L., & van Driel, R. (1992b). Nuclear frameworks: Concepts and operational definitions. Cell Biol. Int. Rep. 16, 837–852.

Sudhakar, K., Sivakumar, N., Behal, A., & Rao, M.R.S. (1992). Evolutionary conservation of a germ cell-specific lamin persisting through mammalian spermiogenesis. Exp. Cell Res. 198, 78–84.

Swedlow, J.R., Agard, D.A., & Sedat, J.W. (1993a). Chromosome structure inside the nucleus. Curr. Op. Cell Biol. 5, 412–416.

Swedlow, J.R., Sedat, J.W., & Agard, D.A. (1993b). Multiple chromosomal populations of topoisomerase II detected *in vitro* by time-lapse, three-dimensional wide field microscopy. Cell 73, 97–108.

Ulitzur, N. & Gruenbaum, Y. (1989). Nuclear envelope assembly around sperm chromatin in cell-free preparations from *Drosophila* embryos. FEBS Lett. 259, 113–116.

van Driel, R., Humbel, B., & de Jong, L. (1991). The nucleus: A black box being opened. J. Cell. Biochem. 47, 311–316.

Verheijen, R., van Venrooij, W.J., & Ramaekers, F. (1988). The nuclear matrix structure and composition. J. Cell Sci. 90, 11–36.

Vester, B., Smith, A., Krohne, G., & Benavente, R. (1993). Presence of a nuclear lamina in pachytene spermatocytes of the rat. J. Cell Sci. 104, 557–563.

von Kries, J.P., Buhrmeister, H., & Strätling, W.H. (1991). A matrix/scaffold attachment region binding protein: Identification, purification, and mode of binding. Cell 64, 123–135.

Vorburger, K., Kitten, G.T., & Nigg, E.A. (1989b). Modification of nuclear lamin proteins by a mevalonic acid derivative occurs in reticulocyte lysates and requires the cysteine residue of the C-terminal CXXM motif. EMBO J. 8, 4007–4013.

Vorburger, K., Lehner, C.F., Kitten, G., Eppenberger, H.M., & Nigg, E.A. (1989a). A second higher vertebrate B-type lamin: cDNA sequence determination and *in vitro* processing of chicken lamin B2. J. Mol. Biol. 208, 405–415.

Wansink, D.G., Schul, W., van der Kraan, I., van Steensel, B., van Driel, R., & de Jong, L. (1993). Fluorescent labeling of nascent RNA reveals transcription by RNA polymerase II in domains scattered throughout the nucleus. J. Cell Biol. 122, 283–293.

Ward, G.E. & Kirschner, M.W. (1990). Identification of cell cycle-regulated phosphorylation sites on nuclear lamin C. Cell 61, 561–577.

Weber, K., Plessmann, U., & Traub, P. (1989). Maturation of nuclear lamin A involves a specific carboxy-terminal trimming, which removes the polyisoprenylation site from the precursor; implications for the structure of the nuclear lamina. FEBS Lett 257, 411–414.

Wiese C. & Wilson K.L. (1993). Nuclear membrane dynamics. Curr. Op. Cell Biol. 5, 387–394.

Wilson, K.L. & Newport, J. (1988). A trypsin-sensitive receptor on membrane vesicles is required for nuclear envelope formation *in vitro*. J. Cell Biol. 107, 57–68.

Wolda, S.L. & Glomset, J.A. (1988). Evidence for modifications of lamin B by a product of mevalonic acid. J. Biol. Chem. 263, 5997–6000.

Wolin, S.L., Krohne, G., & Kirschner, M.W. (1987). A new lamin in *Xenopus* somatic tissues displays strong homology to human lamin A. EMBO J. 6, 3809–3818.

Wood, E. & Earnshaw, W.C. (1990). Mitotic chromatin condensation *in vitro* using somatic cell extracts and nuclei with variable levels of endogenous topoisomerase II. J. Cell Biol. 111, 2839–2850.

Woodcock, C.L., McEwen, B.F., & Frank, J. (1991). Ultrastructure of chromatin II. Three-dimensional reconstruction of isolated fibers. J. Cell Sci. 99, 107–114.

Worman, H.J., Juan, J., Blobel, G., & Georgatos, S.D. (1988). A lamin B receptor in the nuclear envelope. Proc. Natl. Acad. Sci. USA 85, 8531–8534.

Xing, Y., Johnson, C.V., Dobner, P.R., & Lawrence, J.B. (1993). Higher level organization of individual gene transcription and RNA splicing. Science 259, 1326–1330.

Yuan, J., Simos, G., Blobel, G., & Georgatos, S.D. (1991). Binding of lamin A to polynucleosomes. J. Biol. Chem 296, 9211–9215.

Zachar, Z., Kramer, J., Mims, I.P., & Bingham, P.M. (1993). Evidence for channeled diffusion of pre-mRNAs during nuclear RNA transport in metazoans. J. Cell Biol. 121, 729–742.

Zeligs, J.D. & Wollman, S.H. (1979). Mitosis in rat thyroid epithelial cells *in vivo*. I. Ultrastructural changes in cytoplasmic organelles during the mitotic cycle. J. Ultrastruct. Res. 66, 53–77.

Zewe, M., Höger, T.H., Fink, T., Lichter, P., Krohne, G., & Franke, W.W. (1991). Gene structure and chromosomal localization of the murine lamin B2 gene. Eur. J. Cell Biol. 56, 342–350.

ADDITIONAL LITERATURE

Firmbach-Kraft, I. & Stick, R. (1995). Analysis of nuclear lamin isoprenylation in *Xenopus* oocytes: isoprenylation of lamin B3 precedes its uptake into the nucleus. J. Cell Biol. 129, 17–24.

Haas, M. & Jost, E. (1993). Functional analysis of phosphorylation sites in human lamin A controlling lamin disassembly, nuclear transport and assembly. Eur. J. Cell Biol. 62, 237–247.

Osman, M., Paz, M., Landesman, Y., Fainsod, A., & Gruenbaum, Y. (1990). Molecular analysis of the *Drosophila* nuclear lamin gene. Genomics 8, 217–224.

Riemer, D. & Weber. K. (1994). The organization of the gene for *Drosophila* lamin C: Limited homology with vertebrate lamin genes and lack of homology versus the Drosophila lamin Dmo gene. Eur. J. Cell Biol. 63, 299–306.

Stick, R. (1994). The gene structure of B-type nuclear lamins of *Xenopus laevis*: implications for the evolution of the vertebrate lamin family. Chromosome Res. 2, 376–382.

INDEX

ABP (*see* Filamin)
ABP-50, 12, 13, 15
ABP-120, 27, 30, 31
ABP1p, 6
ABPs (*see* Actin-binding proteins)
Acetylcholine receptor, 131
Actin (*see also* Actin-binding proteins; Microfilaments)
  as actin-binding protein, 28-29
  β isoform, 126
  in cell-cell junctions, 143-145
  in cortical networks, 126, 135-136, 150
  in erythrocyte cytoskeleton, 126
  in focal contacts, 136-143
  intermediate filaments and, 150
  as main protein in cell, 2
  microtubules and, 152-153
  myosin I and, 146
  nonfilamentous, 19-21
  in nucleus, 31, 33
  phosphorylated, 28
  polymerization
    bundling proteins and, 16
    caldesmon and, 23
    capping proteins and, 10-11
    in cell locomotion, 183-184
    inhibition of, 5, 19-21, 191
    membrane-associated, 24-25
  ribosylated, 28-29
  synapsins and, 146

transmembrane proteins and, 145-146
viscosity, 16, 20
Actin-binding protein (*see* Filamin)
Actin-binding proteins (ABPs), 2
  (*see also specific proteins*)
  actin as, 28-29
  binding domains, 29, 31-33
  bundling proteins, 4, 12-17
  capping proteins, 4, 6-11, 187
  cofilamentous proteins, 21-24
  cross-linking proteins, 17
  lipid binding, 26, 171-172, 194-195
  membrane-associated proteins, 24-28
  monomer binding proteins, 19-21
  in nucleus, 31, 33
  redundancy, 29-31
  regulation by phosphoinositides, 3, 4, 29, 187-189, 191-192
  research prospects, 33-34
  severing proteins, 2-6, 19, 187-189
α-Actinin, 12, 13, 15, 16, 24, 138
  actin-binding site, 32
  calcium dependence, 182
  at cell-cell junctions, 145
  in cell locomotion, 30
  at focal contacts, 137, 139, 140, 141, 142-143
  lipid binding, 182-183
β-Actinin, 9

J A I P R E S S

Advances in Neural Science

Edited by **Sudarshan Malhotra**, *Department of Zoology, University of Alberta*

Volume 2, 1995, 235 pp. $97.50
ISBN 1-55938-625-8

CONTENTS: Preface, *Sudarshan K. Malhotra.* Phosphorylation of Neurofilament Proteins, *Michael G. Sacher, Eric S. Athlan and Walter W. Mushynski.* Neuronal Development in Embryos of the Mollusc, Helisoma Trivolvis. Multiple Roles of Serotonin, *Jeffrey I. Goldberg.* Opioid Growth Factor and Retinal Morphogenesis, *Ian S. Zagon, Tomoki Isayama and Patricia J. McLaughlin.* The Molecular Bases of Nerve Regeneration, *Joanne K. Daniloff and Laura G. Remsen.* Trophic Actions of Gonadal Steroids on Neuronal Functioning Normalcy and Following Injury, *Kathleen A. Kujawa and Kathryn J. Jones.* Are Epigenetic Factors Involved in the Normal Expression of Neuronal Phenotypes During Spinal Development?, *Eric Philippe and Raymond Marchand.* Plasticity of Descending Spinal Pathways in Developing Mammals, *George F. Martin, Ganesh T. Ghooray, Xian Ming Wang Xiao, and Ming Xu.* Development of the Mammalian Auditory Hindbrain, *Frank H. Willard.* Subject Index.

Also Available:
Volume 1 (1993) $97.50

JAI PRESS INC.
55 Old Post Road # 2 - P.O. Box 1678
Greenwich, Connecticut 06836-1678
Tel: (203) 661- 7602   Fax: (203) 661-0792

Printed and bound by CPI Group (UK) Ltd, Croydon, CR0 4YY

03/10/2024

01040436-0006